TABLE

Standard Normal, Cumulative Pr

(For Negative Values of Z, Area

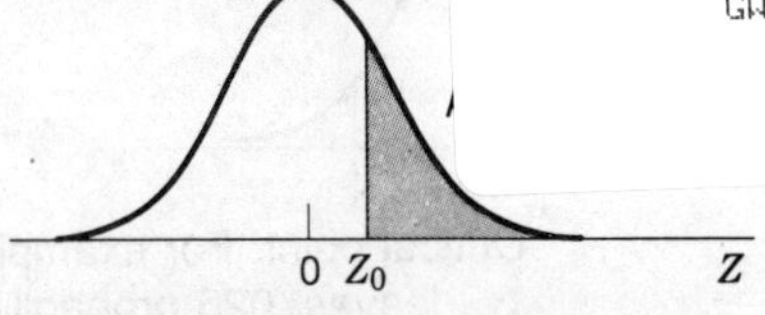

Z_o	.00	.01	.02	.03	.04	.05	.06	.07	.08	.09
	Second Decimal Place of Z_o									
0.0	.5000	.4960	.4920	.4880	.4840	.4801	.4761	.4721	.4681	.4641
0.1	.4602	.4562	.4522	.4483	.4443	.4404	.4364	.4325	.4286	.4247
0.2	.4207	.4168	.4129	.4090	.4052	.4013	.3974	.3936	.3897	.3859
0.3	.3821	.3783	.3745	.3707	.3669	.3632	.3594	.3557	.3520	.3483
0.4	.3446	.3409	.3372	.3336	.3300	.3264	.3228	.3192	.3156	.3121
0.5	.3085	.3050	.3015	.2981	.2946	.2912	.2877	.2843	.2810	.2776
0.6	.2743	.2709	.2676	.2643	.2611	.2578	.2546	.2514	.2483	.2451
0.7	.2420	.2389	.2358	.2327	.2296	.2266	.2236	.2206	.2177	.2148
0.8	.2119	.2090	.2061	.2033	.2005	.1977	.1949	.1922	.1894	.1867
0.9	.1841	.1814	.1788	.1762	.1736	.1711	.1685	.1660	.1635	.1611
1.0	.1587	.1562	.1539	.1515	.1492	.1469	.1446	.1423	.1401	.1379
1.1	.1357	.1335	.1314	.1292	.1271	.1251	.1230	.1210	.1190	.1170
1.2	.1151	.1131	.1112	.1093	.1075	.1056	.1038	.1020	.1003	.0985
1.3	.0968	.0951	.0934	.0918	.0901	.0885	.0869	.0853	.0838	.0823
1.4	.0808	.0793	.0778	.0764	.0749	.0735	.0722	.0708	.0694	.0681
1.5	.0668	.0655	.0643	.0630	.0618	.0606	.0594	.0582	.0571	.0559
1.6	.0548	.0537	.0526	.0516	.0505	.0495	.0485	.0475	.0465	.0455
1.7	.0446	.0436	.0427	.0418	.0409	.0401	.0392	.0384	.0375	.0367
1.8	.0359	.0352	.0344	.0336	.0329	.0322	.0314	.0307	.0301	.0294
1.9	.0287	.0281	.0274	.0268	.0262	.0256	.0250	.0244	.0239	.0233
2.0	.0228	.0222	.0217	.0212	.0207	.0202	.0197	.0192	.0188	.0183
2.1	.0179	.0174	.0170	.0166	.0162	.0158	.0154	.0150	.0146	.0143
2.2	.0139	.0136	.0132	.0129	.0125	.0122	.0119	.0116	.0113	.0110
2.3	.0107	.0104	.0102	.0099	.0096	.0094	.0091	.0089	.0087	.0084
2.4	.0082	.0080	.0078	.0075	.0073	.0071	.0069	.0068	.0066	.0064
2.5	.0062	.0060	.0059	.0057	.0055	.0054	.0052	.0051	.0049	.0048
2.6	.0047	.0045	.0044	.0043	.0041	.0040	.0039	.0038	.0037	.0036
2.7	.0035	.0034	.0033	.0032	.0031	.0030	.0029	.0028	.0027	.0026
2.8	.0026	.0025	.0024	.0023	.0023	.0022	.0021	.0021	.0020	.0019
2.9	.0019	.0018	.0017	.0017	.0016	.0016	.0015	.0015	.0014	.0014
3.0	.00135									
3.5	.000 233									
4.0	.000 031 7									
4.5	.000 003 40									
5.0	.000 000 287									

TABLE V

t Critical Points

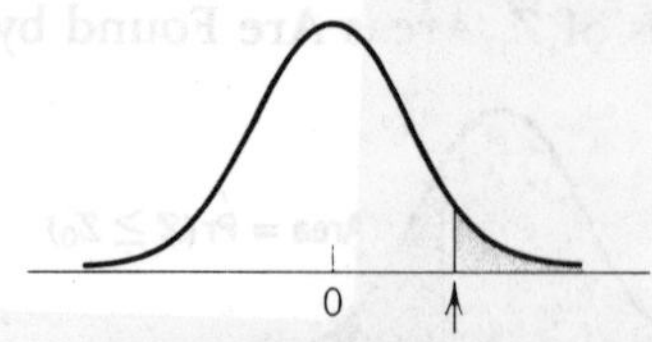

Critical point. For example: $t_{.025}$ leaves .025 probability in the tail.

d.f.	$t_{.25}$	$t_{.10}$	$t_{.05}$	$t_{.025}$	$t_{.010}$	$t_{.005}$	$t_{.0025}$	$t_{.0010}$	$t_{.0005}$
1	1.000	3.078	6.314	12.706	31.821	63.637	127.32	318.31	636.62
2	.816	1.886	2.920	4.303	6.965	9.925	14.089	22.326	31.598
3	.765	1.638	2.353	3.182	4.541	5.841	7.453	10.213	12.924
4	.741	1.533	2.132	2.776	3.747	4.604	5.598	7.173	8.610
5	.727	1.476	2.015	2.571	3.365	4.032	4.773	5.893	6.869
6	.718	1.440	1.943	2.447	3.143	3.707	4.317	5.208	5.959
7	.711	1.415	1.895	2.365	2.998	3.499	4.020	4.785	5.408
8	.706	1.397	1.860	2.306	2.896	3.355	3.833	4.501	5.041
9	.703	1.383	1.833	2.262	2.821	3.250	3.690	4.297	4.781
10	.700	1.372	1.812	2.228	2.764	3.169	3.581	4.144	4.537
11	.697	1.363	1.796	2.201	2.718	3.106	3.497	4.025	4.437
12	.695	1.356	1.782	2.179	2.681	3.055	3.428	3.930	4.318
13	.694	1.350	1.771	2.160	2.650	3.012	3.372	3.852	4.221
14	.692	1.345	1.761	2.145	2.624	2.977	3.326	3.787	4.140
15	.691	1.341	1.753	2.131	2.602	2.947	3.286	3.733	4.073
16	.690	1.337	1.746	2.120	2.583	2.921	3.252	3.686	4.015
17	.689	1.333	1.740	2.110	2.567	2.898	3.222	3.646	3.965
18	.688	1.330	1.734	2.101	2.552	2.878	3.197	3.610	3.922
19	.688	1.328	1.729	2.093	2.539	2.861	3.174	3.579	3.883
20	.687	1.325	1.725	2.086	2.528	2.845	3.153	3.552	3.850
21	.686	1.323	1.721	2.080	2.518	2.831	3.135	3.257	3.189
22	.686	1.321	1.717	2.074	2.508	2.819	3.119	3.505	3.792
23	.685	1.319	1.714	2.069	2.500	2.807	3.104	3.485	3.767
24	.685	1.318	1.711	2.064	2.492	2.797	3.091	3.467	3.745
25	.684	1.316	1.708	2.060	2.485	2.787	3.078	3.450	3.725
26	.684	1.315	1.706	2.056	2.479	2.779	3.067	3.435	3.707
27	.684	1.314	1.703	2.052	2.473	2.771	3.057	3.421	3.690
28	.683	1.313	1.701	2.048	2.467	2.763	3.047	3.408	3.674
29	.683	1.311	1.699	2.045	2.462	2.756	3.038	3.396	3.659
30	.683	1.310	1.697	2.042	2.457	2.750	3.030	3.385	3.646
40	.681	1.303	1.684	2.021	2.423	2.704	2.971	3.307	3.551
60	.679	1.296	.1671	2.000	2.390	2.660	2.915	3.232	3.460
120	.677	1.289	1.658	1.980	2.358	2.617	2.860	3.160	3.373
∞	.674 $= z_{.25}$	1.282 $= z_{.10}$	1.645 $= z_{.05}$	1.960 $= z_{.025}$	2.326 $= z_{.010}$	2.576 $= z_{.005}$	2.807 $= z_{.0025}$	3.090 $= z_{.0010}$	3.291 $= z_{.0005}$

STATISTICS
Discovering Its Power

WILEY SERIES IN PROBABILITY AND MATHEMATICAL STATISTICS

Probability and Mathematical Statistics

ADLER • The Geometry of Random Fields
ANDERSON • The Statistical Analysis of Time Series
ANDERSON • An Introduction to Multivariate Statistical Analysis
ARAUJO and GINE • The Central Limit Theorem for Real and Banach Valued Random Variables
ARNOLD • The Theory of Linear Models and Multivariate Analysis
BARLOW, BARTHOLOMEW, BREMNER, and BRUNK • Statistical Inference Under Order Restrictions
BARNETT • Comparative Statistical Inference
BHATTACHARYYA and JOHNSON • Statistical Concepts and Methods
BILLINGSLEY • Probability and Measure
CASSEL, SARNDAL, and WRETMAN • Foundations of Inference in Survey Sampling
COCHRAN • Contributions to Statistics
DE FINETTI • Theory of Probability, Volumes I and II
DOOB • Stochastic Processes
FELLER • An Introduction to Probability Theory and Its Applications, Volume I, *Third Edition*, Revised; Volume II, Second Edition
FULLER • Introduction to Statistical Time Series
GRENANDER • Abstract Inference
HANNAN • Multiple Time Series
HANSEN, HURWITZ, and MADOW • Sample Survey Methods and Theory, Volumes I and II
HARDING and KENDALL • Stochastic Geometry
HOEL • Introduction to Mathematical Statistics, *Fourth Edition*
HUBER • Robust Statistics
IOSIFESCU • Finite Markov Processes and Applications
ISAACSON and MADSEN • Markov Chains
KAGAN, LINNIK, and RAO • Characterization Problems in Mathematical Statistics
KENDALL and HARDING • Stochastic Analysis
LAHA and ROHATGI • Probability Theory
LARSON • Introduction to Probability Theory and Statistical Inference, *Second Edition*
LARSON • Introduction to the Theory of Statistics
LEHMANN • Testing Statistical Hypotheses
MATTHES, KERSTAN, and MECKE • Infinitely Divisible Point Processes
PARZEN • Modern Probability Theory and Its Applications
PURI and SEN • Nonparametric Methods in Multivariate Analysis
RANDLES and WOLFE • Introduction to the Theory of Nonparametric Statistics
RAO • Linear Statistical Inference and Its Applications, *Second Edition*
ROHATGI • An Introduction to Probability Theory and Mathematical Statistics
RUBINSTEIN • Simulation and The Monte Carlo Method
SCHEFFE • The Analysis of Variance

Probability and Mathematical Statistics (Continued)

SEBER • Linear Regression Analysis
SEN • Sequential Nonparametrics: Invariance Principles and Statistical Inference
SERFLING • Approximation Theorems of Mathematical Statistics
TJUR • Probability Based on Radon Measures
WILLIAMS • Diffusions, Markov Processes, and Martingales, Volume I: Foundations
ZACKS • Theory of Statistical Inference

Applied Probability and Statistics

ANDERSON, AUQUIER, HAUCK, OAKES, VANDAELE, and WEISBERG • Statistical Methods for Comparative Studies
ARTHANARI and DODGE • Mathematical Programming in Statistics
BAILEY • The Elements of Stochastic Processes with Applications to the Natural Sciences
BAILEY • Mathematics, Statistics and Systems for Health
BARNETT • Interpreting Multivariate Data
BARNETT and LEWIS • Outliers in Statistical Data
BARTHOLOMEW • Stochastic Models for Social Processes, *Third Edition*
BARTHOLOMEW and FORBES • Statistical Techniques for Manpower Planning
BECK and ARNOLD • Parameter Estimation in Engineering and Science
BELSLEY, KUH, and WELSCH • Regression Diagnostics: Identifying Influential Data and Sources of Collinearity
BENNETT and FRANKLIN • Statistical Analysis in Chemistry and the Chemical Industry
BHAT • Elements of Applied Stochastic Processes
BLOOMFIELD • Fourier Analysis of Time Series: An Introduction
BOX • R. A. Fisher, The Life of a Scientist
BOX and DRAPER • Evolutionary Operation: A Statistical Method for Process Improvement
BOX, HUNTER, and HUNTER • Statistics for Experimenters: An Introduction to Design, Data Analysis, and Model Building
BROWN and HOLLANDER • Statistics: A Biomedical Introduction
BROWNLEE • Statistical Theory and Methodology in Science and Engineering, *Second Edition*
BURY • Statistical Models in Applied Science
CHAMBERS • Computational Methods for Data Analysis
CHATTERJEE and PRICE • Regression Analysis by Example
CHERNOFF and MOSES • Elementary Decision Theory
CHOW • Analysis and Control of Dynamic Economic Systems
CHOW • Econometric Analysis by Control Methods
CLELLAND, BROWN, and deCANI • Basic Statistics with Business Applications, *Second Edition*
COCHRAN • Sampling Techniques, *Third Edition*
COCHRAN and COX • Experimental Designs, *Second Edition*
CONOVER • Practical Nonparametric Statistics, *Second Edition*
CORNELL • Experiments with Mixtures: Designs, Models and The Analysis of Mixture Data
COX • Planning of Experiments
DANIEL • Biostatistics: A Foundation for Analysis in the Health Sciences, *Second Edition*
DANIEL • Applications of Statistics to Industrial Experimentation
DANIEL and WOOD • Fitting Equations to Data: Computer Analysis of Multifactor Data, *Second Edition*

Applied Probability and Statistics (Continued)

DAVID • Order Statistics, *Second Edition*
DEMING • Sample Design in Business Research
DODGE and ROMIG • Sampling Inspection Tables, *Second Edition*
DRAPER and SMITH • Applied Regression Analysis, *Second Edition*
DUNN • Basic Statistics: A Primer for the Biomedical Sciences, *Second Edition*
DUNN and CLARK • Applied Statistics: Analysis of Variance and Regression
ELANDT-JOHNSON • Probability Models and Statistical Methods in Genetics
ELANDT-JOHNSON and JOHNSON • Survival Models and Data Analysis
FLEISS • Statistical Methods for Rates and Proportions, *Second Edition*
FRANKEN • Queues and Point Processes
GALAMBOS • The Asymptotic Theory of Extreme Order Statistics
GIBBONS, OLKIN, and SOBEL • Selecting and Ordering Populations: A New Statistical Methodology
GNANADESIKAN • Methods for Statistical Data Analysis of Multivariate Observations
GOLDBERGER • Econometric Theory
GOLDSTEIN and DILLON • Discrete Discriminant Analysis
GROSS and CLARK • Survival Distributions: Reliability Applications in the Biomedical Sciences
GROSS and HARRIS • Fundamentals of Queueing Theory
GUPTA and PANCHAPAKESAN • Multiple Decision Procedures: Theory and Methodology of Selecting and Ranking Populations
GUTTMAN, WILKS, and HUNTER • Introductory Engineering Statistics, *Second Edition*
HAHN and SHAPIRO • Statistical Models in Engineering
HALD • Statistical Tables and Formulas
HALD • Statistical Theory with Engineering Applications
HAND • Discrimination and Classification
HARTIGAN • Clustering Algorithms
HILDEBRAND, LAING, and ROSENTHAL • Prediction Analysis of Cross Classifications
HOEL • Elementary Statistics, *Fourth Edition*
HOLLANDER and WOLFE • Nonparametric Statistical Methods
JAGERS • Branching Processes with Biological Applications
JESSEN • Statistical Survey Techniques
JOHNSON and KOTZ • Distributions in Statistics
Discrete Distributions
Continuous Univariate Distributions—1
Continuous Univariate Distributions—2
Continuous Multivariate Distributions
JOHNSON and KOTZ • Urn Models and Their Application: An Approach to Modern Discrete Probability Theory
JOHNSON and LEONE • Statistics and Experimental Design in Engineering and the Physical Sciences, Volumes I and II, *Second Edition*
JUDGE, GRIFFITHS, HILL and LEE • The Theory and Practice of Econometrics
KALBFLEISCH and PRENTICE • The Statistical Analysis of Failure Time Data
KEENEY and RAIFFA • Decisions with Multiple Objectives
LAWLESS • Statistical Models and Methods for Lifetime Data
LEAMER • Specification Searches: Ad Hoc Inference with Nonexperimental Data

Applied Probability and Statistics (Continued)

McNEIL • Interactive Data Analysis
MANN, SCHAFER and SINGPURWALLA • Methods for Statistical Analysis of Reliability and Life Data
MEYER • Data Analysis for Scientists and Engineers
MILLER • Survival Analysis
MILLER, EFRON, BROWN, and MOSES • Biostatistics Casebook
MONTGOMERY and PECK • Introduction to Linear Regression
NELSON • Applied Life Data Analysis
OTNES and ENOCHSON • Applied Time Series Analysis: Volume I, Basic Techniques
OTNES and ENOCHSON • Digital Time Series Analysis
POLLOCK • The Algebra of Econometrics
PRENTER • Splines and Variational Methods
RAO and MITRA • Generalized Inverse of Matrices and Its Applications
RIPLEY • Spatial Statistics
SCHUSS • Theory and Applications of Stochastic Differential Equations
SEAL • Survival Probabilities: The Goal of Risk Theory
SEARLE • Linear Models
SPRINGER • The Algebra of Random Variables
UPTON • The Analysis of Cross-Tabulated Data
WEISBERG • Applied Linear Regression
WHITTLE • Optimization Under Constraints
WILLIAMS • A Sampler on Sampling
WONNACOTT and WONNACOTT • Econometrics, *Second Edition*
WONNACOTT and WONNACOTT • Introductory Statistics, *Third Edition*
WONNACOTT and WONNACOTT • Introductory Statistics for Business and Economics, *Second Edition*
WONNACOTT and WONNACOTT • Regression: A Second Course in Statistics
WONNACOTT and WONNACOTT • Statistics: Discovering Its Power
ZELLNER • An Introduction to Bayesian Inference in Econometrics

Tracts on Probability and Statistics

BARNDORFF-NIELSEN • Information and Exponential Families in Statistical Theory
BHATTACHARYA and RAO • Normal Approximation and Asymptotic Expansions
BIBBY and TOUTENBERG • Prediction and Improved Estimation in Linear Models
BILLINGSLEY • Convergence of Probability Measures
JARDINE and SIBSON • Mathematical Taxonomy
KELLY • Reversibility and Stochastic Networks
RAKTOE, HEDAYAT, and FEDERER • Factorial Designs
TOUTENBERG • Prior Information in Linear Models

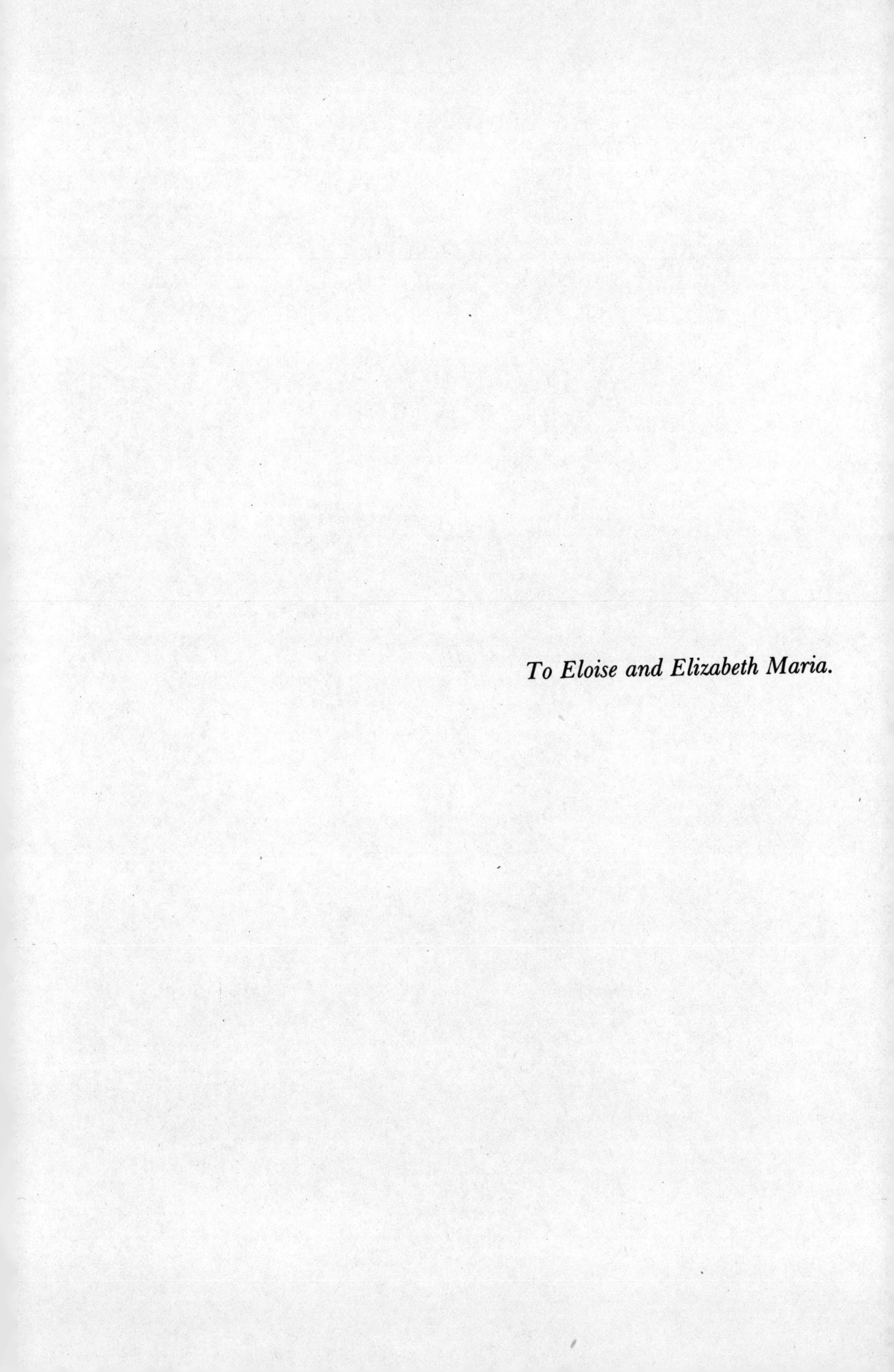

To Eloise and Elizabeth Maria.

Preface

This is an introduction to statistics at an easy and applied level. Written for a one- or two-semester course, the topics are carefully ordered so that students who use the book for only one semester can nevertheless be confident that they have covered the most important subjects.

TO THE STUDENT

Statistics is the intriguing study of how you can describe an unknown world by opening a few windows on it. You will discover the excitement of thinking in a way you have never thought before.

This book is not a novel, and it cannot be read that way. Whenever you come to a numbered example in the text, try first to answer it yourself. Only after you have given it hard thought and, we hope, solved it, should you consult the solution we provide. The same advice holds for the exercise problems at the end of each section. These problems have been kept computationally as simple as possible, so that you can concentrate on insight rather than arithmetic. At the same time, we have tried to make them realistic by the frequent use of real data. ,

The more challenging problems and sections are indicated by a star (*). For example, in Chapters 8 and 9 we give some problems that are best answered by using a computer package: we want students who like computers to see their power; but at the same time we keep these exercises optional, so that other students can fully master the text without using a computer.

Brief answers to all odd-numbered problems are given in the back of the book. Their completely worked-out solutions are available in the student's manual.

TO THE INSTRUCTOR

Our previous textbooks (*Introductory Statistics* and *Introductory Statistics for Business and Economics*) have been aimed at the middle to high end of the first-course market; this one is aimed at the more elementary level. It is substantially easier than our earlier books in two ways: the standard statistical topics covered here are, whenever possible, treated in a simpler way; and the more demanding topics have been dropped altogether. Yet our previous books remain very helpful as references; those students who occasionally want a more advanced treatment, for example, can refer to our *Introductory Statistics* with no roadblocks—it has similar notation, and follows the same order of topics.

With so many books already designed for the elementary course, why another? We felt there are still too few texts that are both intuitively appealing and cover the important topics. In taking on this challenge, we have introduced two distinctive features:

1. *Teaching by example.* New concepts are introduced and illustrated with examples. Many of these examples, formally numbered and set off in blue for easy reference, are formed as questions for students to answer themselves. This type of learning is a real pleasure, in both the classroom and individual reading; students find that working out the answers can be as enjoyable as doing a recreational puzzle.

   In a sense, it's the Socratic method: we try to ask the right question that will start students thinking on their own. In our own classrooms, we have often found the ensuing discussion teaches us as much as the students, and sharing this learning is a very rewarding experience.

2. *Highlighting of important topics.* As you skim the Table of Contents, you will see the topics we feel are important. Some deserve special note here. First, we regard regression as the most powerful tool that can be learned in a basic course; so we get to it early and deal with it thoroughly. We include multiple regression, for example, and emphasize its value in reducing bias in observational studies.

   We stress confidence intervals and p values, because we feel these are much more teachable than classical hypothesis testing. Of course, we do examine classical tests—emphasizing tail probabilities (prob-values) more than testing at a fixed level such as 5%. But the text is structured so you can spend as much or as little time on this subject as you choose. Our subtle shift in emphasis from testing to estimation manifests itself in several ways. For example, nearly all statistical techniques (such as the difference of two means and regression slopes) are first introduced with confidence intervals; tests are only undertaken later. And to further stress estimation, we have a short chapter on shrinkage estimates of the Bayesian or James-Stein type. Such modern topics are not only very useful and interesting, but are also easily learned—they supplement common sense rather than contradict it.

In a one-semester course, the material up to multiple regression in Chapter 8 can easily be covered. Alternatively, the course can cover the basics as far as confidence intervals in Chapter 5, followed by a smorgasbord chosen from the remaining chapters. In a two-semester course, it should be possible to teach all of the book at a relaxed and thoughtful pace.

ACKNOWLEDGMENTS

So many people have contributed so much to this book that it is impossible to thank them adequately. However, we must express our special thanks to Jon Baskerville, Robin Carter, L. K. Chan, A. M. Chaudhry, John Koval, and especially to the students of Stats 137 at the University of Western Ontario during 1976 to 1980, who gave many helpful suggestions on the manuscript as it developed.

London, Ontario, Canada, 1981

Thomas H. Wonnacott
Ronald J. Wonnacott

Contents

PART I BASIC STATISTICS

Chapter 1 **The Nature of Statistics 3**
1-1 Random Sampling 3
1-2 Randomized Experiments 8
1-3 Randomized Experiments in the Social Sciences 13
1-4 Regression 14
1-5 Brief Outline of the Book 15
Chapter Summary 16

Chapter 2 **Descriptive Statistics 19**
2-1 Frequency Tables and Graphs 19
2-2 Center of a Distribution 24
2-3 Spread of a Distribution 33
2-4 Calculations Using Relative Frequencies 37
Chapter Summary 39

Chapter 3 **Probability Distributions 42**
3-1 Probabilities for Discrete Random Variables 42
*3-2 Probability Trees 50
3-3 Mean and Variance 53
3-4 The Binomial Distribution 58
3-5 Continuous Distributions 66
3-6 The Normal Distribution 67
Chapter Summary 76

PART II INFERENCE FOR MEANS AND PROPORTIONS

Chapter 4 **Sampling 83**
4-1 Random Sampling 83
4-2 Monte Carlo 89

4-3 How Reliable Is the Sample Mean? 98
4-4 Proportions 108
Chapter Summary 115

Chapter 5 Confidence Intervals 119
5-1 Introduction: Deduction and Induction 119
5-2 95% Confidence Interval for a Mean μ 121
5-3 Using t, When σ is Estimated by s 129
5-4 Difference in Two Means $(\mu_1 - \mu_2)$ 133
5-5 Proportions 144
5-6 One-Sided Confidence Intervals 148
Chapter Summary 151

Chapter 6 Hypothesis Testing 156
6-1 Hypothesis Testing Using Confidence Intervals 156
6-2 Prob-Value (One-Sided) 161
6-3 Classical Hypothesis Tests 170
*6-4 Classical Tests Reconsidered 176
*6-5 Prob-Value (Two-Sided) 180
Chapter Summary 183

PART III RELATING TWO OR MORE VARIABLES

Chapter 7 Simple Regression 191
7-1 Introduction 191
7-2 Fitting a Least Squares Line 194
7-3 The Regression Model 199
7-4 Sampling Variability 204
7-5 Confidence Intervals and Tests for β 206
Chapter Summary 210

Chapter 8 Multiple Regression 213
8-1 Introduction 213
8-2 The Regression Model 216
8-3 The Least Squares Fitted Plane 217
8-4 Confidence Intervals and Statistical Tests 223
8-5 Regression Coefficients as Magnification Factors 226
Chapter Summary 233

Chapter 9 **Regression Extensions 236**
9-1 Dummy (0-1) Variables 236
*9-2 Analysis of Variance (ANOVA) 244
9-3 Simplest Non-Linear Regression 246
Chapter Summary 249

Chapter 10 **Analysis of Variance (ANOVA) 253**
10-1 One-Factor Analysis of Variance 253
*10-2 Extensions 264
Chapter Summary 266

Chapter 11 **Correlation 268**
11-1 Simple Correlation 268
11-2 Correlation and Regression 276
Chapter Summary 285

PART IV FURTHER TOPICS

Chapter 12 **An Introduction to Bayes Estimation 291**
12-1 Resolving the Classical Dilemma 291
12-2 Bayes Estimates in Other Cases 295
12-3 Conclusions 300
*12-4 Bayes Confidence Intervals 302
Chapter Summary 303

Chapter 13 **Nonparametric Statistics 304**
13-1 The Sign Test 305
*13-2 Confidence Interval for the Median 308
13-3 The *W* Test for Two Independent Samples 312
Chapter Summary 316

Chapter 14 **Chi-Square Tests 318**
14-1 χ^2 Tests for Goodness of Fit 318
14-2 Contingency Tables 323
Chapter Summary 330

Appendix A **Monte Carlo Using Normal Random Numbers 332**

Appendix B **Lines 335**

Appendix C **Solution of Linear Equations 338**

Appendix D **Tables 339**

I Random Digits 341
II Random Normal Numbers 342
III (a) Binomial Coefficients 343
(b) Binomial Probabilities, Individual 344
(c) Binomial Probabilities, Cumulative 346
IV Standard Normal Probabilities, Cumulative 348
V *t* Critical Points 349
VI F Critical Points 350
VII χ^2 Critical Points 352
VIII Wilcoxon-Mann-Whitney Two-Sample Test 353

Bibliography 354
Answers to Odd-Numbered Problems 357
Photo Credits 369
Index 371

PART I

BASIC STATISTICS

CHAPTER 1

The Nature of Statistics

Life is the art of drawing sufficient conclusions from insufficient premises.

Samuel Butler

Statistics, like life, is an art—the art of making wise decisions in the face of uncertainty. Many people think of statistics as simply collecting numbers. Indeed, this was its original meaning: State-istics was the collection of population and economic information vital to the state. But statistics is now much more than this. It has developed into a scientific method of analysis widely applied in business and all the social and natural sciences. To get an idea of what modern statistics is, we will examine a couple of typical applications—a political poll, and an experimental surgical technique.

1-1 RANDOM SAMPLING

Before a presidential election, the Gallup poll tries to pick the winner. It also tries to predict how much support each candidate will get from men and women, whites and blacks, Protestants and Catholics, and so on. To be concrete, consider the problem of predicting the proportion of the population that will support the Democratic candidate in the next

U.S. presidential election. Clearly, canvassing the entire population would be an unrealistic task. All we can do is to take a sample, in the hope that the sample porportion will provide a good estimate of the population proportion.

Just how should the sample be chosen? Some interesting lessons can be learned from history. In 1936, for example, when polling was in its infancy, the *Literary Digest* tried to predict the U.S. vote in the presidential election. They mailed questionnaires to ten million voters chosen from lists such as telephone books and club memberships—lists that tended to be more heavily Republican than the voting population at large. Only a quarter responded—and, as it turned out, they tended to be much more Republican than the nonrespondents. This sample was so mismanaged ("biased") that it pointed to a Republican majority. Election day produced a rude surprise: Less than 40% of the voter population were Republicans, and the Democratic incumbent, Roosevelt, was elected with an historic majority.

Other examples of biased samples are easy to find. Informal polls of people on the street are often biased because the interviewer may select people that seem civil and well dressed; a surly worker or harassed mother is overlooked. Members of congress cannot rely on their mail as an unbiased sample of their constituency, since mail is a sample of people with strong opinions and includes an inordinate number of cranks and members of pressure groups.

From such bitter experience important lessons have been learned: To avoid bias, *every* voter must have a chance to be counted. And to avoid slighting any voter, even unintentionally, the sample should be selected *randomly.* There are various ways of doing this, but the simplest to visualize is the following: Put each voter's name on a chip, stir the chips thoroughly in a large bowl, and draw out a sample of, say, a thousand chips. This gives the names of the thousand voters who make up what is called a *simple random sample* of size $n = 1000$.

Unfortunately, in practice simple random sampling is often very slow and expensive. For example, in polling the population of American voters, it would be very difficult to track down the many isolated voters who would turn up in the sample. Much more efficient is *multistage sampling:* From the nation as a whole, take a random sample of a few cities (and counties); within each of these cities, take a random sample of a few wards; finally, within each ward take a random sample of several individuals. While methods like this are frequently used, in this book we will assume simple random sampling (as in drawing chips from a bowl)—leaving the sophisticated variations to an advanced textbook.

Simple random samples will not reflect the population perfectly, of course. If only a few voters are drawn at random, the luck of the draw will be a factor. For example, how might a sample of just 10 voters turn out, from a population of voters split 50-50 Democrat and Republican? The likeliest result is a sample of 5 Democrats, but the luck of the draw might produce 8 or 9 Democrats—just as 10 flips of a fair coin might produce 8 or 9 heads. That is, the sample proportion of Democrats might be 80 or 90%—a far cry from the population proportion of 50%.

In larger samples, the sample proportion P will be a more reliable estimate of the population proportion of Democrats (which we denote by π, the Greek equivalent of our P.) In fact, the easiest way to show how well π is estimated by P is to give a so-called *confidence interval:*

$$\pi = P \pm \text{a small error} \tag{1-1}$$

with crucial questions being, "How small is this error?" and "How sure are we that we are right?" Since this typifies the very core of the book, we state the answer more precisely, in the language of Chapter 5 (where you will find it fully derived):

For simple random sampling, we can state with 95% confidence that

$$\boxed{\pi = P \pm 1.96\sqrt{\frac{P(1-P)}{n}}} \tag{1-2}$$

where π and P are the population and sample proportions, and n is the sample size.

Before we illustrate this confidence interval, we repeat the warning that we gave in the preface: Every numbered example in this text is an exercise that you should actively work out yourself, rather than passively read. We therefore put each example in the form of a question for you to answer; if you get stuck, then you may read the solution. But in all cases remember that *statistics is not a spectator sport.* You cannot learn it by watching, any more than you can learn to ride a bike by watching. You have to jump on and take a few spills.

Example 1-1

The Gallup poll is a combination of multistage and other kinds of random sampling that provides about the same accuracy as simple random sampling. Throughout this book, therefore, we will do little damage in assuming it actually *is* a simple random sample for purposes of applying equation (1-2).

Just before the 1980 presidential election, a Gallup poll of 1500 voters showed 720 for Carter and the remaining 780 for Reagan (ignoring third-party candidates). Calculate the 95% confidence interval for the population proportion π of all voters who were for Carter.

Solution

The sample size is $n = 1500$ and the sample proportion is

$$P = \frac{720}{1500} = .48$$

Substitute these into equation (1-2):

$$\pi = .48 \pm 1.96 \sqrt{\frac{.48(.52)}{1500}}$$

$$\pi = .48 \pm .03 \qquad (1\text{-}3)$$

That is, with 95% confidence, the proportion for Carter in the whole population of voters was between 45% and 51%.

One of the major objectives of this book will be to construct confidence intervals like equation (1-3)—or, as we will hereafter abbreviate references to equations, "like (1-3)." Another related objective is to *test hypotheses.* For example, suppose a claim is made that only 40% of the population supports Carter. In mathematical terms, this hypothesis may be written $\pi = .40$. On the basis of the information in (1-3), we would reject this hypothesis, of course. In general, there is always this kind of close association between confidence intervals and hypothesis tests.

We can make several other crucial observations about (1-2):

1. The estimate is *not* made with certainty; we are only 95% confident. We must concede the 5% possibility that the draw turned

up a misleading sample—just as in flipping a coin 10 times, it is possible that 8 or 9 heads will occur.

2. As sample size n increases, we note that the error allowance in (1-2) shrinks. For example, if we increased our sample to 15,000 voters, and continued to observe a proportion of .48 for Carter, the 95% confidence interval would shrink to the more precise value:

$$\pi = .48 \pm .01 \tag{1-4}$$

This is also intuitively correct: A larger sample contains more information, and hence allows a more precise conclusion.

In conclusion, what does random sampling accomplish? It allows us to make an *unbiased* estimate of the unknown population—including a confidence interval that shows the uncertainty involved.

PROBLEMS

1-1 Ten days before the 1980 presidential election, a Gallup poll showed the following percentages supporting Carter. (As we mentioned already, treat the sample from each group as a random sample.)

men	49% (n = 600)	under 30	48% (n = 200)
women	58% (n = 600)	over 30	55% (n = 1000)

(a) For each group, calculate a 95% confidence interval for the population proportion supporting Carter.

(b) Star each case where you can conclude with 95% confidence that Carter had a majority (or minority).

1-2 Project yourself back in time to six recent U.S. presidential elections. In parentheses we give the results of the Gallup pre-election poll of 1500 voters (ignoring third-party candidates, as usual).

Year	*Democrat*	*Republican*
1960	Kennedy (51%)	Nixon (49%)
1964	Johnson (64%)	Goldwater (36%)
1968	Humphrey (50%)	Nixon (50%)
1972	McGovern (38%)	Nixon (62%)
1976	Carter (51%)	Ford (49%)
1980	Carter (48%)	Reagan (52%)

(a) In each case, construct a 95% confidence interval for the proportion of Democratic supporters in the whole population.

(b) Mark each case where the interval is wrong, that is, fails to include the true proportion π given in the following list of actual voting results:

1960	Kennedy	50.1%
1964	Johnson	61.3%
1968	Humphrey	49.7%
1972	McGovern	38.2%
1976	Carter	51.1%
1980	Carter	44.7%

1-3 Criticize each of the following sampling plans, pointing out the bias and suggesting how to reduce it.

(a) In order to predict the vote on a municipal bond, a survey selected every corner house and asked the housewife, if she was in, which way she intended to vote.

(b) To estimate the average income of its alumni five years after graduation, a university polled all the alumni who returned to their fifth reunion.

1-2 RANDOMIZED EXPERIMENTS

So far we have seen how randomization frees a sample of bias. In this section we will see how randomization similarly frees an experimental design of bias. We begin with a careful look at what constitutes good scientific evidence.

(a) Use of a Control Group

Medicine provides a good example, because the issues are clear and the state of the art quite advanced. In 1962 a new treatment for ulcers was reported: The patient swallows a balloon, into which a refrigerant liquid is then pumped, in order to freeze the stomach. The objective is to temporarily close down the digestive process and give the stomach a start on healing. The question is, how well does this freezing treatment work in practice? The innovator (Wangensteen, 1962) tried the method on 24 patients, and all of them were cured.

This suggests that the treatment is extraordinarily effective. It is just possible, however, that it has no effect at all; the patients might have

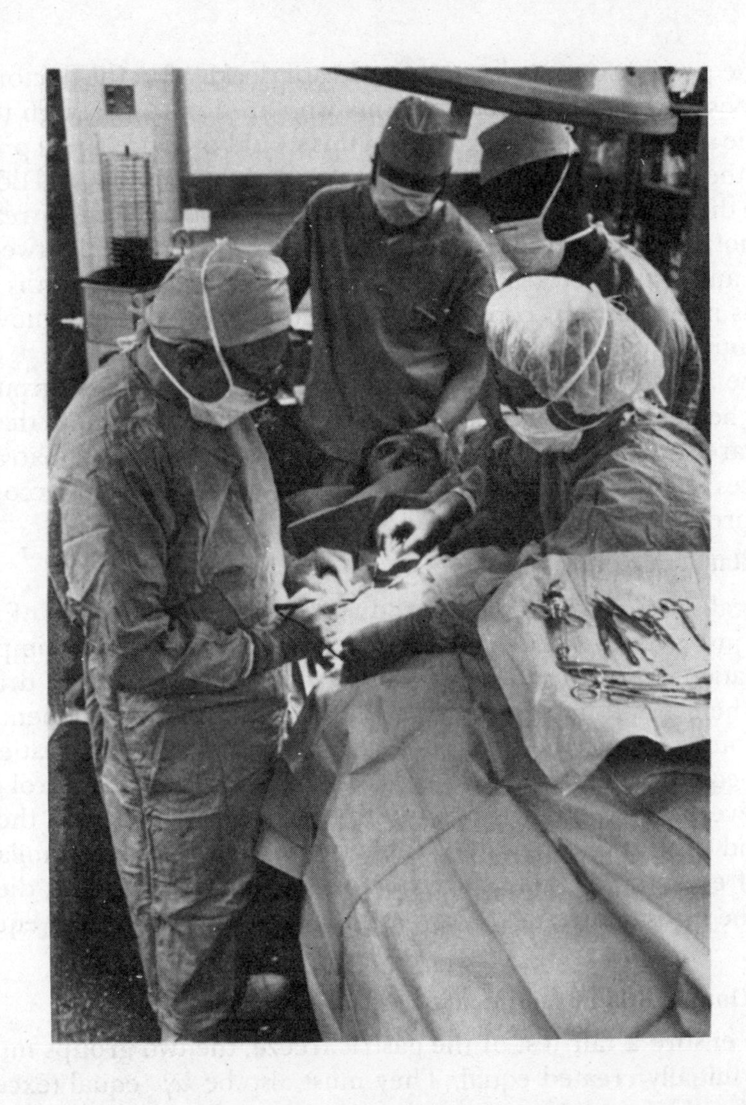

recovered without it. How do we distinguish between these alternative explanations? The answer is to compare the *treated* group of patients with an untreated (or *control*) group; if the treated group does better, then the treatment is effective. This leaves us with an important question that must be considered at the beginning of any such experiment: Given an initial common pool of patients, how does the researcher decide which individuals should be assigned to the treatment group, and which to the control group?

One possible answer is to let the doctor decide. But the doctor might well choose to treat just those patients who are healthy enough to withstand the stress of the treatment. And this would produce a real problem: Even if the treatment *were* worthless, the treatment group would do better because these patients were initially healthier. The effect of the treatment could not, after all, be judged by observing differences between the treated and untreated (control) group. In short, initial health is an extraneous influence that confounds the picture. How do we remove such extraneous influences?

The answer is to ensure that both the control and test groups are equally healthy. At the same time, we would like to ensure that both groups are similar in other respects; in other words, *all* such extraneous influences should be neutralized. The most effective way to accomplish this is through randomization.

(b) Randomization

Recall how randomization produced an unbiased sample of voters. This is just what we need in experimental design, too. We simply put every patient's name on a chip, stir the chips in a bowl, and draw out half of them at random. These patients will be given the treatment, while the rest are put into the control group. Then every healthy patient will have an equal chance of going into the treatment group or control group. So will every weak patient, of course. Thus, in terms of health, the treatment and control groups will be initially equal, on average. Similarly, in terms of every other relevant characteristic—such as age, sex, diet, and so on (the list is endless)—the two groups will again be initially equal, on average.

(c) Double-Blind Experiments

To ensure a fair test of the gastric freeze, the two groups must not only be initially created equal. They must also be *kept* equal (except, of course, for the fact that one is getting the treatment and the other is not). To see how they might miss being kept equal, suppose the doctor who finally evaluates the patients knows who has been treated and who has not; she might tend unconsciously to give a more favorable report to the treated patients (especially if it is a treatment she herself has developed). Consequently, even the doctor should be kept blind about who has been treated. So, of course, should the patients. (A patient who wants to please the doctor may tend to overstate the effectiveness of the cure.) That is, the experiment should be *double blind*.

To keep the patients in the dark, control patients can be given a *placebo* (dummy treatment) that cannot be distinguished from the real treatment. In the case of the gastric freeze, a very ingenious placebo was devised (Ruffin, 1969): Each control patient was subjected to an operation just like the gastric freeze, except that a bypass was put into the balloon to return the refrigerant liquid before it could freeze the stomach. Thus the doctors, as well as the patients, were kept blind. The only person to know was the statistician who flipped the coin and accordingly switched the bypass on or off.

When Ruffin put the gastric freeze to this double-blind and randomized test, the results were very interesting indeed: With 82 patients in the treatment group and 78 in the control group, he found:

> The results of this study demonstrate conclusively that the freezing procedure was no better than the (placebo) in the treatment of duodenal ulcer. . . . It is reasonable to assume that the relief of pain and subjective improvement reported by early investigators was probably due to the (positive, short-run) psychologic effect of the procedure.
>
> The importance of random assignment of patients to treatment and the double-blind method in clinical trials has been emphasized repeatedly, but these features are still too frequently ignored. Only by strict adherence to such principles and resisting the urge to publish until data have been gathered by these rigorous methods will false leads be kept to a minimum and erroneous conclusions avoided.

(d) Summary

Our conclusions so far may be stated in general. Suppose the statistician is unable to randomize, and can only observe the doctor's decision on who will get the treatment. Then the result is an *observational study* that is cluttered up with uncontrolled extraneous factors.

On the other hand, if the decision as to who will get the treatment is made in a deliberate, randomized way (e.g., by drawing names from a bowl, or flipping a coin for each patient), then good control is achieved on all the extraneous factors. If further, the patient and the doctors are kept blind about who is getting the treatment and who is not, the result is a *double-blind randomized experiment.* As shown in Figure 1-1, such experiments are the scientific ideal:

Randomized double-blind experiments ensure that on average the two groups are initially equal, and continue to be treated equally. Thus a fair comparison is possible.	(1-5)

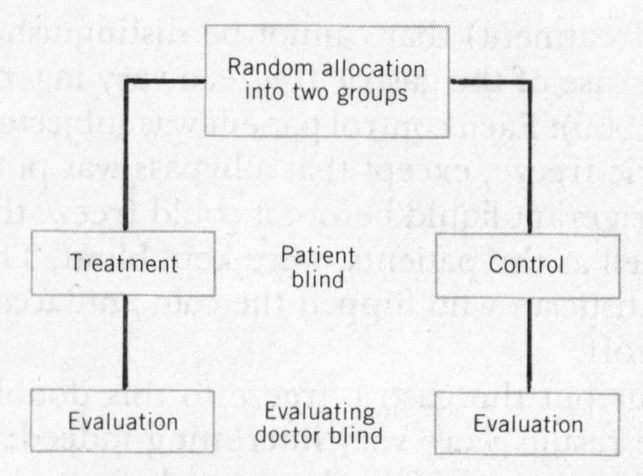

FIGURE 1-1 The logic of the double-blind randomized experiment.

(e) Some Ethical Issues

Is it ethical to experiment with people? Although this is often the way the question is stated, it can be very misleading. *Every* time a new medical procedure or social program is introduced, at some stage it *has* to be tried on people for the first time, that is, experimentally. So the real question is: Do we experiment carefully, or haphazardly? Do we experiment with randomized control and learn quickly, or do we continue to run poor experiments and thus subject people to unnecessary risks?

In fact, haphazard studies not only involve an opportunity lost, with unnecessary risks imposed on their subjects; in addition they may build up a store of misinformation that leads to a harmful treatment being given to many people in the future. Perhaps the ethical dilemma can best be summarized by quoting a surgeon (Peacock, 1972; via Tufte, 1974):

> One day when I was a junior medical student, a very important Boston surgeon visited the school and delivered a great treatise on a large number of patients who had undergone successful operations for vascular reconstruction. At the end of the lecture, a young student at the back of the room timidly asked, "Do you have any controls?" Well, the great surgeon drew himself up to his full height, hit the desk, and said, "Do you mean did I not operate on half of the patients?" The hall grew very quiet then. The voice at the back of the room very hesitantly replied, "Yes, that's what I had in mind." Then the visitor's fist really came down as he thundered, "Of course not. That would have doomed half of them to their death." God, it was quiet then, and one could scarcely hear the small voice ask, "Which half?"

1-3 RANDOMIZED EXPERIMENTS IN THE SOCIAL SCIENCES

(a) Sometimes Not Possible

So far we have concluded that a randomized experiment is strongly preferred to an observational study. But sometimes a randomized experiment is not possible. Illustrations abound, especially in the social sciences. For example, suppose we wish to determine whether sex affects college faculty salaries. Specifically, do women earn less, simply because they are women?

Suppose, for a sample of professors, that we could assign a sex to each at random (heads you're a women, tails you're a man), and then watch what happens to the salaries of women compared to men over the next 10 years. In this way we could remove the effect on salaries of extraneous influences, such as years an individual has been teaching. But, of course, we can't randomly assign sex. So we must fall back on an observational study, taking professors as they come.

(b) Sometimes Not Feasible

How might we determine the deterrent effect of capital punishment? We might get a conclusive answer if we could choose 25 of the 50 states at random and enforce capital punishment there; the remaining 25 states would be the control group, with a maximum penalty of life imprisonment, say. But would states ever agree to surrender their power over life and death for the sake of experimental science?

As another example, we might find the effect of interest rates on investment, if we could select 5 of the next 10 years at random, and assign a high interest rate. (The other five years would be the low-interest control years.) But could we, or should we, persuade the Federal Reserve to give up its control over interest rates (and thus over inflation and unemployment) simply to expand our experimental knowledge of the interest rate/investment relation?

How does a college education affect a person's life, on average? This question might be answered if we could find a volunteer sample of high school graduates about to enter college and randomly send half to college, while keeping out the other half as the control group. But who would volunteer, in the interests of furthering knowledge, to risk such a large part of their lives?

(c) Sometimes Not Done, even when Feasible

Experiments on some other educational issues may nevertheless be possible. For example, suppose we wish to evaluate the educational benefits of a free educational program for prekindergarten children (such as Headstart). Since there are many more applicants than places available (due to limited funds), selecting individuals at random would be a fair way of deciding which children would be allowed to participate. At the same time, such randomization would admirably suit the needs of a fair and valid scientific experiment. Unfortunately, randomization is seldom done, and so the value of such programs remains in dispute.

It is interesting to speculate on why randomization isn't done more often, even when it costs relatively little. Is it because some investigators just don't appreciate its importance? Or is it because some administrators cannot admit that a mere coin does a better job of assignment than they do?

Whatever the reason, randomization could be done more frequently; indeed, whenever practical, it should be undertaken. We cannot repeat this point often enough, since it is one of the most important ones we make in this book: *Randomized assignment ensures that an experiment is free of bias.*

1-4 REGRESSION

As already noted, the difficulty of randomization in the social sciences often means that there is no alternative but to take individuals as they come, in an observational study. But how can we reduce the bias in such a study, due to uncontrolled extraneous factors?

> We observe and record the extraneous factors. Then, instead of holding them constant by design (as we ideally would have done if we could have randomized), we analyze our data in a compensating way that gives us, insofar as possible, the same answer *as if we had held the extraneous factors constant.* The technical method used to accomplish this is called *multiple regression analysis,* or just *regression.* (1-6)

Although it does the best job possible under the circumstances, regression still cannot do a perfect job—it is just not possible to record or even identify the endless list of extraneous influences. So we must recognize that no method of analyzing an observational study—not even regression—can *completely* compensate for a lack of randomized control.

Actually, regression is very useful in randomized experiments as well as in observational studies—because it very powerfully describes, in a single equation, how one variable is related to others. For example, it can show how lung performance is related to a person's age, sex, smoking habits, and environmental hazards. In fact, regression is probably the single most important tool you will ever use, and so we will place heavy emphasis on it.

1-5 BRIEF OUTLINE OF THE BOOK

The poll we discussed was typical of the basic statistical analysis that will be covered in Chapters 2 to 6. This will allow us to scientifically estimate a wide variety of phenomena: the proportion of voters in a congressional district who are Democratic, the average income of workers in an industry, or the crime rate in a certain neighborhood. Each of these studies is relatively simple, since it measures just one variable.

In Chapters 7 to 11 we take up the even more interesting question of how one variable is related to several others. (To use our previous example, how is lung performance related to age, sex, etc?) It is in these chapters that we develop regression analysis, along with numerous applications.

The last three chapters, Chapters 12 to 14, are devoted to shorter, but nonetheless important, topics.

PROBLEMS

1-4 Give some historical examples of useless or even harmful "treatments" that persisted for many years, because they were not evaluated properly (e.g., drilling holes in people's heads to let the demons out).

1-5 Give some present-day examples of "treatments" that are perhaps useless or harmful, but still persist because they have not been evaluated well and nobody really knows their true effect (e.g., life imprisonment versus capital punishment).

In which cases would it be relatively easy to evaluate the treatment properly? How?

In which cases would it be extremely difficult to evaluate the treatment properly? How?

1-6 "The possession of such a degree as the MBA will double lifetime earnings as compared to those of a high school graduate." (From Bostwick, 1977, p. 7.)

Granted that MBA graduates really do earn twice as much as high school graduates on average, is this statement accurate? If so, elaborate. If not, correct it.

CHAPTER 1 SUMMARY

1-1 Statistics is a way to describe a whole population just by looking at a sample that is properly drawn from it. To avoid bias, the sample must be random. Then a confidence interval can be constructed, with an error allowance that shows the sampling uncertainty.

1-2 In an experiment to determine whether a treatment is effective, how can we avoid bias? We must use a random process to decide who gets the treatment and who gets left as a control. And anyone who might prejudice the results must be kept blind about who has received the treatment and who has not.

1-3 Randomized experiments are becoming more common in the social sciences now as well as in the life sciences. Yet much more could be done.

1-4 When there are many influences present, the appropriate statistical technique is regression (multiple regression). It is especially valuable when randomized experiments are not possible, when we have to be satisfied with an observational study that takes the data as it comes.

REVIEW PROBLEMS

These review problems included at the end of each chapter give an overview of the material so far. Because they do not fit neatly into a pigeonhole, they require more thought. Consequently, they provide the best preparation for meeting real life—and exams.

1-7 On the basis of a psychological test, a sample of 80 school children were classified as having high or low self-esteem. In addition, the parents were asked, among other things, whether or not the child was breast-fed. The 80 children were then classified as follows (Coopersmith, 1967):

	Bottle-fed	*Breast-fed*
High self-esteem	11	22
Low self-esteem	25	22

To what extent do you think this data shows that "breast-feeding has a negative effect on a child's later self-esteem"?

1-8 George Box, one of the world's foremost statisticians, has said, "To find out what happens to a system when you interfere with it, you have to actually interfere with it, not just passively observe it." In the context of this chapter, what does this mean?

CHAPTER 2

Descriptive Statistics

The average statistician is married to 1.75 wives who try their level best to drag him out of the house 2¼ nights a week with only 50 percent success.

He has a sloping forehead with a 2 percent grade (denoting mental strength), ⅝ of a bank account, and 3.06 children who drive him ½ crazy; 1.65 of the children are male.

Only .07 percent of all statisticians are ¼ awake at the breakfast table where they consume 1.68 cups of coffee—the remaining .32 dribbling down their shirt fronts. . . . On Saturday nights, he engages ⅓ of a baby-sitter for his 3.06 kiddies, unless he happens to have ⅝ of a mother-in-law living with him who will sit for ½ the price. . . .

W. F. Miksch (1950)

In Chapter 1 we discussed how a random sample allows us to make a scientific inference about the underlying population. The sample must first be simplified, however, and described by one or two summary numbers; each is called a *statistic.*

For instance, in the Gallup poll of Example 1-1, the polltaker would record the answers of the 1500 people in the sample, obtaining a sequence such as C,R,R,C,C,R, . . . , where C and R represent Carter and Reagan supporters. An appropriate summary of this sample is the statistic P, the sample proportion of Carter supporters; this can be used to make an inference about π, the population proportion. Admittedly, this statistic P is trivial to compute. It required merely counting the number of C's and then dividing by the sample size, obtaining $P = 720/1500 = .48$.

Now we will take a look at some other samples, and calculate the appropriate statistics to summarize them.

2-1 FREQUENCY TABLES AND GRAPHS

(a) Discrete Example

In a sample of 50 completed American families, let us record the number of children X in each family. We call X a *discrete* variable because its possible values progress in steps (0, 1, 2, 3, . . .) rather than continuously. Suppose the 50 values of X turn out to be

$$3, 2, 2, 0, 5, 1, 4, 0, \ldots, 7, 2$$

To simplify, we keep a running tally of the possible outcomes in Table 2-1. In the third column we record, for example, that a zero-child family was observed 9 times; the *relative* frequency is $9/50 = .18$ (or 18%), and is recorded in the final column.

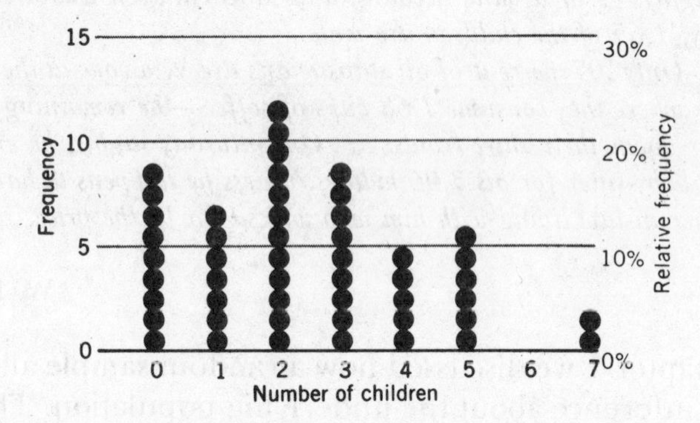

FIGURE 2-1 Frequency and relative frequency distribution of the number of children (sample of 50 completed American families).

TABLE 2-1

Calculation of the Frequency and Relative Frequency of the Number of Children (Sample of 50 completed American Families)

Number of Children	*Tally of 50 Families*	*Frequency* (f)	*Relative Frequency* $\left(\frac{f}{n}\right)$
0	𝍸 ||||	9	.18
1	𝍸 ||	7	.14
2	𝍸 𝍸 ||	12	.24
3	𝍸 ||||	9	.18
4	𝍸	5	.10
5	𝍸 |	6	.12
6		0	0
7	||	2	.04
		Sum = 50 = n	Sum = 1.00 ✓

The information in the third column is called the *frequency distribution,* and is graphed in Figure 2-1. The *relative frequency distribution* in the last column can be shown on the same graph, using a different vertical scale on the right. To emphasize the meaning of this graph, we have represented each family by a dot. Thus the leftmost 9 dots represent the 9 families having zero children, and so on.

(b) Continuous Example

Suppose we sample 200 American men's height. We call height X a *continuous* variable, since its possible values vary continuously (X could be 64, 65, 66, . . . , or *anywhere in between,* such as 64.328··· inches). It no longer makes sense to talk about the frequency of a specific value of X, since we will never again observe anyone exactly 64.328··· inches tall. Instead we can tally the frequency of heights within a cell (such as 58.5 to 61.5 inches) as in Table 2-2.

The cells have been chosen somewhat arbitrarily, but with the following conveniences in mind:

1. The number of cells is a reasonable compromise between too much detail and too little. Usually 5 to 15 cells is appropriate.
2. Each cell midpoint, which hereafter will represent all observations in the cell, is a convenient whole number.

TABLE 2-2

Frequency and Relative Frequency of the Heights of 200 Men

Cell Boundaries	*Cell Midpoint*	*Tally*	*Frequency* (f)	*Relative Frequency* $\left(\frac{f}{n}\right)$
58.5–61.5	60	||||	4	.02
61.5–64.5	63	~~||||~~ ~~||||~~ ||	12	.06
·	66	·	44	.22
·	69	·	64	.32
·	72	·	56	.28
	75		16	.08
76.5–79.5	78	||||	4	.02
			Sum = 200 = n	Sum = 1.00

Figure 2-2 illustrates the grouping of the 200 heights into cells. In panel (a), the 200 men are represented by 200 tiny dots strung out along the X axis. The grouped data can then be graphed in panel (b). (Note how this is simply a graph of the frequency distribution in the second-to-last column of Table 2-2.) Bars instead of lines are used to represent frequencies as a reminder that the observations occurred throughout the cell, and not just at the midpoint. Such a graph is called a *bar diagram* or *histogram*.

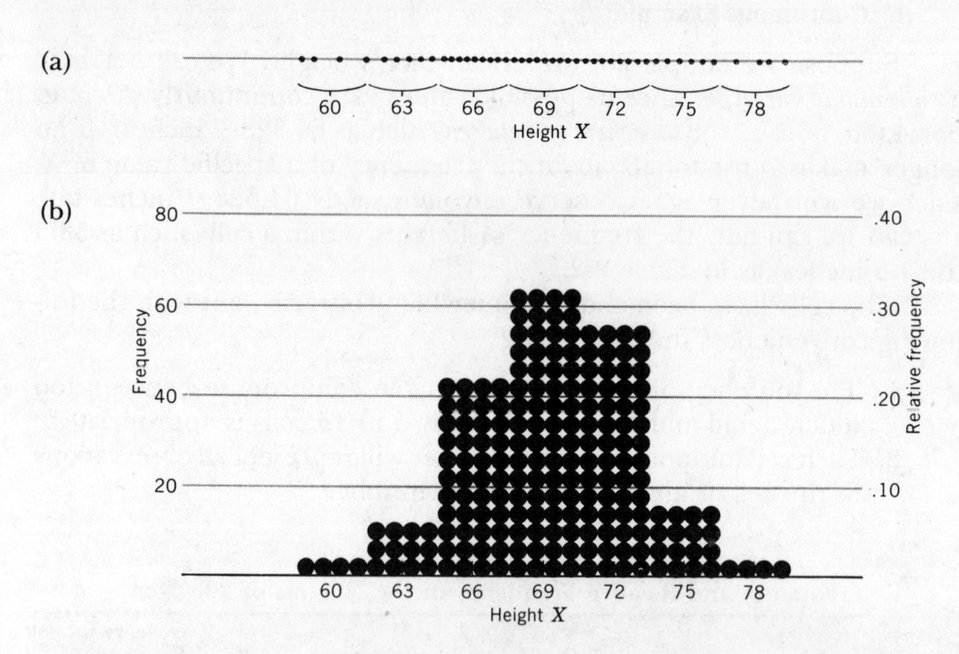

FIGURE 2-2 *(a)* The grouping of observations into cells, illustrating the first two columns of Table 2-2. *(b)* The bar graph for the grouped data.

Once more, to emphasize the meaning of this graph, we have represented each observation (man's height) by a dot. Thus the rightmost 4 dots represent the 4 tallest men, and so on. To be even more concrete, Figure 2-3 illustrates how a bar graph can be constructed by lining up actual people in a photograph, from shortest to tallest. This emphasizes how *the size of each bar represents the number of individuals.*

A bar graph can be extremely useful in showing relative position. For example, returning to Figure 2-2b, what can we say about a man 64.5 inches tall (on the right-hand boundary of the second cell)? He is relatively small, with only a few men smaller than he. To be specific, only 8% are

FIGURE 2-3 A living bar graph (from Joiner, 1975).

smaller. (From the last column of Table 2-2, we add up the first two cells: 2% + 6% = 8%.) His height is therefore said to be the 8th *percentile.* At the other end, a height of 73.5 inches would be the 90th percentile.

We next turn to the question of how to summarize the frequency distribution with one or two simple numbers (statistics) that measure the center and the spread of the distribution. These concepts will be illustrated with the continuous distribution of men's heights. For discrete distributions—such as family size—these concepts remain just the same.

PROBLEMS

In many of the exercises based on real data, such as Problem 2-1 or 2-2, we have tried to preserve the reality yet keep the arithmetic manageable by using the following device: The available data usually is a whole population, or at least a very large sample. We construct a small sample that reflects the population as closely as possible.

2-1 In a large American university in 1969, a random sample of 5 women professors gave the following annual salaries (in thousands of dollars, rounded. From Katz, 1973):

9, 12, 8, 10, 16

Without sorting into cells, graph the salaries as dots on the X-axis.

2-2 At the same university, a random sample of 25 men professors gave the following annual salaries (in thousands of dollars, rounded):

12	11	19	11	22	22	13	11	17	12
20	14	17	14	24	26	9	12	19	18
21	21	11	9	15					

(a) Without sorting into cells, graph the salaries as dots on the X axis. Be sure to use the same scale as in Problem 2-1, so you can compare the graphs of the men and women. In your view how good is the evidence that, over the whole university, men tend to earn more than women? (This issue will be answered more precisely in Chapter 5.)

(b) Now sort the data into cells with midpoints of 10, 15, 20, and 25, and graph it.

2-3 In Problem 2-2:

(a) What percentile is a salary of 17.5 thousand dollars? And a salary of 25 thousand dollars?

(b) What salary is the 50th percentile, that is, the middle observation, with 12 below and 12 above?

2-2 CENTER OF A DISTRIBUTION

There are many different ways to define the center of a distribution. We will discuss the three most popular—the mode, the median, and the mean—starting with the simplest.

(a) The Mode

Since *mode* is the French word for fashion. the mode of a distribution is defined as the most frequent (fashionable) value. In the example of men's heights, the mode is 69 inches, since this cell has the greatest frequency, or longest bar in Figure 2-2. That is, the mode is where the distribution peaks.

Although the mode is very easy to obtain—at a glance—it is not a very good measure of central tendency, since it often depends on the arbitrary grouping of the data. It is also possible to draw a sample where the largest frequency occurs twice (or more). Then there are two peaks, and the distribution is called *bimodal*.

(b) The Median

The *median* is the 50th percentile, that is, the value below which 50% of the observations fall. Since it splits the distribution into halves, it is sometimes called the middle value. In the sample of 200 detailed heights shown in Figure 2-2*a*, the median may be estimated by reading off the 100th value from the left, 69.4 inches, say. (Actually, there are two middle values, the 100th and 101st. So the median is defined halfway between them. In a sample with an odd number of observations, this ambiguity does not arise. For example, in a sample of 5 observations, the median would be the 3rd highest, leaving 2 below and 2 above).

If the only information available is the grouped frequency distribution in Figure 2-2*b*, the median can only be approximated by choosing an appropriate value within the median cell. The simplest solution is to use the cell midpoint, 69 inches. More accurate approximations can be found elsewhere (Wonnacott, 1977, for example).

(c) The Average or Mean, $\overline{X}$

The word average is derived from the Arabic root *awar* meaning damaged goods. Even today, in marine law, average means the equitable division of loss among the interested parties. For example, suppose 4 shippers suffered losses of \$90, \$50, \$0, and \$20. Then their average loss is found by dividing the total loss among them equally:

$$\text{Average} = \frac{90 + 50 + 0 + 20}{4} = \frac{160}{4} = 40 \qquad \text{(2-1)}$$

To generalize this, suppose a sample of n observations is denoted by $X_1, X_2, \ldots, X_n$. Then the average $\overline{X}$ is found by summing them and dividing by the sample size n:

$$\overline{X} \equiv \frac{1}{n}(X_1 + X_2 + \cdots + X_n) \qquad \text{(2-2)}$$

where $\equiv$ is a symbol meaning *equals, by definition.* We customarily abbreviate the sum of all the X values by ΣX, where Σ is sigma, the Greek equivalent of our S (as in Sum). Thus (2-2) can be written briefly as

$$\boxed{\overline{X} \equiv \frac{1}{n} \Sigma X} \tag{2-3}$$

The average for the sample of heights could be computed by summing all 200 observations and dividing by 200. However, this tedious calculation can be greatly simplified by using the grouped data in Table 2-3. Let us denote the first cell midpoint by X_1 and use it[1] to approximate all the observations in the first cell (f_1 in number). Similar approximations hold for all the other cells too, so that

$$\overline{X} \simeq \frac{1}{n}\left[\underbrace{(X_1 + X_1 + \cdots + X_1)}_{f_1 \text{ times}} + \underbrace{(X_2 + X_2 + \cdots + X_2)}_{f_2 \text{ times}} + \cdots\right] \tag{2-4}$$

$$\overline{X} \simeq \frac{1}{n}\left[X_1 f_1 + X_2 f_2 + \cdots\right] \tag{2-5}$$

where $\simeq$ means "approximately equals". In brief notation,

$$\boxed{\text{For grouped data, } \overline{X} \simeq \frac{1}{n} \Sigma X f} \tag{2-6}$$

Note that this formula is just like (2-3), except that we are careful to count every X with the frequency f that it occurs. Using (2-6), we calculate the mean height in Table 2-3, in the first three columns (the last three columns will be explained later).

---

[1]Note that X_1 is now used in a different sense than in equation (2-2), where it meant the first observation instead of the lowest cell midpoint. This ambiguity will cause no difficulty, since the context will make clear which meaning is intended.

TABLE 2-3
Calculation of Mean and Standard Deviation of Men's Heights

Given Data		*Calculation of $\overline{X}$*	*Calculation of s (Using $\overline{X}$ Rounded to 69)*		
X	f	Xf	$(X-\overline{X})$	$(X-\overline{X})^2$	$(X-\overline{X})^2 f$
60	4	240	−9	81	324
63	12	756	−6	36	432
66	44	2904	−3	9	396
69	64	4416	0	0	0
72	56	4032	3	9	504
75	16	1200	6	36	576
78	4	312	9	81	324
$n = 200$		$\overline{X} = \frac{13{,}860}{200} = 69.3$			$s^2 \simeq \frac{2556}{199} \simeq 12.84$; $s \simeq \sqrt{12.84} \simeq 3.58$

(d) The Mean Interpreted as the Balancing Point

The 200 heights appeared in Figure 2-2*a* as points along the X axis. If we think of each observation as a one-pound mass, and the X axis as a weightless supporting rod, we might ask where this rod balances. Our intuition suggests "the center."

The precise balancing point, also called the center of gravity, is given by the physics formula:

$$\frac{1}{n} \sum X \tag{2-7}$$

which is exactly the same as the formula for the mean. Thus we may think of the sample mean as the "balancing point" of the data, symbolized by ▲ in our graphs.

(e) Comparison of Mean, Median, and Mode

The three measures of center are compared in Figure 2-4*a*, where we show a distribution that has a single peak and is symmetric (i.e., one half is the mirror image of the other). In this case the mode, median, and mean all coincide.

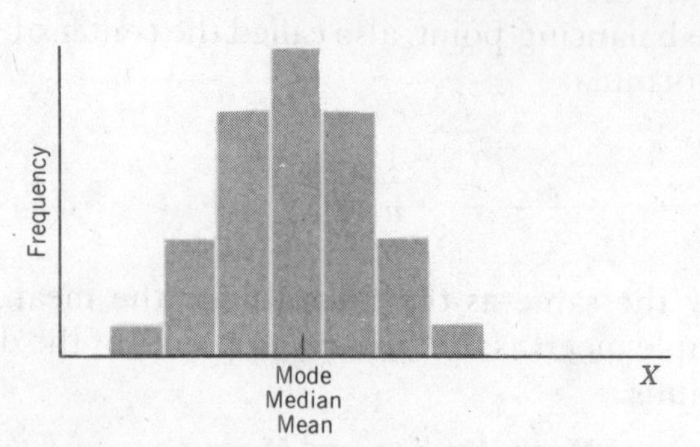

FIGURE 2-4 *(a)* A symmetric distribution with a single peak. The mode, median, and mean coincide at the point of symmetry.

But what if the distribution is skewed? For example, Figure 2-4*b* shows a long tail to the right. Then the three measures differ. To see why, let us ask whether the median, for example, will coincide with the

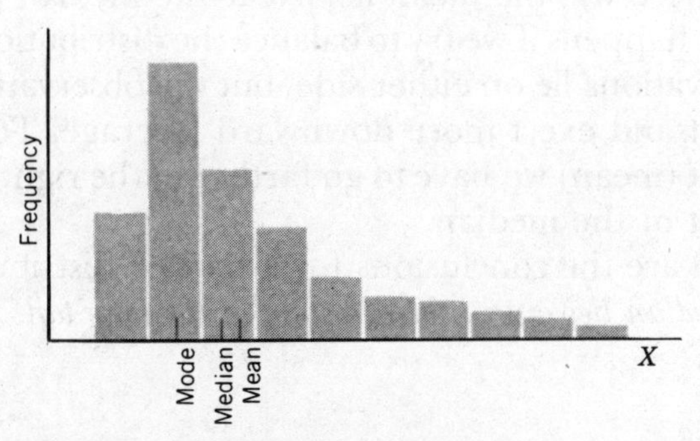

FIGURE 2-4 *(b)* A right-skewed distribution, showing the median and the mean out in the direction of the long tail.

mode? With so many observations strung out in the right-hand tail, we have to move from the peak value toward the right in order to pick up half the observations. Thus the median is to the right of the mode.

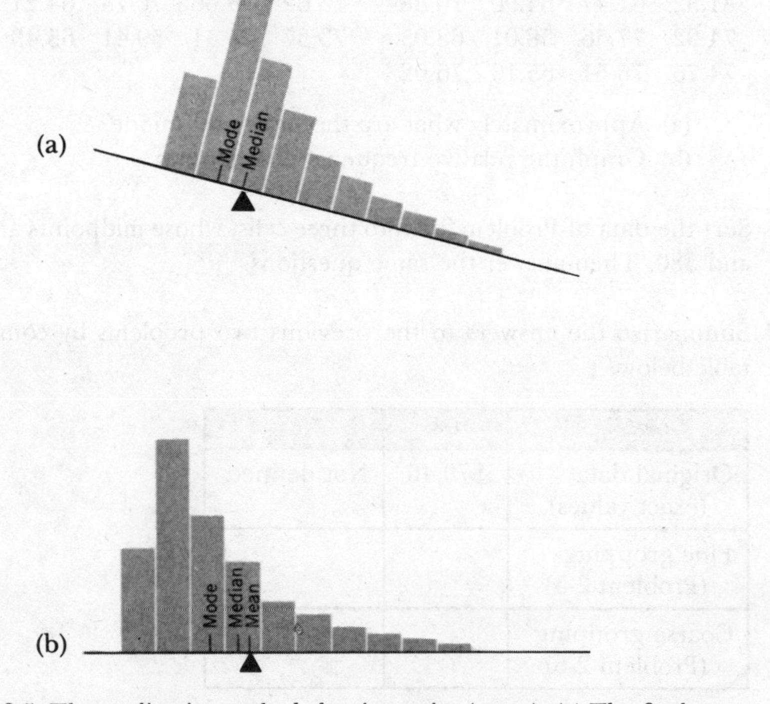

FIGURE 2-5 The median is not the balancing point (mean). *(a)* The farther-out observations tip the attempted balance at the median. *(b)* The actual balancing point (mean) is therefore to the right.

Finally, where will the mean lie? Near the median perhaps? Figure 2-5 shows what happens if we try to balance the distribution at the median. Half the observations lie on either side, but the observations on the right are farther out and exert more downward leverage. To find the actual balancing point (mean) we have to go farther to the right. Thus the mean lies to the right of the median.

What then are the conclusions for a skewed distribution? *Relative to the mode, the median lies out in the direction of the long tail. And the mean lies even farther out.*

PROBLEMS

2-4 For Problem 2-1 and Problem 2-2(*a*) and (*b*), calculate the mean and mark it on the graph.

2-5 Sort the following daily profit figures of 20 newsstands into five cells, whose midpoints are $60, $65, $70, $75, and $80:

$81.32	61.47	64.90	70.88	76.02	75.06	76.73	64.21
74.92	77.56	58.01	68.05	73.37	75.41	59.41	65.43
74.76	76.51	65.10	76.02				

(a) Approximately what are the mean and mode?
(b) Graph the relative frequency distribution.

2-6 Sort the data of Problem 2-5 into three cells, whose midpoints are $60, $70, and $80. Then answer the same questions.

2-7 Summarize the answers to the previous two problems by completing the table below.

Grouping	*Mean*	*Mode*
Original data (exact values)	$70.46	Not defined
Fine grouping (Problem 2-5)		
Coarse grouping (Problem 2-6)		

(a) Why is the mode not a good measure?

(b) Which gives a closer approximation to the true ungrouped mean: the coarse or the fine grouping?

2-8 In Problem 2-5:

(a) How good is a profit of $72.50? Answer by stating what percentile it is.

*(b)[2] What is the median profit?

2-9 In the following distribution, the mean, median, and mode are among the five little marks on the X axis. Pick them out—without resorting to calculation.

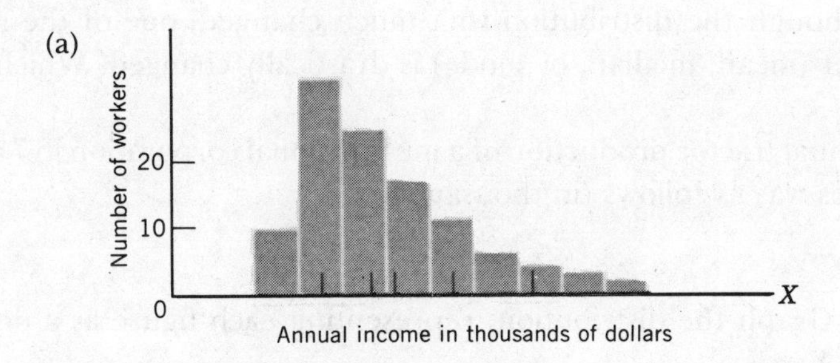

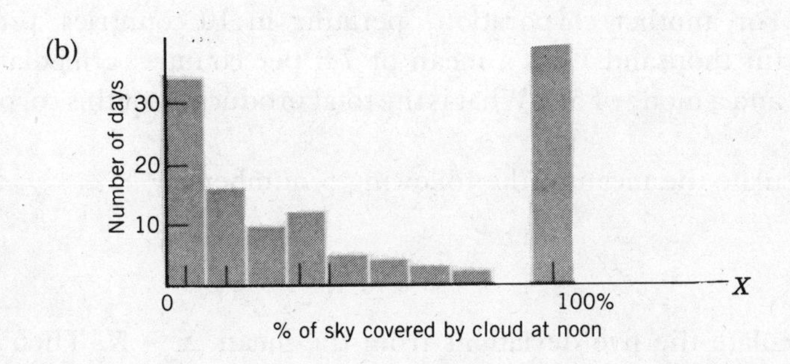

[2]As we mentioned in the preface, a star next to a problem (or section) indicates that it is more difficult. Although it provides an interesting challenge for the advanced student, it is optional.

2-10 In Problem 2-9(*a*), suppose 2 zeros were mistakenly added to one of the incomes, making it 100 times too large. How would this affect the mean and median? Underline the correct choice in each of the following square brackets:

(a) The [mean, median] would erroneously be increased substantially.

(b) The [mean, median] would be changed very little, or not at all, depending on whether the one mutilated observation was originally below or above the [mean, median].

(c) If a measure of center is not sensitive to outlier observations, we call it *robust.* That is, the [mean, median] is robust.

2-11 In Problem 2-9, suppose the rightmost bar in part (b) was reduced slightly, to 3/4 of its present value.

Although the distribution isn't much changed, one of the measures of center (mean, median, or mode) is drastically changed. Which one?

2-12 The annual tractor production of a multinational corporation in 7 different countries was as follows (in thousands):

6, 8, 6, 9, 11, 5, 60

(a) Graph the distribution, representing each figure as a dot on the X axis.

(b) What is the total production? The mean? The median? The mode? Mark these on the graph.

(c) For another corporation operating in 10 countries, production (in thousands) has a mean of 7.8 per country, a median of 6.5, and a mode of 5.0. What is the total production of this corporation?

2-13 (a) Calculate the mean of the following 5 numbers:

3, 7, 8, 12, 15

(b) Calculate the five deviations from the mean, $X - \overline{X}$. Then calculate the average of these deviations.

(c) Write down any set of numbers. Calculate their mean, and then the average deviation from the mean.

*(d) Prove that, for every possible sample, the average deviation from the mean is exactly zero. Is this equally true for deviations from the median?

2-3 SPREAD OF A DISTRIBUTION

Although average height may be the most important single statistic, it is also important to know how spread out or varied the observations are. As with measures of center, we find that there are several measures of spread. We start again with the simplest.

(a) The Range

The range is simply the distance between the largest and smallest value:

$$\text{Range} \equiv \text{largest} - \text{smallest observation}$$

For men's heights in Figure 2-2, the range is about 79.5 − 58.5 = 21. It may be fairly criticized on the grounds that it tells us nothing about the distribution except where it ends. And using only these two end observations may be unreliable. We therefore turn to measures of spread that take account of all the observations.

(b) Mean Squared Deviation (MSD)

To get an overall measure of variability, let us look at all the deviations from the mean, $X - \overline{X}$. Some are positive, and some negative, as Figure 2-6 illustrates. On the whole, their average is 0, which is a useless measure of spread.

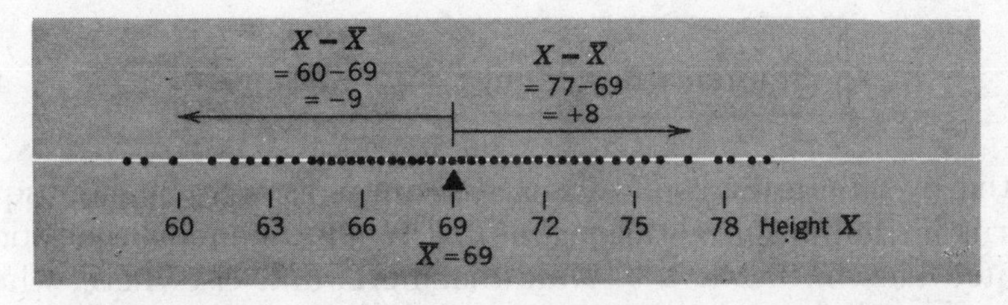

FIGURE 2-6 Two typical deviations from the mean $(X - \overline{X})$.

In order to get a worthwhile measure of spread, therefore, we first *square* the deviations to make them all positive numbers, so that it then makes sense to average. When the deviations are squared and averaged this way, the result is the average (or mean) squared deviation:

$$\text{Mean squared deviation, MSD} \equiv \frac{1}{n}\sum (X - \overline{X})^2 \tag{2-8}$$

For grouped data, we must as usual modify this formula, by counting each squared deviation with the frequency f that it occurs. Thus (2-8) becomes

$$\text{For grouped data, MSD} \simeq \frac{1}{n}\sum (X - \overline{X})^2 f \tag{2-9}$$

(c) Variance and Standard Deviation

For certain technical reasons, it is customary in (2-8) to use the divisor $n - 1$ instead of n. This gives a slightly different measure of spread, the variance:

$$\boxed{\text{Variance, } s^2 \equiv \frac{1}{n-1}\sum (X - \overline{X})^2} \tag{2-10}$$

$$\boxed{\text{For grouped data, } s^2 \simeq \frac{1}{n-1}\sum (X - \overline{X})^2 f} \tag{2-11}$$

Finally we define

$$\boxed{\text{Standard deviation, } s \equiv \sqrt{\text{variance}}} \tag{2-12}$$

Note that by taking the square root, we compensate for having squared the terms in defining the variance in (2-10). The standard deviation s is calculated back in Table 2-3. Note that here, and elsewhere, when we calculate $\overline{X}$ or s from grouped data, the divisor n is the *sample size,* not the number of rows in the table.

We note that the various deviations $(X - \overline{X})$ in Table 2-3 vary in magnitude from 0 up to 9. The standard deviation s is 3.58, which is smaller than some of the deviations and larger than others. Thus s is, in a sense, a typical deviation. Since this provides a rough check on our arithmetic, as well as an intuitive meaning for s, we state it in general:[3]

The standard deviation s lies somewhere between the smallest and largest deviations $(X - \overline{X})$.	(2-13)

In conclusion, the sample mean $\overline{X}$ is the most common measure of center, and the sample standard deviation s is the most common measure of spread. We refer to $\overline{X}$ and s^2 as the first and second *moments* of the sample.

(d) Degrees of Freedom (d.f.)

The MSD was a good measure of spread, provided we only want to describe the sample. But typically we want to go one step further, and make a statistical inference about the population. For this purpose the sample variance is better, as the following intuitive argument indicates.

If only $n = 1$ observation were available, we could calculate the mean; but there would be no way of calculating the spread. Only to the extent that n exceeds 1 can we get information about the variance. That is, there are essentially only $(n - 1)$ pieces of information for the variance, and this is the appropriate divisor.

Customarily, pieces of information are called *degrees of freedom* (d.f.), and our argument is summarized as

One d.f. is used for the mean, leaving $(n - 1)$ d.f. for the variance.	(2-14)

***(e) Rounding Error**

In Table 2-3, when we set out to calculate s^2, we rounded $\overline{X}$ from 69.3 down to 69, in order to avoid decimals (and thus introduced a rounding error $e = .3$). Rounding makes good sense for two reasons:

1. In practice, samples are usually large and calculations are done by computer. The point of going through a hand calculation such

[3]Both (2-13) and (2-15) would be exactly true (theorems) if s^2 had a divisor of n. (Even with s^2 having a divisor of $n - 1$, however, they are still valid for every distribution we shall encounter.)

as Table 2-3 is *not* to make you a calculating wizard, but instead, to give you a feeling for what the variance means. And for this, rough figures will do.

2. The rounding error in $\overline{X}$ produces an error in s^2 that is surprisingly small. Specifically, if we let e denote the error in rounding off the mean, then

$$\text{The variance is overestimated by only } e^2 \tag{2-15}$$

This overestimation is usually so small that we will ignore it. For example, in rounding $\overline{X}$ from 69.3 to 69 in Table 2-3, the rounding error e is .3, and e^2 is only .09. Thus the variance $s^2 = 12.84$ is overestimated by only .09, which doesn't really matter. (And if it is ever required, the correct value is easily obtained by subtraction: $12.84 - .09 = 12.75$.)

PROBLEMS

In each of the following problems, check that the calculated value of s satisfies the "typical deviation" rule (2-13).

2-14 Compute the standard deviation of the following sample of 5 women professors' salaries (same data as Problem 2-1):

9, 12, 8, 10, 16

2-15 Compute the standard deviation of the following sample of 25 men professors' salaries (same grouped data as Problem 2-2):

X	f
10	9
15	6
20	8
25	2

2-16 For the sample below of 20 daily profit figures (same grouped data as Problem 2-6), calculate:

(a) The mean squared deviation.
(b) The variance and standard deviation.

X	f
60	5
70	7
80	8

2-4 CALCULATIONS USING RELATIVE FREQUENCIES

Sometimes original data is not available and only a summary is given, in the form of the relative frequency distribution. From this, how can we calculate $\overline{X}$ and s?

To derive the appropriate formula for $\overline{X}$, recall that for grouped data,

$$\overline{X} \simeq \frac{1}{n}(X_1 f_1 + X_2 f_2 + \dots) \qquad \text{(2-16)}$$

(2-5) repeated

$$= X_1\left(\frac{f_1}{n}\right) + X_2\left(\frac{f_2}{n}\right) + \dots$$

$$\overline{X} \simeq \Sigma X\left(\frac{f}{n}\right) \qquad \text{(2-17)}$$

like (2-6)

That is, $\overline{X}$ is just the sum of the cell midpoints X, weighted according to the relative frequencies f/n. In the same way, it would be easy to show that the MSD can also be calculated from the relative frequencies:

$$\text{MSD} \simeq \Sigma (X - \overline{X})^2 \left(\frac{f}{n}\right) \qquad \text{(2-18)}$$

like (2-9)

Finally, by comparing (2-11) with (2-9), we can easily calculate s^2 from the MSD:

$$\text{Variance, } s^2 = \left(\frac{n}{n-1}\right)\text{MSD} \qquad \text{(2-19)}$$

To show how these formulas work, in Table 2-4 we use them to calculate $\overline{X}$ and s. Note that the answers agree, of course, with the previous answers in Table 2-3.

TABLE 2-4

Calculation of Mean and Standard Deviation from the Relative Frequency Distribution (n = 200 observations. Compare to Table 2-3)

Given Relative Frequency		*Calculation of $\overline{X}$ Using (2-17)*	*Calculation of MSD Using (2-18) ($\overline{X}$ Rounded to 69)*		
X	$\frac{f}{n}$	$X\left(\frac{f}{n}\right)$	$(X - \overline{X})$	$(X - \overline{X})^2$	$(X - \overline{X})^2\left(\frac{f}{n}\right)$
60	.02	1.20	−9	81	1.62
63	.06	3.78	−6	36	2.16
66	.22	14.52	−3	9	1.98
69	.32	22.08	0	0	0
72	.28	20.16	3	9	2.52
75	.08	6.00	6	36	2.88
78	.02	1.56	9	81	1.62
	1.00✓	$\overline{X} = 69.30$	MSD $\simeq$ 12.78 by (2-19), $s^2 \simeq \left(\frac{200}{199}\right) 12.78 \simeq 12.84$ $s \simeq \sqrt{12.84} \simeq 3.58$		

PROBLEM

2-17 Compute the mean and standard deviation of the following sample of 25 salaries (same grouped data as in Problem 2-15)

Salary (Midpoint of Cell)	*Relative Frequency*
10	.36
15	.24
20	.32
25	.08
	1.00

CHAPTER 2 SUMMARY

2-1 Masses of data can be summarized in a frequency distribution, shown as a table or graph.

2-2 To define the center of a distribution, the most common single measure is the mean:

$$\overline{X} = \frac{1}{n} \Sigma X f \qquad \text{like (2-6)}$$

2-3 To define the spread of a distribution, the most common single measure is the standard deviation:

$$s = \sqrt{\frac{1}{n-1} \Sigma (X - \overline{X})^2 f} \qquad \text{like (2-11)}$$

2-4 The mean and standard deviation can equally well be calculated from the *relative* frequency distribution.

REVIEW PROBLEMS

2-18 The 92 supreme court justices whose terms ended before 1976 had the following lengths of service on the bench:

Length of Service	*Midpoint*	*Frequency*
0 to 5 years	2.5	12
5 to 10 years	7.5	22
10 to 15 years	12.5	11
15 to 20 years	17.5	15
20 to 25 years	22.5	12
25 to 30 years	27.5	10
30 to 35 years	32.5	9
35 to 40 years	37.5	1
	Total	92

Calculate the average length of service, and the standard deviation.

2-19 Suppose that the disposable annual incomes of the 5 million residents of a certain country had a mean of $4800 and a median of $3400.

(a) What is the disposable income of the whole country?

(b) What can you say about the shape of the distribution of incomes?

2-20 In the 1970s, the U.S. unemployment rate was as follows:[4]

Year	*Unemployment*
1970	4.9%
1971	5.9
1972	5.6
1973	4.9
1974	5.6

Year	*Unemployment*
1975	8.5%
1976	7.7
1977	7.0
1978	6.0
1979	5.7

Calculate the average unemployment rate, and the standard deviation:

(a) For first 5 years, 1970–1974.

(b) For the last 5 years, 1975–1979.

(c) For all 10 years. How is this answer related to the answers in (a) and (b)?

2-21 The U.S. annual inflation rate in the 1970s was as follows:

Year	*Inflation*
1970	5.9%
1971	4.3
1972	3.3
1973	6.2

Year	*Inflation*
1974	11.0%
1975	9.1
1976	5.8
1977	6.5
1978	7.7
1979	10.9

Calculate the average annual inflation rate:

(a) For the first 4 years, 1970–1973 (before OPEC quadrupled oil prices).

(b) For the next 6 years, 1974–1979.

(c) For all 10 years.

*(d) Suppose the original data was lost, and only the 4- and 6-year averages in (a) and (b) were available. Would it still be possible to calculate the overall 10-year average? How?

---

[4]The data for Problems 2-20, 2-21, and 2-22 is taken from the Statistical Abstract of the United States (Bureau of the Census, U.S. Dept. of Commerce). This is also the source for many other Problems throughout the book that quote a specific year.

2-22 In 1975 the world's population was 4 billion, with the following approximate breakdown. What was the growth rate for the world as a whole?

Region	*Population (millions)*	*Annual Growth Rate (%)*
Europe	500	.5%
N. America	240	.8
Russia	260	.9
Asia	2300	2.1
S. America	300	2.6
Africa	400	2.8
World	4000	?

CHAPTER 3

Probability Distributions

Let us see if there is justice upon the earth, or if we are ruled by chance.

Sir Arthur Conyn Doyle (Sherlock Holmes)

3-1 PROBABILITIES FOR DISCRETE RANDOM VARIABLES

To introduce random variables, we will use some familiar examples. As always, these are meant to be *active* learning devices: Solve them yourself first, and only then should you look at the solutions to see what further you can learn.

Example 3-1

Throw a fair die 50 times. Or simulate this by consulting the random numbers in Appendix Table I at the back of the book, disregarding the digits 0, 7, 8, and 9. Since the remaining digits 1, 2, . . . , 6 will of course still be equally likely, they will provide an accurate simulation of a die.

Graph the relative frequency distribution:

(a) After 10 throws.

(b) After 50 throws.

(c) After zillions of throws (guess).

Solution

Since we don't have a die at hand, we will simulate (it's also faster). Although it would be best to start at a random spot in Table I, we will start at the beginning so that our work will be easy to follow. The first few digits in Table I are

3~~9~~65~~7~~64545 1~~9~~~~9~~~~0~~6~~9~~ . . .

with the irrelevant digits stroked out.

The information in the first sample ($n = 10$ throws) is assembled in Table 3-1, with the relative frequencies (f/n) calculated in the last column and reproduced in Table 3-2 in column (a). The second sample (with n increased to 50 throws) is shown in column (b). The emerging pattern begins to confirm our expectations: In the long run, a fair die will come up equally often on every face (in fact, this is the very definition of fairness), and so we don't have to actually carry out the zillions of throws for part (c). These three relative frequency distributions are displayed in graph form in Figure 3-1.

TABLE 3-1

Relative Frequency Distribution for a Die, using a Sample of $n = 10$ Observations

Number of Dots (X)	*Tally*	*Frequency* (f)	*Relative Frequency* $\left(\frac{f}{n}\right)$
1	|	1	.10
2		0	0
3	|	1	.10
4	||	2	.20
5	|||	3	.30
6	|||	3	.30

TABLE 3-2

Relative Frequency Distributions for a Die, Using Various Sample Sizes

X Number of Dots	$\frac{f}{n}$ (a) $n = 10$	(b) $n = 50$	(c) $n = \infty$
1	.10	.22	1/6 = .167
2	0	.12	1/6 = .167
3	.10	.14	1/6 = .167
4	.20	.14	1/6 = .167
5	.30	.14	1/6 = .167
6	.30	.24	1/6 = .167
	1.00√	1.00√	1.00√

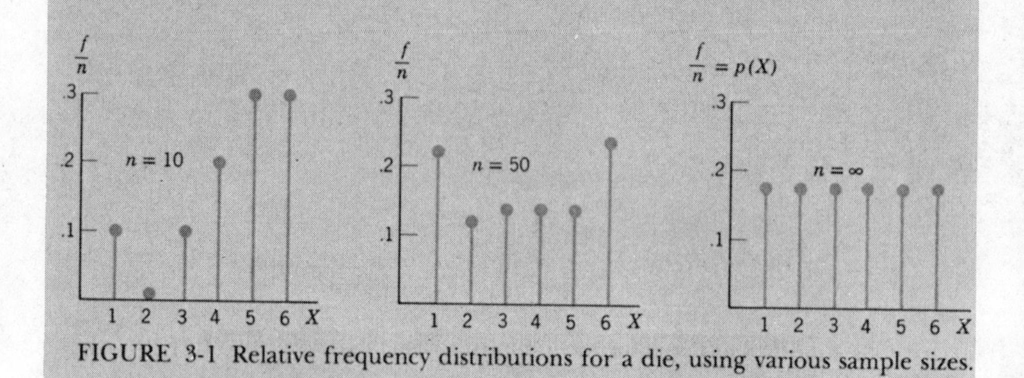

FIGURE 3-1 Relative frequency distributions for a die, using various sample sizes.

The number of dots that come up when a die is thrown is an example of a *random variable,* and is customarily denoted by a capital letter:

$$X = \text{number of dots when a fair die is thrown} \qquad (3\text{-}1)$$

In general, a random variable may be defined as a numerical outcome of a random experiment.

In Table 3-2, we noted in column (a) how the relative frequencies fluctuate wildly when there are few observations in the sample. Yet eventually, in column (c), the relative frequencies settle down to a limiting distribution, which we call the *probability distribution* of X, and denote by $p(X)$:

$$\text{Probability} = \text{limiting relative frequency} \tag{3-2}$$

Just as all the relative frequencies must add to 1 (as in Table 3-2, for example) so must all the probabilities add to 1:

$$\Sigma\, p(X) = 1 \tag{3-3}$$

Since probability is often called *chance,* this equation may be informally stated as, *all the chances add up to 100%.*

In Example 3-1, because the die was fair, we could deduce the limiting distribution even before throwing the die; we therefore call it a *prior* probability distribution. We must admit that in practice, however, no die is perfectly fair. So a prior probability distribution like this must be taken not as the absolute truth, but as *a good approximation* to how an actual die would behave.

Now let's use these concepts of probability on some more practical examples.

Example 3-2

Assuming boys and girls are equally likely, graph the probability distribution of the random variable:

$$X = \text{number of girls that will occur if you have 3 children}$$

This problem is mathematically equivalent to tossing 3 fair coins, of course.

Solution

Clearly, the values of X are 0, 1, 2, and 3. But they are not all equally probable. For example, it is a common observation that families with one girl are more prevalent than families with no girls. Just how much more prevalent? On the one hand, there are three ways to get a family with one girl: The first, second, or third birth could be a girl. On the other hand, there is only one way to get a family with no girls: Every birth must be a boy. Thus the probability for $X = 1$ is three times as great as for $X = 0$.

By symmetry, the probability for one boy ($X = 2$) must be the same as for one girl ($X = 1$). Similarly, the probability for no boys ($X = 3$) must be the same as for no girls ($X = 0$).

Thus the probability distribution must have the following shape:

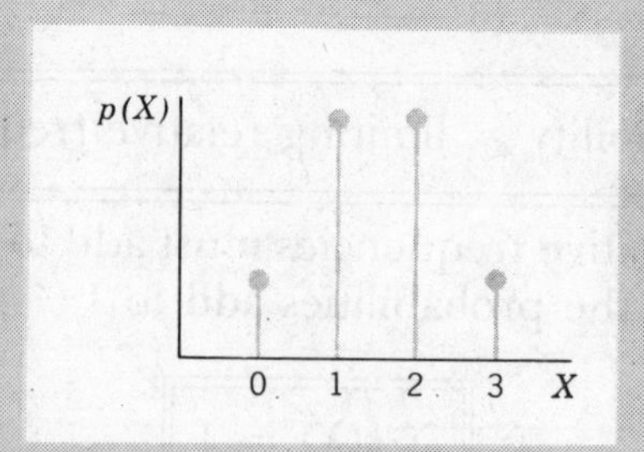

That is, the probabilities are in the ratio 1, 3, 3, 1. For them to add up to 1, as specified by (3-3), the probabilities must therefore be 1/8, 3/8, 3/8, 1/8. In other words, the probability distribution is

X	$p(X)$
0	1/8
1	3/8
2	3/8
3	1/8
	8/8✓

Example 3-3

Let everyone in the class simulate the experiment of having 3 children and, as in Example 3-2,

$$\text{let } X = \text{the number of girls}$$

You can simulate by using 3 coins, with heads representing a girl, and tails a boy. Or, you can simulate by using 3 digits from Table I, with an even digit representing a girl, and an odd digit a boy. Let the instructor then tabulate and graph all the values of X in a relative frequency distribution. (If the class is small, have everyone repeat the experiment a few times, so that altogether there will be at least a hundred repetitions of the experiment.) Note how this relative frequency distribution approximates the probability distribution in Example 3-2.

Next we will see how a probability distribution can be used to calculate the chance of some interesting event.

Example 3-4

Consider the random variable:

X = the number of girls that will occur if you have 6 children

Its probability distribution (as will be shown in Section 3-4) turns out to be

X	$p(X)$
0	.02
1	.09
2	.23
3	.32
4	.23
5	.09
6	.02
	1.00✓

(a) Graph this distribution.

(b) What is the chance that girls will be a small minority, specifically, the probability of less than 2 girls:

$$\Pr(X < 2) = ?$$

(c) What is the chance that girls will not be such a small minority, that is, the probability of at least 2 girls:

$$\Pr(X \geq 2) = ?$$

Solution

(a)

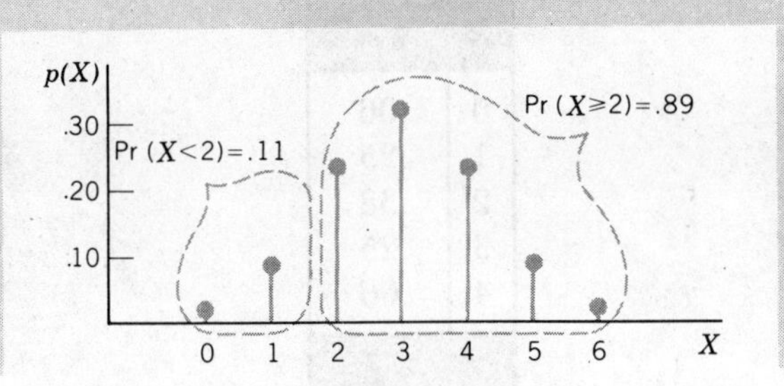

(b) "Less than 2 girls" occurs if either $X = 0$ or $X = 1$. These events occur 2% and 9% of the time, respectively; that is, 11% of the time altogether. Thus

$$\Pr(X < 2) = .11$$

(c) "At least 2 girls" occurs in many ways: $X = 2, 3, 4, 5$, or 6. Just as in part (a), we could add up the respective probabilities:

$$\Pr(X \geq 2) = .23 + .32 + .23 + .09 + .02 = .89 \qquad (3\text{-}4)$$

Remarks The probabilities derived in (b) and (c) are shown in the figure. Notice that the event $X \geq 2$ consists of exactly those cases *not* included in $X < 2$, and is therefore called the *complementary* event.

For complementary events, there is an easy way to calculate probability: Since the total probability must add up to 1 [according to (3-3)], this means that

$$\Pr(\text{complementary event}) + \Pr(\text{event}) = 1$$

$$\boxed{\Pr(\text{complementary event}) = 1 - \Pr(\text{event})} \qquad (3\text{-}5)$$

Thus, in Example 3-4, we could have given an easier answer to question (c):

$$\Pr(X \geq 2) = 1 - \Pr(X < 2)$$
$$= 1 - .11 = .89 \qquad (3\text{-}4)\text{confirmed}$$

PROBLEMS

3-1 In a family of 4 children the number of girls X has the following probability distribution (the *derivation* of the probability distributions in these exercises will be given in Section 3-2).

X	$p(X)$
0	.06
1	.25
2	.38
3	.25
4	.06
	1.00✓

(a) Let everyone in the class conduct the experiment, and record X. (Or simulate with Table I. In this case, let an even digit represent a girl, and an odd digit a boy.) Let the instructor then array all the values of X into a relative frequency distribution. Does this distribution resemble the given probability distribution above?

(b) Find:

(i) $\Pr(X < 3)$

(ii) $\Pr(X \geq 3)$

(c) Graph $p(X)$, and show the probabilities in (b).

3-2 When two dice are thrown, let X denote the larger number. For example, if the two dice turn up 5 and 2, then $X = 5$. The probability distribution of X is the following:

X	$p(X)$
1	.03
2	.08
3	.14
4	.19
5	.25
6	.31
	1.00✓

Answer the same questions as in Problem 3-1.

3-3 When three dice are thrown, the largest number X has the following probability distribution.

X	$p(X)$
1	.01
2	.03
3	.09
4	.17
5	.28
6	.42
	1.00✓

Answer the same questions as in Problem 3-1.

*3-2 PROBABILITY TREES

The star on this (or any other) section indicates it may be skipped, since it is not absolutely essential for understanding the rest of the book. Nevertheless, this section provides a very useful technique for deriving probability distributions.

For example, consider once again the birth of 3 children. We make the same assumption as in Example 3-2, that boys and girls are equally likely. We will show how a probability tree can be used as an alternative method to derive the probability distribution of

$$X = \text{the number of girls} \tag{3-6}$$

In Figure 3-2, in column 4 we list the 8 possible outcomes; each involves a particular sequence—for example, boy/girl/boy (BGB). What is its probability? It can be calculated using the concept of relative frequency: imagine millions of couples, each performing the "experiment" of having 3 children. If we summarize the results after they have all had their first child, one half will report "boy," the other half "girl." This is shown by the first branching of the probability tree, in column 1.

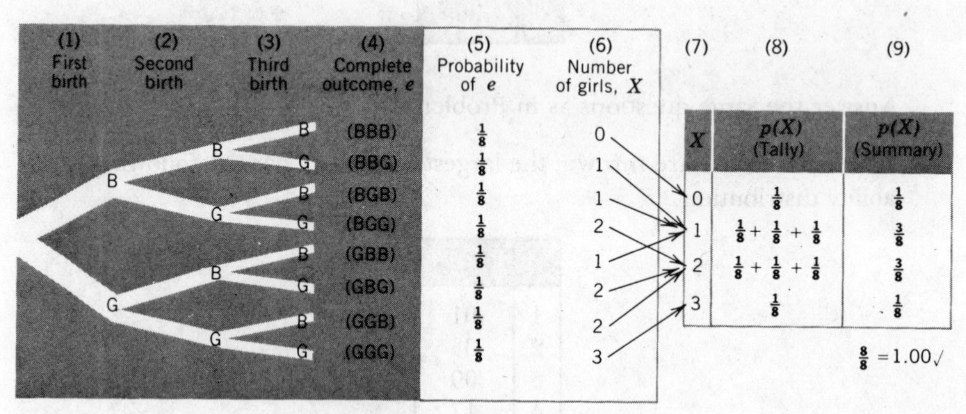

FIGURE 3-2 The probability tree for a family of 3 children.

Now of those couples who initially had B, one-half would report G at the second birth, and of these, one-half would finally report B at the third birth. Thus the proportion of the couples who would report the complete outcome (BGB) is

$$\frac{1}{2} \text{ of } \frac{1}{2} \text{ of } \frac{1}{2} = \frac{1}{8}$$

This relative frequency, or probability, is duly recorded in Figure 3-2 as the third entry in column 5. Similarly, we calculate and record the probability for every one of the 8 possible outcomes. Since boys and girls are equally likely, each probability in this example is the same (1/8).

In column 6, for each outcome we indicate the number of girls X. Then in the last three columns we accumulate the probabilities for each of the possible values of X. Thus, for example, we note that the probability of exactly two girls ($X = 2$) is 3/8—because of all the couples who performed this experiment, 1/8 had (BGG), 1/8 had (GBG), and another 1/8 had (GGB), making a total of 3/8 altogether.

We note with satisfaction that we have obtained the same distribution for X as in Example 3-2. In the next example we will show how a tree can still be used in the same straightforward way even though the experiment is asymmetrical (i.e., the initial probabilities now are *not* .50-.50).

Example 3-5

(a) Suppose on every birth the probability of a boy is 52% and a girl is 48%. (According to U.S. birth statistics, this is a more realistic assumption than our earlier .50-.50 assumption.) Calculate the probability tree for a couple having 3 children.

(b) Then tabulate the probability distribution of X = number of girls.

Solution

As examples,

$$\Pr(\text{BBB}) = 52\% \text{ of } 52\% \text{ of } 52\% = 14\%$$
$$\Pr(\text{GBG}) = 48\% \text{ of } 52\% \text{ of } 48\% = 12\%$$

Continuing in this way, we obtain the tree shown in Figure 3-3.

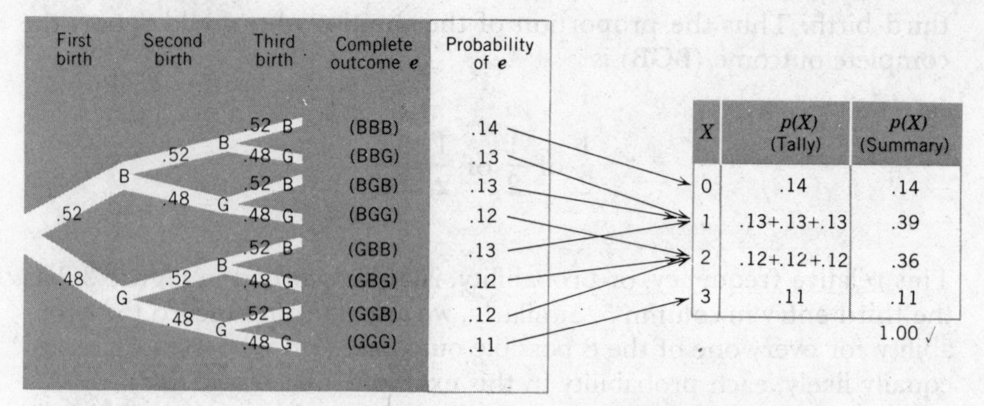

FIGURE 3-3 The probability tree for a family of three children if the probability of a boy on each birth is 52%.

Experiments can be more complicated. For example, there may be several outcomes (branchings) at each trial—as in tossing a die, where 6 possible faces can occur. In any case, an outcome tree can still be drawn up to derive the probability distribution.

PROBLEMS

3-4 In a learning experiment, a subject attempts a certain task several times in a row. Each time his chance of failure is .40. Let X = total number of failures. Find the distribution of X, if he makes:

(a) Two attempts.
(b) Three attempts.

3-5 Repeat Problem 3-4, if the subject learns from his previous trials, especially his previous successes, as follows:

His chance of failure is still .40 at the first trial. However, for later trials his chance of failure drops to .30 if his previous trial was a failure, and drops way down to .20 if his previous trial was a success.

3-6 Derive the probability distribution table in

(a) Problem 3-1.
(b) Problem 3-2.
*(c) Problem 3-3.

3-7 (a) (Acceptance sampling). The manager of a small hardware store buys electric clocks in cartons of 12 clocks each. To see whether each carton is acceptable, 3 clocks are randomly selected and thoroughly tested. If all 3 are of acceptable quality, then the carton of 12 is accepted.

Suppose in a certain carton, unknown to the hardware manager, only 8 of the 12 are of acceptable quality. What is the chance that the sampling scheme will inadvertently accept the carton?

(b) Repeat part (a), if the given carton of 12 clocks has only 6 of acceptable quality.

3-3 MEAN AND VARIANCE

(a) Mean

Just as for samples, we can calculate the mean and variance for probability distributions. (After all, they can be viewed as samples in the limit). An example will illustrate.

Example 3-6

Calculate the mean for each of the frequency distributions tabulated on page 44.

Solution

Since relative frequencies are given, it will be appropriate to use the method of Table 2-4.

	(a) $n = 10$		(b) $n = 50$		(c) $n = \infty$	
X	$\frac{f}{n}$	$X\left(\frac{f}{n}\right)$	$\frac{f}{n}$	$X\left(\frac{f}{n}\right)$	$\frac{f}{n} = p(X)$	$X\,p(X)$
1	.10	.10	.22	.22	1/6	1/6
2	0	0	.12	.24	1/6	2/6
3	.10	.30	.14	.42	1/6	3/6
4	.20	.80	.14	.56	1/6	4/6
5	.30	1.50	.14	.70	1/6	5/6
6	.30	1.80	.24	1.44	1/6	6/6
		$\bar{X} = 4.50$		$\bar{X} = 3.58$		$\mu = 21/6 = 3.50$

In other words, in tossing a fair die, the mean (average) number of dots is 3½.

In Example 3-1 we already noted how the relative frequencies settled down to limiting values $p(X)$, called the probability distribution of X. Similarly, in Example 3-6 we note how the sample mean settles down to a limiting value, which we call the *mean of the probability distribution.* We also call it the *population mean*—because it is based on the conceptual population of all possible throws. It is denoted by the Greek letter μ (*mu*, which rhymes with *new* and is the Greek equivalent of *m* for mean). We shall use a Greek letter for any such limiting or ideal number, in honor of the Greek philosopher Plato who wrote so much about an ideal world.

The mean has several aliases. Historically, meteorologists have used the term *mean* annual rainfall, economists the term *average* profit, and gamblers the term *expected* loss. Yet they all are talking about the same concept. So we must recognize that for any random variable X, all the following terms have exactly the same meaning:

$$\begin{aligned} \mu &= \text{mean of } X \\ &= \text{average of } X \\ &= \text{the expected value of } X \end{aligned}$$

Of course, no gambler actually expects to get 3 heads when he tosses 6 coins, just as no statistician actually expects to have 3.06 children. In any single experiment—whether it be gambling or creating a family—the expected value μ is *not* what you will get. Instead, it is what you can expect, on average, if you repeat the experiment many many times. Specifically, the mean μ has a precise mathematical meaning: As the last column of the table in Example 3-6 shows, to get μ we weight the values of X according to their probabilities $p(X)$, and sum. That is,

$$\boxed{\mu \equiv \Sigma\, X\, p(X)} \qquad (3\text{-}7)$$

(b) Variance

What we have done with the mean is naturally extended to the variance, as the next example shows.

Example 3-7

In Example 3-6, calculate the MSD and variance for the probability distribution $p(X)$ in the second last column of the table.

Solution

First we calculate the MSD, just as we did in Table 2-4. Note that the limiting relative frequencies $p(X)$ now replace the relative frequencies (f/n), and the population mean μ replaces the sample mean $\bar{X}$.

X	$p(X)$	$(X - \mu)$	$(X - \mu)^2$	$(X - \mu)^2 p(X)$
1	1/6	−2.5	6.25	1.04
2	1/6	−1.5	2.25	.38
3	1/6	− .5	.25	.04
4	1/6	.5	.25	.04
5	1/6	1.5	2.25	.38
6	1/6	2.5	6.25	1.04
				MSD = 2.92

Now the variance differs from the MSD only because its divisor is $n - 1$ instead of n. But the two are the same when we are describing the population of all possible throws (i.e., when n is infinitely large). Thus

Variance = MSD = 2.92

Standard deviation = $\sqrt{2.92}$ = 1.71

In Example 3-7, we calculated the standard deviation for the limiting case (the whole population of all possible throws). We therefore call it the *population standard deviation* and denote it by the Greek letter σ (*sigma*, which is the Greek equivalent of s for standard deviation). As the last column of the table in Example 3-7 shows, to get the MSD or variance σ^2, we weight the squared deviations $(X - \mu)^2$ according to their probabilities $p(X)$, and sum:

$$\boxed{\sigma^2 \equiv \text{MSD} \equiv \Sigma\,(X - \mu)^2 p(X)} \tag{3-8}$$

(c) Summary

We repeat the formulas for the population moments μ and σ^2 in Table 3-3, where they may be compared with the corresponding sample moments defined earlier. Note that each population moment is represented with a Greek letter, and is obtained by simply replacing relative frequencies (f/n) by *limiting* relative frequencies $p(X)$. To review these formulas, in Table 3-4 we calculate μ and σ, for X = the number of girls in a 3-child family.

TABLE 3-3

Sample Moments	*Population Moments (Greek letters)*
Sample mean $\bar{X} \equiv \Sigma X \left(\frac{f}{n}\right)$ (2-17)	Population mean $\mu \equiv \Sigma X\, p(X)$ (3-7)
Sample variance $s^2 \simeq \text{MSD} \equiv \Sigma (X - \bar{X})^2 \left(\frac{f}{n}\right)$ (2-18)	Population variance $\sigma^2 \equiv \text{MSD} \equiv \Sigma (X - \mu)^2\, p(X)$ (3-8)

TABLE 3-4

Calculation of the Mean and Variance of X = Number of Girls

Given Probability Distribution		*Calculation of μ using (3-7)*	*Calculation of σ^2 Using (3-8)*		
X	$p(X)$	$X\, p(X)$	$(X - \mu)$	$(X - \mu)^2$	$(X - \mu)^2\, p(X)$
0	1/8	0	−3/2	9/4	9/32
1	3/8	3/8	−1/2	1/4	3/32
2	3/8	6/8	+1/2	1/4	3/32
3	1/8	3/8	+3/2	9/4	9/32
		$\mu = 12/8$ $= 3/2$			$\sigma^2 = 24/32$ $= .75$ $\sigma = \sqrt{.75} = .87$

To summarize, we again emphasize the distinction between sample and population values. We call μ the population mean, since it is based on the whole population. On the other hand, we call $\bar{X}$ the sample mean since it is based on a mere sample drawn from the population. Similarly, σ^2 and s^2 represent population and sample variance, respectively. [Remember: *Greek* letters represent *population* moments that are ideal.]

Since the definitions of μ and σ are similar to those of $\overline{X}$ and s, we find similar interpretations. We continue to think of the mean μ as a weighted average (using probability weights rather than relative frequency weights), which can be interpreted as the balancing point. And the standard deviation σ is a measure of typical spread.

PROBLEMS

3-8 Calculate the mean and standard deviation in Problems 3-1 to 3-3.

3-9 The following table classifies the 28.2 million white American women, aged 45 or more in 1970 who were ever married, according to the number of children they had borne. (At the end of the table, 4.3 million women had more than 4 children, and they are counted as having 6 children—a roughly representative figure).

Children Ever Born to White American Women
(ever married, aged 45 or over, 1970 Census)

Family Size (children)	*Frequency (millions)*
0	4.3
1	4.9
2	7.0
3	4.8
4	2.9
6	4.3
Total	28.2

(a) Calculate the relative frequencies (probability distribution).
(b) Calculate the mean family size.
(c) Calculate the standard deviation.
(d) Graph the probability distribution. Mark the mean as the balancing point on the graph.

3-4 THE BINOMIAL DISTRIBUTION

There are many types of discrete random variables. In this section we study in detail the commonest type—the binomial. The classical example of a binomial variable is

$$S = \text{number of heads in } n \text{ tosses of a coin} \qquad (3\text{-}9)$$

There are many random variables of this type, a few of which are listed in Table 3-5. (We have already encountered not only the coin tossing example, but also another: the number of girls in a family of n children.) To handle all such cases, some general notation is helpful: Each *trial* (each toss of the coin) results in one of two possible outcomes, arbitrarily called *success* or *failure* (heads or tails). Their respective probabilities are π and $(1 - \pi)$, say. Then S, the total number of successes in n trials, has a binomial distribution, which can be calculated with an easy formula. The formula of course will include n and π, which are called the binomial *parameters*. First studied in depth 250 years ago by the French mathematician De Moivre (and reproduced in most modern statistics texts, such as Wonnacott, 1977), the formula is customarily written:

Binomial distribution

$$p(S) = \binom{n}{S} \pi^S (1 - \pi)^{n-S} \qquad (3\text{-}10)$$

where $\binom{n}{S}$ is called the *binomial coefficient,* and is tabulated in Appendix Table IIIa. Alternatively, it can be calculated with a formula; for example, for $n = 8$ and $S = 3$, the formula is

$$\binom{8}{3} = \frac{8(7)(6)}{3(2)(1)} \qquad (3\text{-}11)$$

The formula simply started with 8 and 3 (in the numerator and denominator, respectively) and successively reduced them by 1 until there

TABLE 3-5

Examples of Binomial Variables

Trial	*Success*	*Failure*	π	n	S
Tossing a fair coin	Head	Tail	1/2	Number of tosses	Number of heads altogether
Birth of a child	Girl	Boy	Practically 1/2	Family size	Number of girls in family
Pure guessing on a multiple choice questions (with 5 choices, say)	Correct	Wrong	1/5	Exam size	Number of correct answers
Drawing a voter at random in a poll	Republican	Democrat	Proportion of Republicans in the population	Sample size of voters	Number of Republicans in the sample
Drawing a woman at random in a fertility survey	Pregnant	Not pregnant	Pregnancy rate in the population of women	Sample size	Number of pregnant women in the sample

are altogether 3 factors (in both numerator and denominator). In general, this can be written[1]

$$\binom{n}{S} \equiv \frac{n(n-1)(n-2)\cdots \text{to } S \text{ factors}}{S(S-1)(S-2)\cdots 1} \tag{3-12}$$

Before using (3-10), we must be aware of an important assumption: The binomial distribution is appropriate only if the trials are *independent* (more formally called *statistically independent*). By this we mean:

Trials are *independent* if the probabilities for each trial are not affected by knowing the outcome of the other trials.	(3-13)

As an example of independence, knowing that the first toss of a coin has come up heads will not affect the probability on the second toss, if the coin is properly tossed. However, if the coin is poorly tossed (e.g., tossed so that it is likely to turn over just once), then knowing that the first toss is heads will reduce the probability of heads on the second toss; this is an example of dependence.

As another example of independence, suppose you are drawing two cards from a deck and you *replace* the first before drawing the second. Then the chance of, say, an ace on the second draw is 4/52, independent of whatever the card on the first draw happened to be.

On the other hand, if you *keep out* the first card, your chances on the second draw are altered, since you can't get that first card again. Thus if the first card is an ace, the chance of the second card being an ace is reduced to 3/51; if the first card is a non-ace, the chance of the second card being an ace is increased to 4/51. This is another example of dependence, where the binomial formula (3-10) cannot be applied.

---

[1]The special case $\binom{n}{n}$ is easily derived in (3-12) and equals 1. At the other extreme, the case $\binom{n}{0}$ is not covered by (3-12); but it is defined to equal 1 also. This makes Appendix Table IIIa symmetrical, with a border of 1's along the left as well as the right side. It is known as the *triangle of Pascal* (French, 1623–1662) or the *triangle of Omar Khayyam* (Persian, 1043?–1123?).

Example 3-8

In a family of 8 children, what is the probability of getting exactly 3 girls? First list the assumptions you find it necessary to make.

Solution

(1) We assume parents do not have any control over the sex of their children. (For example, they have no means of increasing the probability of a girl, even after a run of 7 boys.) That is, each birth is a repeated random event like the toss of a coin. This assumption assures statistical independence, so that the binomial formula may be applied.

(2) We also assume that girls and boys are equally likely, so that the probability of a "success" (girl) is $\pi = .50$. (And unless stated otherwise, it is the assumption we will continue to make throughout the book.) This assumption keeps computations simple, and is accurate enough for most purposes.

The probability of exactly 3 girls, then, is found by substituting $n = 8$, $S = 3$, and $\pi = .5$ into (3-10):

$$p(3) = \binom{8}{3} .5^3 \, .5^5$$

$$= \frac{8 \cdot 7 \cdot 6}{3 \cdot 2 \cdot 1} .5^3 \, .5^5$$

$$= 56(.5)^8 = .2188 \simeq .22 \qquad (3\text{-}14)$$

In practice, it is a nuisance calculating binomial probabilities like this by hand. Why not let the computer do them once and for all? In Appendix Table IIIb we summarize the computer printout for $p(S)$ in (3-10), for a whole array of values of π and n. For instance, we could find the answer in Example 3-8 very easily. In Table IIIb we simple look up $n = 8$, $\pi = .50$, and $S = 3$. This immediately gives the answer .2188, which confirms (3-14).

Another example will illustrate Table IIIb clearly.

Example 3-9

A poll of 5 voters is to be randomly drawn from a large population, 60% of whom are Democrats (as in the 1964 U.S. presidential election, for example, when Johnson defeated Goldwater).

(a) The number of Democrats in the sample can vary anywhere from 0 to 5. Tabulate its probability distribution.
(b) What is the probability of exactly 3 Democrats in the sample?
(c) What is the probability of at least 3 Democrats in the sample?
(d) Graph your answers to (a), (b), and (c).

Solution

(a) Each voter that is drawn constitutes a trial. On each such trial, the probability of a Democratic voter (success) is $\pi = .60$. In a total of $n = 5$ trials, we want the probability of S successes, where $S = 0, 1, \ldots, 5$. So we simple look up Appendix Table IIIb and copy down the distribution for $n = 5$ and $\pi = .60$.

S	$p(S)$
0	.0102
1	.0768
2	.2304
3	.3456
4	.2592
5	.0778
	1.000✓

(c) Sum $= .6826 \simeq .68$ (for $S = 3, 4, 5$)

(b) From the table, we find:

$$p(3) \simeq .35 \tag{3-15}$$

(c) From the table, we add up:

$$\begin{aligned} p(3) + p(4) + p(5) &= .3456 + .2592 + .0778 \\ &= .6826 \simeq .68 \end{aligned} \tag{3-16}$$

(d)

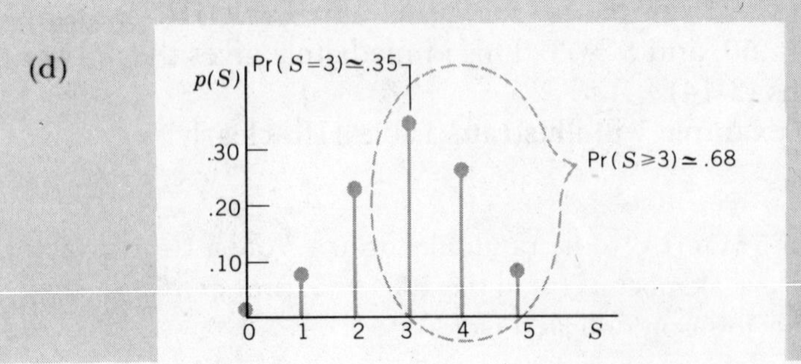

In practice, we often want to add up a string of probabilities as in Example 3-9(c). Again, why not let the computer do it once and for all? In the next table, Appendix Table IIIc, we summarize the computer output for the sum of the probabilities in the right-hand tail—called the *cumulative* binomial probability. For instance, we could find the answer in Example 3-9(c) very easily. In Table IIIc we simply look up $n = 5$, $\pi = .60$, and $S_0 = 3$. This immediately gives the answer .6826, which confirms (3-16).

Another Example will illustrate Table IIIc once more.

Example 3-10

(a) Suppose a warship takes 10 shots at a target, and it takes at least 4 hits to sink it. If the warship has a record of hitting on 20% of its shots in the long run, what is the chance of sinking the target?
(b) What crucial assumption did you make in part (a)? Why may it be questionable?
(c) To appreciate this crucial assumption, put yourself in the position of the captain of the British battleship *Prince of Wales* in 1941. The gunners on the German *Bismark* have just homed in on the British *Hood,* sinking it with their second shot. They now turn their fire on you, and after an initial miss they make a direct hit. Do you leave the probability they will hit you on the next shot unchanged, or do you revise it?

Solution

(a) For $n = 10$ and $\pi = .20$, we look up $S_0 = 4$ in the cumulative binomial, Table IIIc. It immediately gives the answer, $.1209 \simeq .12$.
(b) The gunners generally adjust each shot to allow for previous mistakes. Thus their probability of hitting the target improves over time (rather than being a constant $\pi = .20$, say, as the binomial assumes). Also, once they succeed in hitting the target, the gunners will be more likely to stay on target thereafter; that is, the shots may be dependent (rather than independent, as the binomial assumes).
(c) We agree with the captain of the *Prince of Wales*—he revised upward the probability that he would be hit on the next shot, and prudently broke off the action. (Other British ships picked up the trail of the *Bismark,* however, and sank it a few days later.)

Example 3-10 clearly illustrated some of the limitations of the binomial model. We therefore end with a word of caution. The binomial is only valid for *independent* trials with a *constant* probability of success.

PROBLEMS

3-10 Assume boys and girls are equally likely on every birth. Of all families with 6 children,

(a) What proportion have 3 boys and 3 girls?
(b) What proportion have 4 or more girls?
(c) What proportion have a girl as both the oldest and youngest?

3-11 A multiple choice exam consists of 10 questions, each having 5 possible answers to choose among. To pass, a mark of 50% (5 out of 10 questions correct) is required. What is the chance of passing if:

(a) You go into the exam without knowing a thing, and have to resort to pure guessing?
(b) You have studied enough so that on each question, 3 choices can be eliminated. Then you have to make a pure guess between the remaining 2 choices.

3-12 In one of Mendel's genetics experiments, the three possible colors for the offspring's flower (with their probabilities) were as follows:

red 25%
pink 50%
white 25%

Also, the offspring were statistically independent, so that knowing how the first turned out would not affect the odds on the next. Among 8 offspring, what is the probability that:

(a) At least half would be pink?

(b) None would be red?

3-13 A bomber pilot in World War II was estimated to have a 2% chance of being shot down on each mission. If he flies 50 missions, is he sure to be shot down? If so, why? If not, what is the correct probability that he would survive?

3-14 (a) One hundred coins are spilled at random on the table, and the total number of heads S is counted. The distribution of S is binomial, with $n =$ ____ and $\pi =$ ____. Although this distribution would be tedious to tabulate, the *average* (mean) of S is easily guessed to be ____.

(b) Repeat (a) for $S =$ the number of aces when 30 dice are spilled at random.

(c) Repeat (a) for $S =$ the number of correct answers when 50 true—false questions are answered by pure guessing.

3-15 (a) On the basis of Problem 3-14, guess the formula for the mean of a general binomial variable S in terms of n and π.

(b) It can be proved that the formula for the variance is $n\pi(1 - \pi)$. Use this to calculate the variance of S in Problem 3-14, all parts.

3-16 (a) Using the binomial tables, verify the distribution in Example 3-4 (for $X =$ the number of girls in a 6 child family).

(b) From the table in (a), calculate the mean and variance of X.

(c) Calculate the mean and variance of X using the formulas in Problem 3-15:

$$\mu = n\pi$$
$$\sigma^2 = n\pi(1-\pi) \tag{3-17}$$

(d) Do you find (b) or (c) easier? Now that you understand it, you should use the easier way.

3-5 CONTINUOUS DISTRIBUTIONS

In Figure 2-2 we saw how a continuous variable such as men's height could be represented by a bar graph showing relative frequencies. This graph is reproduced in Figure 3-4*a* (with men's height now measured in feet rather than inches; furthermore, the Y axis has been shrunk to the same scale as the X axis). The sum of all the relative frequencies (i.e., the sum of all the *heights* of the bars) in Figure 3-4*a* is of course 1, as first noted in Table 2-2.

We now find it convenient to change the vertical scale to relative frequency *density* in Figure 3-4b, a rescaling that makes the total area (i.e., the sum of all the *areas* of the bars) equal to 1. We accomplish this by defining

$$\boxed{\text{Relative frequency density} \equiv \frac{\text{relative frequency}}{\text{cell width}}} \tag{3-18}$$

$$= \frac{\text{relative frequency}}{1/4} \tag{3-19}$$

$$= 4 \text{ (relative frequency)}$$

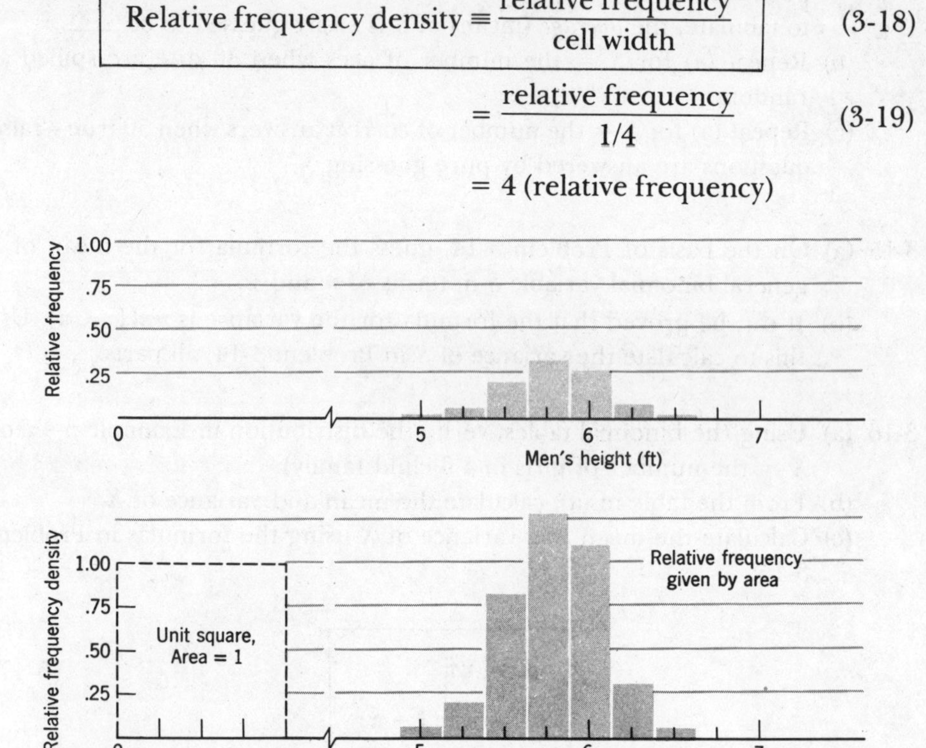

FIGURE 3-4 (*a*) Relative frequency histogram with total heights = 1. (*b*) Rescaled into relative frequency density, making total *area* = 1.

Thus in Figure 3-4, panel (*b*) is 4 times as high as panel (*a*); we also see that panel (*b*) now has an area equal to 1.

In Figure 3-5 we show what happens to the relative frequency density of a continuous random variable as

1. Sample size increases.
2. Cell size decreases.

With a small sample, chance fluctuations influence the picture. But as sample size increases, chance is averaged out, and relative frequencies settle down to probabilities. At the same time, the increase in sample size allows a finer definition of cells. While the area remains fixed at 1, the relative frequency density becomes approximately a curve, the *probability density function,* which we simply call the *probability distribution,* $p(X)$.

3-6 THE NORMAL DISTRIBUTION

For many random variables, the probability distribution is a specific bell-shaped curve, called the *normal* curve, or *Gaussian* curve (in honor of the German Karl Friedrick Gauss, 1777–1855, the greatest mathematical genius of his time). This is the most common and useful distribution in statistics. For example, errors made in measuring physical and economic phenomena often are normally distributed. In addition, many other probability distributions (such as the binomial) often can be approximated by the normal curve.

(a) Standard Normal Distribution

The simplest of the normal distributions is the *standard normal* distribution shown in Figure 3-6, and called simply the Z distribution. It is distributed around a mean $\mu = 0$ with a standard deviation $\sigma = 1$. Thus

Each Z value is *the number of standard deviations* away from the mean.	(3-20)

We often want to calculate the probability (i.e., the area under the curve) beyond a given value of Z, like the value 1.4 in Figure 3-6*b*. This, and all other such tail probabilities, have already been calculated by statisticians and are set out in Appendix Table IV (which is very similar to the binomial tail probabilities in Table IIIc). The following example illustrates how to use this table.

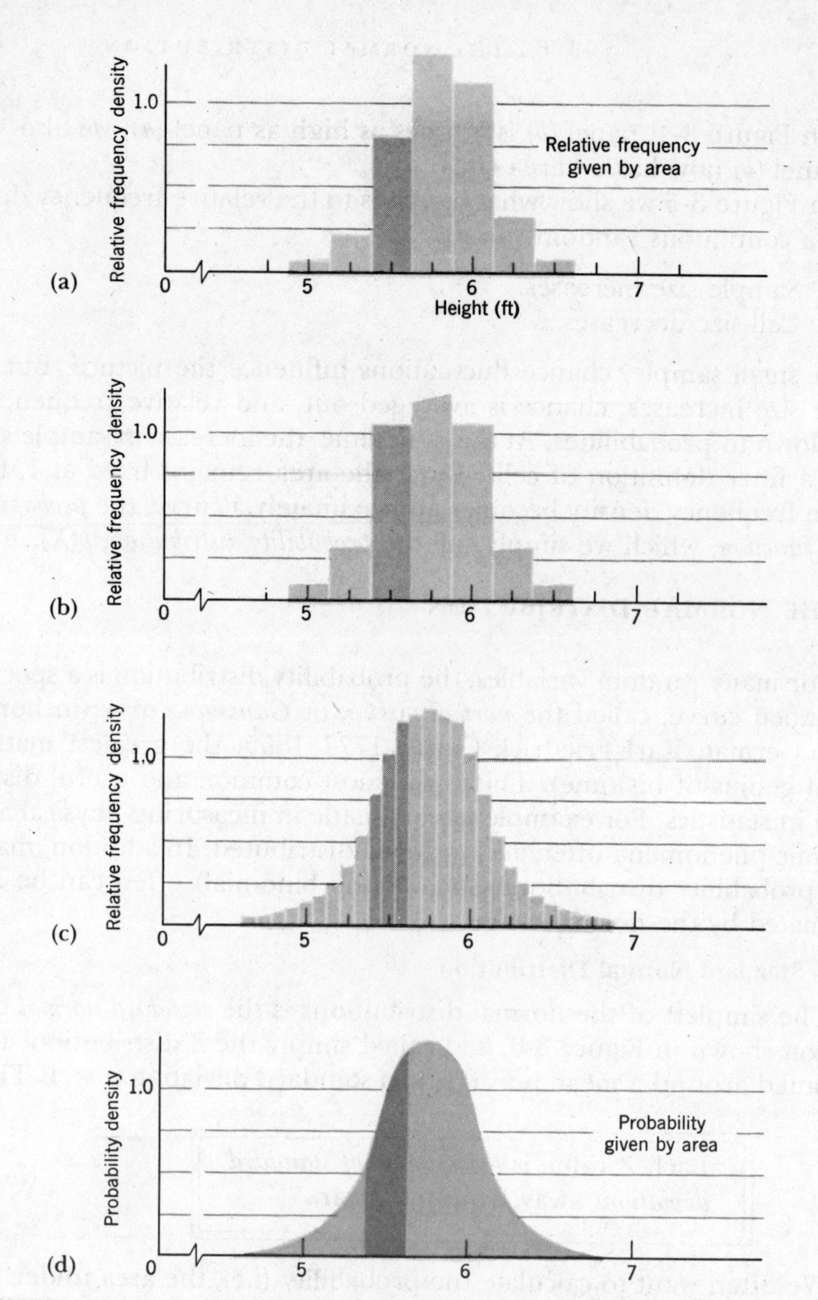

FIGURE 3-5 How relative frequency density may be approximated by a probability density as sample size increases, and cell size decreases. (*a*) Small n, as in Figure 3-4*b*. (*b*) Large enough n to stabilize relative frequencies. (*c*) Even larger n, to permit finer cells while keeping relative frequencies stable. (*d*) For very large n, this becomes (approximately) a smooth probability density curve.

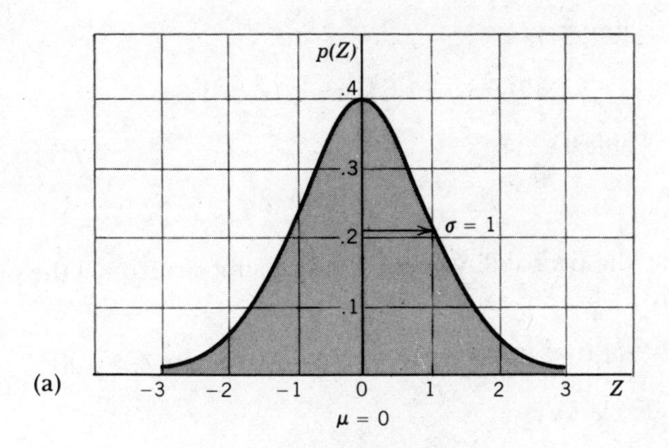

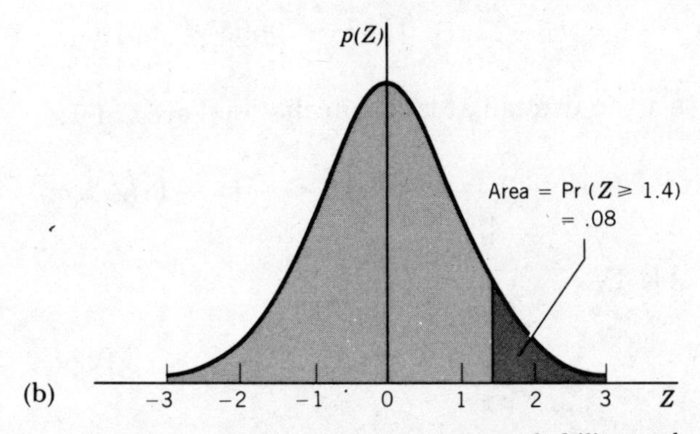

FIGURE 3-6 (*a*) Standard normal distribution. (*b*) Probability enclosed by the standard normal curve beyond a given point, say $Z_0 = 1.4$.

Example 3-11

If Z has a standard normal distribution, find

(a) $\Pr(Z > 1.64)$
(b) $\Pr(Z < -1.64)$
(c) $\Pr(1.0 < Z < 1.5)$
(d) $\Pr(-1 < Z < 2)$
(e) $\Pr(-2 < Z < 2)$

Solution

Each calculation is set out below, and is illustrated in the corresponding panel of Figure 3-7.

(a) By Appendix Table IV,

$$\Pr(Z > 1.64) = .0505$$

(b) By symmetry,

$$Pr(Z < -1.64) = Pr(Z > 1.64)$$

By Table IV,

$$= .0505$$

(c) Take the probability above 1.0, and subtract from it the probability above 1.5:

$$Pr(1.0 < Z < 1.5) = Pr(Z > 1.0) - Pr(Z > 1.5)$$

By Table IV,

$$= .1587 - .0668 = .0919$$

(d) Subtract the two tail areas from the total area of 1:

$$Pr(-1 < Z < 2) = 1 - Pr(Z < -1) - Pr(Z > 2)$$

By Table IV,

$$= 1 - .1587 - .0228 = .8185$$

(e)
$$Pr(-2 < Z < 2) = 1 - Pr(Z < -2) - Pr(Z > 2)$$
$$= 1 - 2(.0228) = .9544$$

(b) General Normal Distribution

In general, normal random variables have various different means μ and standard deviations σ. But because they all have the same bell shape, *any* normal variable can be easily transformed into the standard normal, as shown in the examples that follow. Then probabilities can be read off Table IV as before.

Example 3-12

When the population of American men have their heights X arrayed into a frequency distribution, it looks like Figure 3-8—a normal distribution with mean $\mu = 69$ inches, and standard deviation $\sigma = 3$ inches. What proportion of these men exceed 75 inches (6 ft, 3 in.)? Or, equivalently,

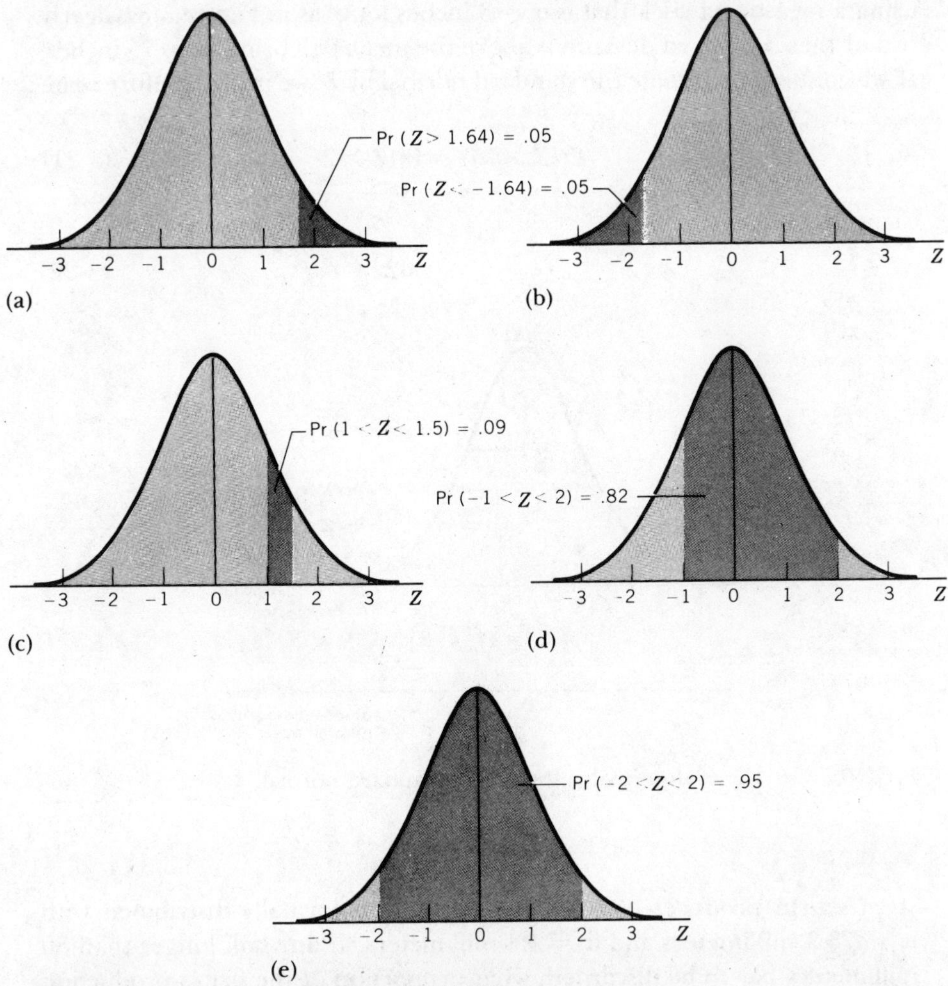

FIGURE 3-7 Standard normal probabilities.

what is the probability that a man drawn at random will exceed 75 inches? In symbols,

$$\Pr(X > 75) = ?$$

Solution

To derive the probability that X is more than 75 inches, we first ask: How many standard deviations above the mean is that? We may think of using a measuring stick that is $\sigma = 3$ inches long, as in Figure 3-8. Clearly two of these standard deviations above the mean will bring us to 75 inches. If we continue to denote the standard normal by Z, we may therefore write

$$\Pr(X > 75) = \Pr(Z > 2) \tag{3-21}$$

Using Table IV,

$$= .0228 \tag{3-22}$$

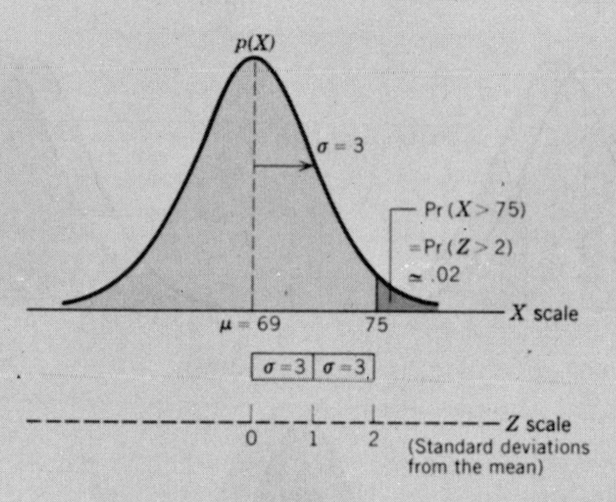

FIGURE 3-8 General normal rescaled to a standard normal.

Example 3-13

A firm produces bolts whose length X is normally distributed with $\mu = 78.3$ millimeters and $\sigma = 1.4$ millimeters. If any bolt longer than 80 millimeters has to be discarded, what proportion of the firm's production will be lost?

Solution

As Figure 3-9 shows, in this case the geometry does not make clear exactly what Z value corresponds to the critical X value of 80. So we must solve for the critical Z value algebraically, as follows:

(i) The critical X value of 80 differs from its mean by

$$80 - 78.3 = 1.7$$

(ii) How many standard deviations is this? Since there are 1.4 millimeters in each standard deviation, 1.7 millimeters is

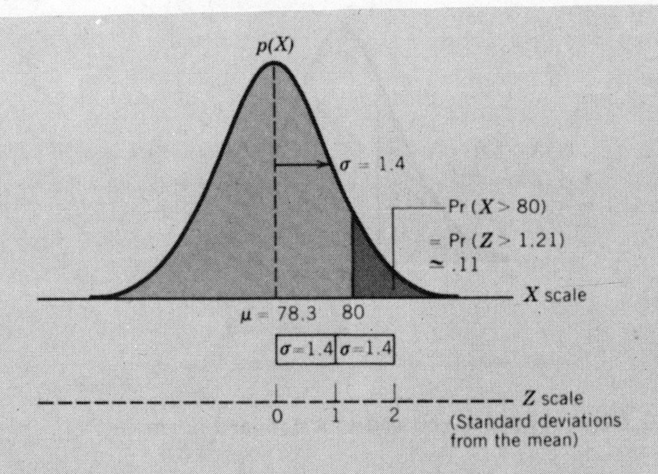

FIGURE 3-9 General normal rescaled to a standard normal

$$\frac{1.7}{1.4} = 1.21 \text{ standard deviations} \tag{3-23}$$

Thus the critical X value is 1.21 standard deviations above its mean; in other words, the critical Z value is 1.21. Then the answer is, from Table IV,

$$\Pr(Z > 1.21) = .1131 \simeq 11\%$$

To generalize how we derived the Z value of 1.21 in (3-23), we note that we first calculated the deviation $(X - \mu)$, and then compared it to the standard deviation σ. That is,

$$Z = \frac{X - \mu}{\sigma} \tag{3-24}$$

Thus the Z value gives the number of standard deviations away from the mean—as we first saw in (3-20).

Example 3-14

Suppose that scores on an aptitude test are normally distributed about a mean $\mu = 60$ with a standard deviation $\sigma = 20$. What proportion of the scores

(a) Exceed 85?
(b) Fall below 50?

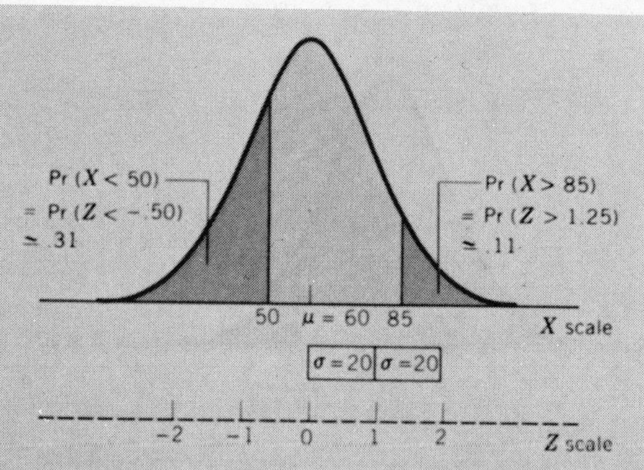

FIGURE 3-10 General normal rescaled to a standard normal.

Solution

(a) As shown in Figure 3-10, we first must *standardize* the score $X = 85$ (that is, express it as a standard normal value):

$$Z = \frac{X-\mu}{\sigma} \qquad \text{(3-24) repeated}$$

$$= \frac{85-60}{20} = \frac{25}{20} = 1.25$$

Thus $$\Pr(X > 85) = \Pr(Z > 1.25)$$
$$= .1056 \simeq 11\%$$

(b) $$Z = \frac{X-\mu}{\sigma} \qquad \text{(3-24) repeated}$$

$$= \frac{50-60}{20} = \frac{-10}{20} = -.50$$

Thus
$$\Pr(X < 50) = \Pr(Z < -.50)$$

By symmetry, $$= \Pr(Z > +.50)$$
$$= .3085 \simeq 31\%$$

PROBLEMS

Since Table IV is so important, it is repeated inside the front cover where it is easy to find.

3-17 If Z is a standard normal variable, calculate:

(a) $\Pr(Z > 1.60)$
(b) $\Pr(1.6 < Z < 2.3)$
(c) $\Pr(Z < 1.64)$
(d) $\Pr(-1.64 < Z < -1.02)$
(e) $\Pr(0 < Z < 1.96)$
(f) $\Pr(-1.96 < Z < 1.96)$
(g) $\Pr(-1.50 < Z < .67)$
(h) $\Pr(Z < -2.50)$

3-18 (a) How far above the mean of the Z distribution must we go so that only 1% of the probability remains in the right-hand tail? That is:

$$\text{if } \Pr(Z \geq Z_0) = .01, \text{ what is } Z_0?$$

Then what percentile is Z_0 called?

(b) How far on either side of the mean of the Z distribution must we go to include 95% of the probability? That is:

$$\text{if } \Pr(-Z_0 < Z < Z_0) = .95, \text{ what is } Z_0?$$

3-19 If X is normal, calculate:

(a) $\Pr(X > 6.8)$, if $\mu = 5$ and $\sigma = 3$
(b) $\Pr(X < 800)$, if $\mu = 500$ and $\sigma = 200$
(c) $\Pr(9.9 < X < 10.1)$, if $\mu = 10.0$ and $\sigma = 0.2$
(d) $\Pr(10 < X < 11)$, if $\mu = 10.73$ and $\sigma = .213$

3-20 The U.S. population of men's heights is approximately normally distributed with a mean of 69 inches and standard deviation of 3 inches. Find the proportion of the men who are:

(a) Under 5 feet (60 inches).
(b) Over 6 feet (72 inches).
(c) Between 5 feet and 6 feet.

3-21 The mathematical Scholastic Aptitude Test (SAT, math) gives scores that range from 200 to 800. They are approximately normally distributed about a mean of 470, with a standard deviation of 120 (for the population of U.S. college-bound seniors in 1978).

Graph this distribution, and then calculate and illustrate the following:

(a) What percentage of students scored between 500 and 600?
(b) What score is the 75th percentile (upper quartile)?
(c) What score is the 25th percentile (lower quartile)?
(d) What is the range between the two quartiles (*interquartile range*)?

CHAPTER 3 SUMMARY

3-1 Probability is the limiting relative frequency, as the sample size n grows larger and larger.

*3-2 To derive probability distributions, probability trees are a very useful technique when there are successive stages (trials) in the experiment.

3-3 The mean μ and standard deviation σ are defined for a probability distribution in the same way as $\bar{X}$ and s were defined for a relative frequency distribution in Chapter 2. The Greek letters emphasize their theoretical (ideal) nature.

3-4 When an experiment consists of independent trials with the same chance of success on each trial, the total number of successes S is called a binomial variable. Its probability distribution can be easily found from the standard formula (3-10), or Table IIIb.

3-5 For random variables that vary continuously, probabilities are given by areas under a continuous distribution (probability density curve).

3-6 The commonest continuous distribution is the bell-shaped normal (Gaussian) distribution, whose tail areas are given in Table IV inside the front cover.

REVIEW PROBLEMS

3-22 IQ scores are distributed approximately normally, with a mean of 100 and standard deviation of 15.

(a) What proportion of IQs are over 120?
(b) Suppose Fred Miller has an IQ of 120. What percentile is this?
(c) Suppose Mary Tibbs has an IQ of 135. What percentile is this?

3-23 Hawaii contains 770,000 people, 60% of whom are Asian, 39% white, and 1% black. If a random sample of 7 persons is drawn:

(a) What is the chance that a majority will be Asians?
(b) What is the chance that none will be black?
(c) For the number of Asians in the sample, what is the expected value? And the standard deviation?

3-24 The following table classifies the 31 million American women ever married, according to family size, in the 1970 census. (At the end of the table, 17% of the women had more than 4 children, and they are counted as having 6 children—a roughly representative figure.)

Children Ever Born to American Women (ever married, aged 45 or over, 1970 Census)

Family Size (children)	*Relative Frequency (%)*
0	16
1	17
2	24
3	16
4	10
6	17
Total	100%

(a) Find:
 (i) mean family size
 (ii) modal family size
 (iv) expected family size
 (iv) standard deviation of family size.

(b) Graph the relative frequency distribution. Mark the mean as the balancing point. Show the standard deviation as the unit of scale as in Figure 3-8.

(c) What family size is commonest? What proportion of the families are this size? How many million families is this?

(d) What proportion of the families have 4 or more children? Less than 4?

3-25 To see how useful means can be, let us consider some questions based on Problem 3-24.

(a) How many Americans would there be a generation later if American women continued to reproduce at the rate shown in the given table? The answer depends not only on the mean figure of 2.55; it also depends on many other factors such as whether unmarried women reproduce at the same rate, what proportion of children live to maturity, and what proportion of babies are female.

A more meaningful figure therefore is the mean number of female babies born to all women, who live to maturity. This average (with a few minor changes, described in Haupt and Kane, 1978) is called the net reproduction rate (NRR) and, to the extent that the NRR stays above 1.00, there will be long-run population growth.

In America in 1970, the NRR rate was 1.17 and there were about 200 million people. If we ignore immigration, changes in life expectancy, and other changes in age distribution, and if we also assume the NRR stays at 1.17, we can get a *projection* of how the population would grow.

What will the projected population size be one generation later? Two generations later? Ten generations later (about 250 years)? This is called *exponential growth.*

(b) In fact, the NRR does not stay constant at all. By 1975 it had dropped to an astonishing .83, less than half its postwar peak of 1.76 in 1957.

Answer (a) assuming the NRR stays at .83. This is called *exponential decline.*

(c) How useful do you think the projections are in (a) and (b)?

3-26 In the 1964 presidential election, 60% voted Democratic and 40% voted Republican. Calculate the probability that a random sample would correctly forecast the election winner; i.e., that a majority of the sample would be Democrats, if the sample size were:

(a) $n = 1$
(b) $n = 3$
(c) $n = 9$

Note how the larger sample increases the probability of a correct forecast.

3-27 The time required to complete a college achievement test was found to be normally distributed, with a mean of 90 minutes and a standard deviation of 15 minutes.

(a) What proportion of the students will finish in 2 hours (120 minutes)?

(b) When should the test be terminated to allow just enough time for 90% of the students to complete the test?

3-28 Suppose that of all patients diagnosed with a rare tropical disease, 40% survive for a year. In a random sample of 8 such patients, what is the probability that 6 or more would survive?

*3-29 (The "Sign Test.") Eight volunteers are to have their breathing capacity measured before and after a certain treatment, and recorded in a layout like the following:

Person \ *Breathing Capacity*	*Before*	*After*	*Improvement*
H.J.	2250	2350	+100
K.L.	2300	2380	+80
M.M.	2830	2800	−30
.	.	.	.
.	.	.	.
.	.	.	.

(a) Suppose that the treatment has no effect whatever, so that the "improvements" represent random fluctuations (resulting from measurement error or minor variation in a person's performance). Also assume that measurement is so precise that an improvement of exactly zero is never observed.

What is the probability that seven or more signs will be +?

(b) If it actually turned out that seven of the eight signs were +, would you question the hypothesis in (a) that the treatment has no effect whatever?

3-30 In the 1972 presidential election, 60% voted for Nixon (Republican). A small random sample of 4 voters was selected, and the number who voted Republican (R) turned out to be 3. This sampling experiment was repeated fifty times; R turned out to be usually 2 or 3, but occasionally as extreme as 4 or 1 or even 0. The results were arrayed in the following table:

R	*Frequency*
0	1
1	1
2	9
3	25
4	14
	50

(a) What is the mean of R?
(b) What is the standard deviation of R?
(c) Graph the relative frequency distribution of R.

*3-31 If the sampling experiment in Problem 3-30 were repeated millions of times (not merely fifty), what would be your answers?

PART II

BASIC INFERENCE: MEANS AND PROPORTIONS

CHAPTER 4

Sampling

I know of scarcely anything so apt to impress the imagination as the wonderful form of cosmic order expressed by the "Law of Frequency of Error (Normal Distribution)." The law would have been personified by the Greeks and deified, if they had known it.

Sir Francis Galton

4-1 RANDOM SAMPLING

(a) Populations and Samples

In the last two chapters we have developed the tools to work out the problem posed in Chapter 1: How do we make an inference about a population on the basis of a relatively small random sample?

The word population has a very specific meaning in statistics: It is the total collection of objects or people to be studied, from which a sample is to be drawn. For example, the population of interest might be all the American voters (if we wished to predict an election). Or it might be the population of all the students at a certain university (if we wished to estimate how many would attend a rock concert).

The population can be any size. For example, it might be only 100 students in a certain men's physical education class (if we wished to estimate their average height). To be specific, suppose this population of 100 heights has the frequency distribution shown in Table 4-1. When we draw our first random observation, what is the chance it will be 63 inches, for example? From the second line of Table 4-1, we find the probability is 6 in 100, or .06. Similarly, the probabilities for all the outcomes can be found in the relative frequency column, which we therefore label $p(X)$. From these probabilities, we calculate the population mean μ and standard deviation σ. Then in Figure 4-1*a* we graph this population distribution, sketched also as a smooth curve.

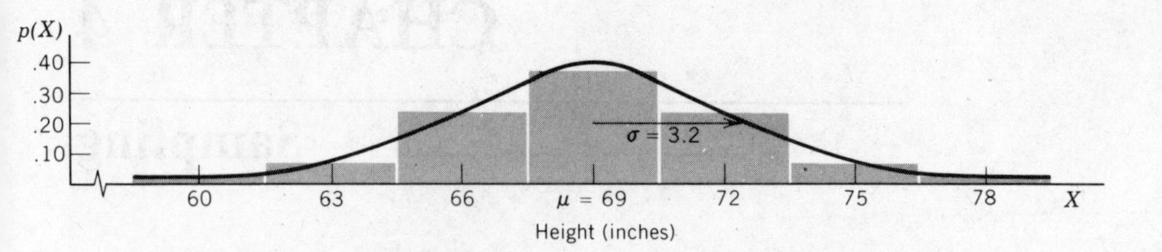

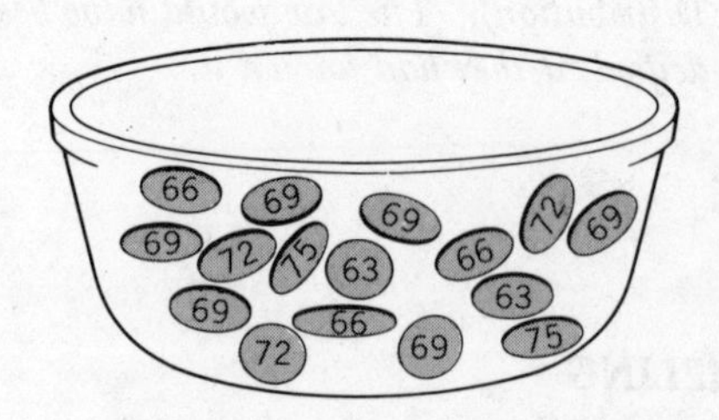

FIGURE 4-1 Population of 100 students' heights (*a*) Bar graph of distribution in Table 4-1. (*b*) Bowl-of-chips equivalent.

TABLE 4-1
A Population of 100 Student's Heights, and the Calculation of μ and σ^2

Population Distribution			*Calculation of mean* μ	*Calculation of variance* σ^2
Height X	*Frequency*	*Relative Frequency, also p(X)*	$X\,p(X)$	$(X - \mu)^2 p(X)$
60	1	.01	.60	.81
63	6	.06	3.78	2.16
66	24	.24	15.84	2.16
69	38	.38	26.22	0
72	24	.24	17.28	2.16
75	6	.06	4.50	2.16
78	1	.01	.78	.81
	N = 100√	1.00√	$\mu = 69.00$	$\sigma^2 = 10.26$ $\sigma = 3.20$

Now let us draw a sample from this population, a sample of $n = 4$ students, for example. As we mentioned in Chapter 1, a sample is called a *simple random sample* (SRS) if each individual in the population is equally likely to be chosen every time we draw an observation. For example, as suggested in Chapter 1, two ways we could take a random sample of 4 students in the physical education class are:

1. *Draw chips from a bowl.* As illustrated in Figure 4-1*b*, the most graphic method is to record each student's height on a chip, mix all these 100 chips in a large bowl, and then draw the sample of $n = 4$ chips.
2. *Assign each student a serial number, and select numbers at random.* For example, let the class assign itself numbers by counting off from 00 to 99. Then we could get our sample by reading pairs of random digits from Appendix Table I. For example, if we start at the beginning, we would select the four students with numbers 39, 65, 76, and 45.

These two sampling methods are mathematically equivalent. Since the second is simpler to employ, it is commonly used in practical sampling. However, the bowlful of chips is conceptually easier; consequently in our theoretical development of random sampling, we will often visualize drawing chips from a bowl.

(b) Sampling With or Without Replacement

In large populations, such as all the American voters, the "population bowl" in Figure 4-1 would contain millions of chips, and it would make practically no difference whether or not we replace each chip before drawing the next. After all, what is one chip in millions? It cannot substantially change the relative frequencies, $p(X)$.

However, in small populations such as the 100 heights in Figure 4-1, replacement of each sampled chip becomes an important issue. If each chip drawn is recorded and then replaced, it restores the population to exactly its original state. Thus later chips are completely independent of the chips drawn earlier. On the other hand, if it is *not* replaced, the probabilities involved in the draw of later chips *will* change. (For example, if the first chip drawn happens to be the only 78″ chip in the bowl, then the probability of getting that chip in succeeding draws becomes zero; it's no longer in the bowl.) In this case, later chips *are* dependent on the earlier chips drawn.

In conclusion, the n observations in a sample will be independent if we sample with replacement. And in large populations, even if we sample without replacement, it is practically the same as with replacement, so that we still essentially have independence. All these cases where the observations are independent are easy to analyze, and lead to very simple formulas. We therefore call them *very simple random samples:*

> For a *very simple random sample (VSRS),* all individuals in the population are *equally* likely to be observed, and the n observations are *independently* drawn. (4-1)

The exception to this independence occurs when the chips are drawn from a small population, *and* are not replaced. This procedure does offer one advantage. Because it ensures that no chip can be repeated, each observation must bring fresh information, and the sample is more valuable. However, this method of keeping out the chips is more difficult to analyze, and so is deferred to a starred section. Everywhere else, we will assume a VSRS, and often refer to it simply as a *random sample.*

(c) Sampling Simulation

Let us illustrate how random sampling (VSRS) works for the phys-ed class shown in Figure 4-1. We use the more practical of the two sampling methods mentioned earlier—namely, method 2 in which every student is assigned a serial number. This can be done geographically, alphabetically, or in any other convenient way. For the phys-ed class of 100 men, we assume they have been numbered from smallest to largest, as shown in the last column of Table 4-2.

TABLE 4-2

A Population of 100 Students' Heights, Serially Numbered for Sampling

Population Distribution			
Height (inches) X	*Frequency*	*Relative Frequency* $p(X)$	*Serial Number*
60	1	.01	00
63	6	.06	01–06
66	24	.24	07–30
69	38	.38	31–68
72	24	.24	69–92
75	6	.06	93–98
78	1	.01	99

The next step in this *simulated sampling* is to read off pairs of digits in Appendix Table I. We can start anywhere, since they are purely random. If we start at the beginning of the sixth row, for example, the first pair of random digits is 02, which is found in Table 4-2 to be the serial number of a man 63 inches tall. Continuing to draw the remaining observations in the same way, we obtain the sample in Table 4-3. If, by the luck of the draw, a serial number had been repeated, we would have included it both times in our sample. This is because we defined our sampling (VSRS) to be with replacement. Thus we randomly pick a man, then replace him in the population, and by the luck of the draw we may pick him again.

To summarize the sample, the sample mean $\overline{X}$ = 66.8 is calculated in Table 4-3.

TABLE 4-3

A Simulated Sample of n = 4 Observations drawn from the Population in Table 4-2

Serial Number	*Height X*
02	63
89	72
08	66
16	66
	$\overline{X}$ = 267/4 = 66.8

In Appendix A at the end of the book we set out an alternative simulation that can be used when the population is normal. Although this alternative is a bit more difficult to understand, it is easier for the computer to use—and hence is often the preferred technique.

(d) Conclusion

The population of 100 heights in Figure 4-1 is graphed again in Figure 4-2. As a reminder that this is more than abstract art, the circles indicate actual people, ordered according to their height, as in Figure 2-2. Moreover, the 4 blue circles show the sample that we drew. The sample mean $\overline{X}$, also shown in blue, is not so extreme as the individual observations and is closer to the population mean μ. This is because an

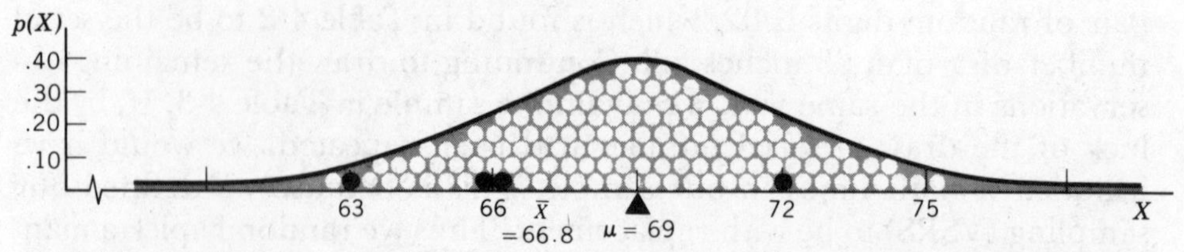

FIGURE 4-2 Random sample (blue) drawn from a population (gray), showing how the sample mean averages out the extremities of the individual observations.

extreme observation such as $X = 63$ tends to be offset by an observation at the other end like $X = 72$, or diluted by more typical observations like $X = 66$. (Notice this is exactly what happened in our sample in Table 4-3.) We therefore conclude:

> Because of averaging, the sample mean $\overline{X}$ is not so extreme (doesn't vary so widely) as the individuals in the population. (4-2)

PROBLEMS

4-1 (a) Draw your own sample of $n = 4$ observations from the population of 100 students' heights in Table 4-2. (To get the random serial numbers, start at a different place in Appendix Table I—by closing your eyes and hitting it randomly with your pencil, for example.) Then calculate $\overline{X}$.
(b) Graph the sample and $\overline{X}$, as in Figure 4-2. Then note that $\overline{X}$ does not vary so widely as the individual observations.

4-2 MONTE CARLO

The standard statistical problem is this: We don't know what the population looks like; specifically, we don't know that the population mean μ is 69 in Figure 4-2. So we draw from this population a relatively small, inexpensive random sample, and calculate its mean $\overline{X}$. In general, the sample mean $\overline{X}$ will be slightly off the population target μ. The crucial question is: How reliable is $\overline{X}$ as an estimate of μ?

In the specific sample we drew, the sample mean ($\overline{X} = 66.8$) turned out to be a pretty good estimate of the population mean ($\mu = 69$). Was this just the luck of the draw, or does a sample mean usually do this well? To answer this, we try our luck again: We draw a second sample of 4 observations, and calculate its sample mean $\overline{X}$; suppose this is 70.5. Now we are beginning to get a picture of the reliability of $\overline{X}$. We just repeat the process over and over, each time calculating a fresh sample mean. When we have accumulated a lot of these, we can indeed answer the question: How good is $\overline{X}$ as an estimate of μ, *in general?* An example will illustrate this *Monte Carlo* analysis.

Example 4-1

(a) Let everyone in your class draw their own sample of n = 4 observations from the population of men's heights in Table 4-2 (starting at a different place in Appendix Table I, of course). Then everyone calculate their sample mean $\bar{X}$.

If your class is small, each student should repeat the experiment several times, so that the class obtains at least 50 values of $\bar{X}$. Then let the instructor tabulate and graph their frequency distribution. (We suggest grouping the values of $\bar{X}$ into cells of width 1, that is, round $\bar{X}$ to the nearest whole number.)

(b) Calculate the mean and standard deviation of the distribution of $\bar{X}$. How are they approximately related to $\mu = 69$ and $\sigma = 3.2$ for the parent population (found in Table 4-1)?

Solution

(a) In our classroom, the following 50 values of $\bar{X}$ were obtained. (In your classroom, you will get a roughly similar result):

$\bar{X}$	*Frequency*	*Relative Frequency*
66	3	.06
67	2	.04
68	13	.26
69	14	.28
70	9	.18
71	4	.08
72	4	.08
73	0	0
74	1	.02
	50	1.00✓

By repeating the sampling experiment over and over like this, we begin to understand the luck of the draw, just as in repeated spinning of a roulette wheel. This is why repeated sampling is often called *Monte Carlo*. If we had repeated millions of values of $\bar{X}$ (rather than just 50), then the relative frequencies in the table above would settle down to the probabilities given in the table below, and graphed in Figure 4-3.

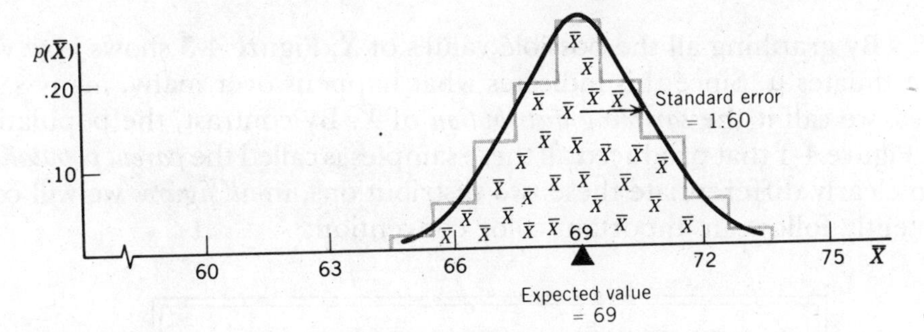

FIGURE 4-3 Sampling distribution of $\overline{X}$, for $n = 4$.

Sampling Distribution		*Calculation of the Expected value*	*Calculation of the Standard error*	
$\overline{X}$	$p(\overline{X})$	$\overline{X}\,p(\overline{X})$	$(\overline{X} - 69)^2$	$(\overline{X} - 69)^2 p(\overline{X})$
65	.01	.65	16	.16
66	.05	3.30	9	.45
67	.12	8.04	4	.48
68	.19	12.92	1	.19
69	.26	17.94	0	0
70	.19	13.30	1	.19
71	.12	8.52	4	.48
72	.05	3.60	9	.45
73	.01	.73	16	.16
	1.00✓	Expected value = 69.00		Variance = 2.56 Standard error = $\sqrt{2.56}$ = 1.60

(b) The expected value of $\overline{X}$ is calculated from the sampling distribution in the table above, and turns out to be 69, exactly the same as the population mean ($\mu = 69$). The standard deviation of $\overline{X}$ is also calculated, and turns out to be 1.60, exactly 1/2 the standard deviation of the population ($\sigma = 3.20$).

To help distinguish these two different standard deviations, the standard deviation of $\overline{X}$ is commonly called the *standard error* of $\overline{X}$, or simply SE. In other words,

$$\begin{aligned} \text{SE} &= \text{standard error of } \overline{X} \\ &= \text{standard deviation of } \overline{X} \end{aligned}$$

By graphing all the possible values of $\overline{X}$, Figure 4-3 shows how well $\overline{X}$ estimates μ. Since this indicates what happens over many, many samples, we call it the *sampling distribution* of $\overline{X}$. By contrast, the population in Figure 4-1 that produced all these samples is called the *parent population.* To clearly differentiate these two distributions, in *all figures* we will consistently follow an important color convention:

Parent populations are gray. Samples and sampling distributions are blue.

(4-3)

From the blue sampling distribution in Figure 4-3, we see that the sample mean $\overline{X}$ will sometimes be above the target (the population mean $\mu = 69$), and sometimes below. Since $\overline{X}$ has no tendency to consistently overestimate or underestimate we call it *unbiased.* That is, its expected value is neither too high nor too low; it is just right. Furthermore, $\overline{X}$ tends to be fairly close to its target: The sample mean fluctuates only 1/2 as much as the individuals within the population. (The standard error of $\overline{X}$ is 1.60 compared to the population standard deviation $\sigma = 3.20$).

Why does the sample mean $\overline{X}$ fluctuate less than the individual observation X? *Because of the averaging out.* We might get a 7-foot man as an individual observation from the population; but we would be far less likely to get a 7-foot average in a sample of 4 men. This is because any 7-foot man who appears in the sample will likely be partially offset by a short man, or at least his effect will be diluted, because he is averaged in with other more typical men. With a larger sample, there should be even more averaging out, as the next example shows.

Example 4-2

Repeat Example 4-1, with the class simulating many random samples of $n = 9$ observations each, instead of $n = 4$.

Solution

(a) The distribution of your sample means should look something like Figure 4-4, which shows how relative frequencies for $\overline{X}$ would settle down to probabilities if we were to take millions of samples.

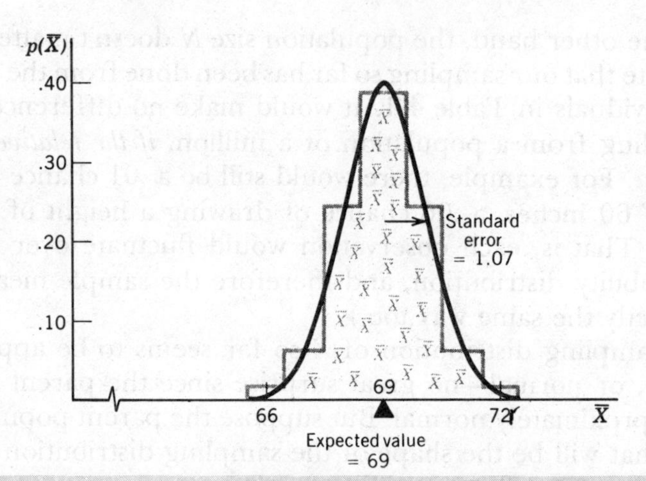

FIGURE 4-4 Sampling distribution of $\overline{X}$, for $n = 9$.

Sampling Distribution		*Calculation of the Expected value*	*Calculation of the Standard error*	
$\overline{X}$	$p(\overline{X})$	$\overline{X}\,p(\overline{X})$	$(\overline{X} - 69)^2$	$(\overline{X} - 69)^2 p(\overline{X})$
66	.01	.66	9	.09
67	.06	4.02	4	.24
68	.24	16.32	1	.24
69	.38	26.22	0	0
70	.24	16.80	1	.24
71	.06	4.26	4	.24
72	.01	.72	9	.09
	1.00√	Expected value = 69.00	Variance =	1.14
			Standard error =	$\sqrt{1.14}$ = 1.07

(b) The expected value of $\overline{X}$ is again 69, the population mean (μ = 69). The standard error of $\overline{X}$ is 1.07, which is only 1/3 the population standard deviation (σ = 3.20).

Comparing Figures 4-4 and 4-3, we see that the larger sample makes $\overline{X}$ a more reliable estimate of μ: $\overline{X}$ fluctuates around μ even less (standard error of $\overline{X}$ = 1.07, rather than 1.60). This then establishes an important conclusion: *The sample size n is critical in determining how much $\overline{X}$ fluctuates.*

(On the other hand, the population size N doesn't matter at all. To see why, note that our sampling so far has been done from the population of 100 individuals in Table 4-1. It would make no difference if we had been sampling from a population of a million, *if the relative frequencies were the same*. For example, there would still be a .01 chance of drawing a height of 60 inches, a .06 chance of drawing a height of .63 inches, and so on. That is, each observation would fluctuate over exactly the same probability distribution, and therefore the sample mean $\overline{X}$ would behave exactly the same way too.)

The sampling distribution of $\overline{X}$ so far seems to be approximately bell-shaped, or normal—no great surprise since the parent population was also approximately normal. But suppose the parent population is *not* normal. What will be the shape of the sampling distribution of $\overline{X}$ then? Our next example will answer this question.

Example 4-3

A class (population) of 80 students was asked how many novels X they each had read in the past three months. The highly skewed distribution of X is presented in the graph that follows. (It might be more interesting to actually use your own class as the population. It simply requires the instructor to gather the population information by a show of hands.)

(a) Calculate the mean and standard deviation of the population.

(b) From the population, simulate drawing of sample of $n = 9$ observations, and calculate $\overline{X}$ (rounded to the nearest half novel.)

(c) Let the instructor tabulate and graph the relative frequency distribution for all the $\overline{X}$ sampled in (b).

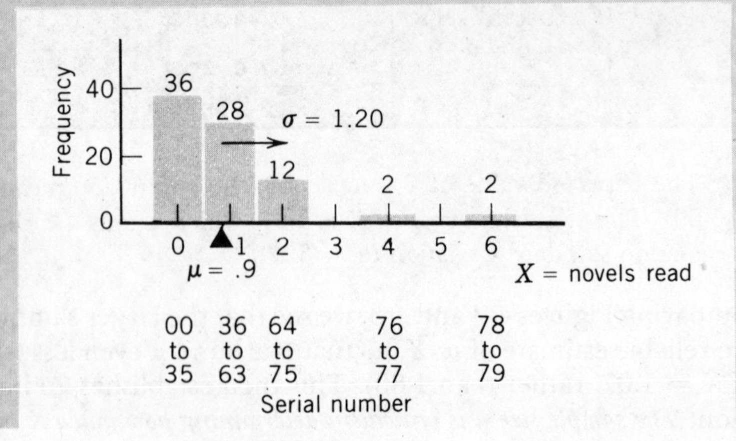

(d) Calculate the expected value and standard error of the sampling distribution of $\overline{X}$ found in (c). How are they related to the population μ and σ found in (a)?

Solution

(a) To describe the parent population (i.e., the distribution of X), we first transcribe the information from the figure into the first two columns below. From this we can calculate the population moments:

Population Distribution			*Calculation of Mean* μ	*Calculation of Variance* σ^2		
X	f	$p(X) = \frac{f}{N}$	$X\,p(X)$	$(X - \mu)$	$(X - \mu)^2$	$(X - \mu)^2 p(X)$
0	36	.45	0	−.9	.81	.365
1	28	.35	.35	.1	.01	.004
2	12	.15	.30	1.1	1.21	.181
3	0	0	0	2.1	4.41	0
4	2	.025	.10	3.1	9.61	.240
5	0	0	0	4.1	16.81	0
6	2	.025	.15	5.1	26.01	.650
	$N = 80$	1.00✓	$\mu = .90$			$\sigma^2 = 1.440$ $\sigma = \sqrt{1.440}$ $= 1.20$

(b) The 80 students in the population are given two-digit serial numbers from 00 to 79, as shown below the graph. To sample, we draw our first observation by selecting a pair of random digits. Again suppose we start at the sixth row of Appendix Table I. Then the first pair of random digits is 02, which is the serial number of a student who read 0 novels. The next pair of random digits is 89, for which there is no student's serial number. So we simply ignore it, and go to the next pair of random digits. (In ignoring the numbers 80 to 99, we do not, of course, change the fact that the numbers 00, 01, 02, . . . , 79 all remain equally likely—

the key idea in random sampling.) We thus obtain the following random sample:

Random Serial Number	*Observed X*
02	0
Ignore 89	
08	0
16	0
Ignore 94	
Ignore 85	
53	1
Ignore 83	
29	0
Ignore 95	
56	1
27	0
09	0
24	0
	$\overline{X} = 2/9 = .22$

(c) If you were to take millions of such samples and calculate $\overline{X}$ for each, the relative frequencies for $\overline{X}$ would settle down to the following probabilities. (In your classroom, with only a few values of $\overline{X}$, you will get a roughly similar result):

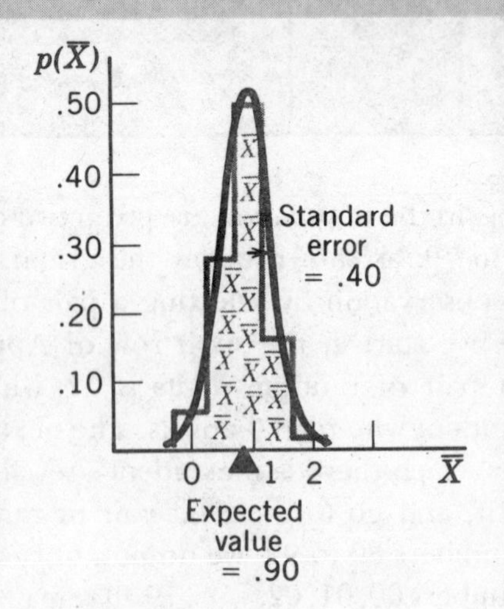

(c) (d)

Sampling Distribution		*Calculation of the Expected value*	*Calculation of the Standard error*	
$\bar{X}$	$p(\bar{X})$	$\bar{X}\, p(\bar{X})$	$(\bar{X} - .90)^2$	$(\bar{X} - .90)^2 p(\bar{X})$
0	.05	.00	.81	.0405
.5	.28	.14	.16	.0448
1.0	.50	.50	.01	.0050
1.5	.16	.24	.36	.0576
2.0	.01	.02	1.21	.0121
	1.00✓	Expected value =.90	Variance =	.1600
			Standard error =	$\sqrt{.1600}$
			=	.40

(d) The distribution of $\bar{X}$ is centered at .90, the same as the population mean (μ = .90). The standard error of $\bar{X}$ is .40, only 1/3 the population standard deviation (σ = 1.20).

As well as being less spread out than the population, the sampling distribution of $\bar{X}$ in Example 4-3 displays a remarkable feature: It is nearly a symmetric normal curve, even though the parent population distribution is very skewed. And this phenomenon is no fluke. As the Problems below will show, $\bar{X}$ regularly has an approximately normal distribution—whether or not the population itself is normal.

PROBLEMS

For each of the following problems, start in a different place in Appendix Table I to generate a random sample. Calculate the sample mean $\bar{X}$. Then have the instructor graph the frequency distribution of all the values of $\bar{X}$ obtained by the class.

4-2 When asked how many children X they intended to have, a population of 20-year-olds gave the following distribution of answers:

X	$p(X)$
0	.05
1	.05
2	.35
3	.25
4	.15
5	.10
6	.05

A demographer, who does not know this population, wants to estimate the population mean μ. Suppose she takes a random sample of $n = 10$ answers, and calculates the sample mean $\overline{X}$. Simulate this sample.

4-3 The daily demand X for fan belts ordered from a large auto supplier has the following distribution (X is rounded to the nearest multiple of 5):

X	$p(X)$
15	.01
20	.05
25	.22
30	.33
35	.26
40	.10
45	.03

A management consultant, who does not know this population, wants to estimate the mean daily demand μ. So he takes a random sample of $n =$ 5 days, and calculates the sample mean $\overline{X}$. Simulate this sample.

4-3 HOW RELIABLE IS THE SAMPLE MEAN?

(a) The Distribution of $\overline{X}$ for Random Samples

In Figure 4-5 we review the Monte Carlo results from the previous section. In column (*a*) we show what happened when we sampled from a normal population: The sample mean also had a normal distribution centered on μ, but with a reduced standard error because of averaging out.

Let us express this with some solid formulas. To say the distribution of $\overline{X}$ is centered at the target (population mean μ) we write

$$\text{Expected Value of } \overline{X} = \mu \tag{4-4}$$

What about the standard error of $\overline{X}$? When the sample size is $n = 4$, then the standard error of $\overline{X}$ is 1/2 of σ, the population standard deviation. When $n = 9$, the standard error of $\overline{X}$ is 1/3 of σ. A simple formula that fits these cases is

$$\text{Standard error of } \overline{X} = \frac{\sigma}{\sqrt{n}} \tag{4-5}$$

These formulas (4-4) and (4-5) not only fit the specific cases shown in Figure 4-5 but also work for *every* case of simple random sampling (as shown, for example, in Wonnacott, 1977, p. 147).

With the expected value and standard error of $\overline{X}$ established, let us finally review the shape of its distribution. In Figure 4-5, column (*a*) shows that a normal population yields a normal sampling distribution for $\overline{X}$. Column (*b*) shows the surprise noted earlier: Even though the population is highly skewed, the sampling distribution of $\overline{X}$ becomes approximately symmetric and normal as the sample size increases. This remarkable shape phenomenon can be proved to be generally true:

Suppose the parent population is nearly normal, *or* the sample size is fairly large (often $n = 5$ or 10 will be large enough). *In either case* the sampling distribution of $\overline{X}$ has an approximately normal shape.	(4-6)

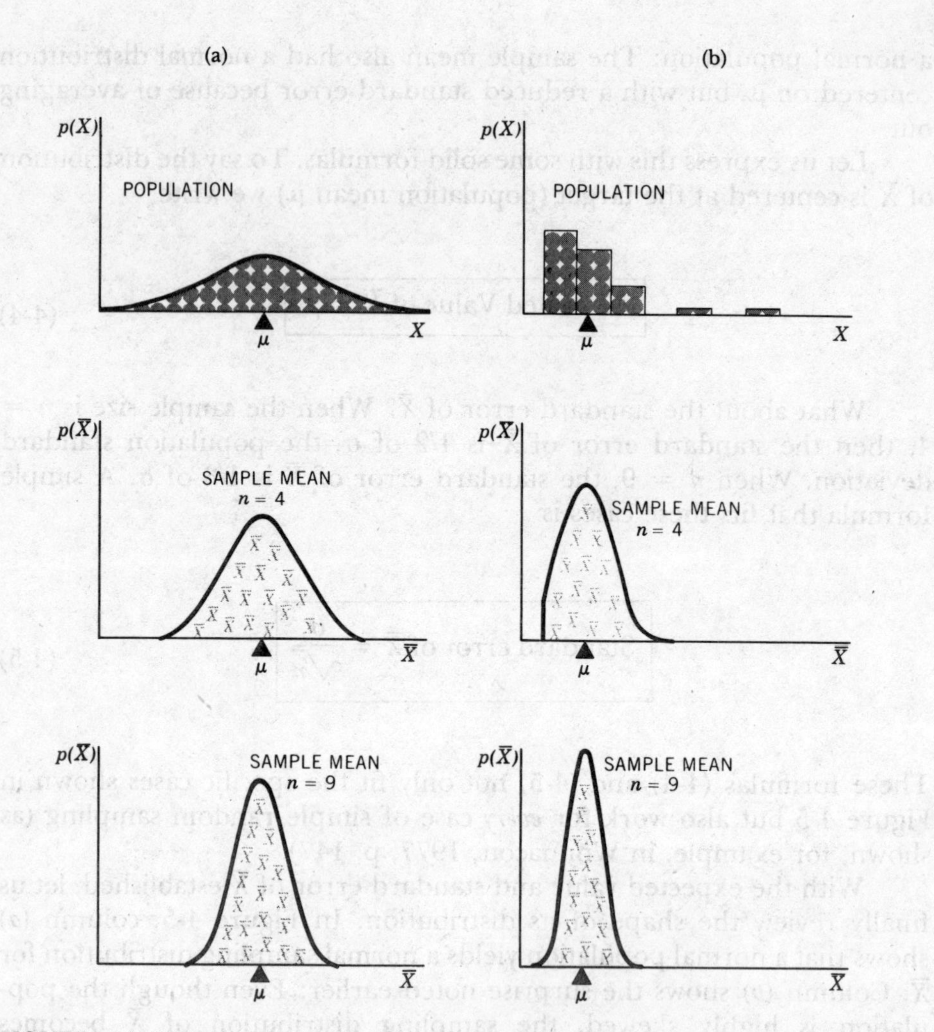

FIGURE 4-5 (*a*) Column (a) shows sampling from a normal population. As sample size n increases, the standard error of $\overline{X}$ decreases. (*b*) Even though the population is not normal, this column shows the sample mean $\overline{X}$ is still approximately normal. And the standard error of $\overline{X}$ is again small.

Our conclusions so far may be summarized into one statement, which nicely sums up the most remarkable and important issue in statistics:[1]

[1]Mathematicians call (4-7) the *Central Limit Theorem*. It is true provided the population has a finite standard deviation, that is, terribly extreme values are not too probable. (In practice, this condition is nearly always met.)

> **The Normal Approximation Theorem.** In random samples (VSRS) of size n, the sample mean $\overline{X}$ fluctuates around the population mean μ with a standard error of $\sigma/\sqrt{n}$ (where σ is the population standard deviation). As n increases, the distribution of $\overline{X}$ therefore fluctuates less and less around its target μ. It also gets closer and closer to normal (bell-shaped). (4-7)

This theorem allows us to answer a great many sampling questions using the familiar normal tables, as the following examples illustrate.

Example 4-4

(a) Suppose a large class in statistics has marks normally distributed around a mean of 72 with a standard deviation of 9. Find the probability that an individual student drawn at random will have a mark over 80.

(b) Find the probability that a random sample of 10 students will have an average mark over 80.

(c) If the population were not normal, what would be your answer to part (b)?

Solution

(a) The distribution for an individual student (i.e., the population distribution) is shown below as the flat gray distribution $p(X)$. The score $X = 80$ is first standardized with its mean $\mu = 72$ and standard deviation $\sigma = 9$:

$$Z = \frac{X - \mu}{\sigma} = \frac{80 - 72}{9} = .89 \qquad \text{(3-24) repeated}$$

Thus,

$$\begin{aligned} \Pr(X > 80) &= \Pr(Z > .89) \\ &= .1867 \simeq .19 \end{aligned}$$

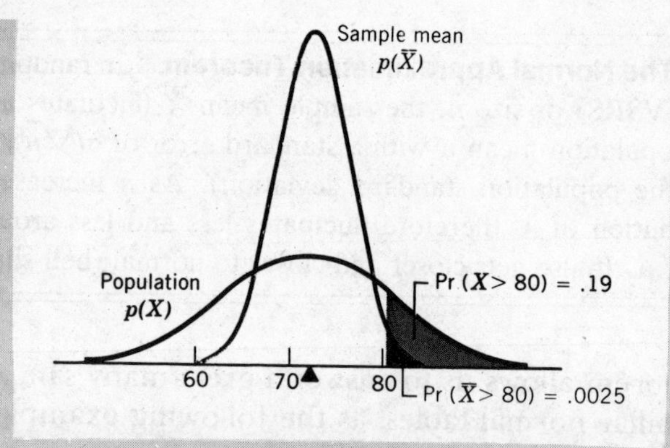

(b) The Normal Approximation Theorem (4-7) assures us that $\bar{X}$ has an approximately normal distribution as shown by the blue curve above, with

$$\text{Expected value} = \mu \quad = 72$$

$$\text{Standard error} = \frac{\sigma}{\sqrt{n}} = \frac{9}{\sqrt{10}} = 2.85$$

We use these to standardize the score $\bar{X} = 80$,

$$Z = \frac{\bar{X} - \mu}{\text{standard error}} \tag{4-8}$$

$$= \frac{80 - 72}{2.85} = 2.81 \qquad \text{like (3-24)}$$

Thus

$$\Pr(\bar{X} > 80) = \Pr(Z > 2.81)$$
$$= .0025$$

These two probability calculations may be easily compared in the figure. Although there is a reasonable chance (about 19%) that a single student will get over 80, there is very little chance (less than 1%) that a sample average of 10 students will perform this well. Once again we see how "averaging out" tends to reduce the extremes.

(c) According to the Normal Approximation Theorem (4-7), $\bar{X}$ has an approximately normal shape, no matter what the shape of the parent population. We would therefore get the same answer.

Often a sample mean is used to estimate a population mean. How likely is it that it will be close? The next example shows how the Normal Approximation Theorem will answer this question.

Example 4-5

A population of men on a large midwestern campus has a mean height $\mu = 69$ inches, and a standard deviation $\sigma = 5.1$ inches. If a random sample of $n = 100$ men is drawn, what is the chance the sample mean $\overline{X}$ will be within an inch of the population mean μ?

Solution

$\overline{X}$ has a normal distribution, according to the Normal Approximation Theorem (4-7), with

$$\text{Expected value} = \mu = 69$$

$$\text{Standard error} = \frac{\sigma}{\sqrt{n}} = \frac{5.1}{\sqrt{100}} = .51$$

We want to find the probability that $\overline{X}$ is within 1 inch of $\mu = 69$, that is, between 68 and 70. So we first calculate the probability above 70, beginning with its standardization:

$$Z = \frac{\overline{X} - \mu}{\text{standard error}} = \frac{70 - 69}{.51} = 1.96$$

Thus,

$$\Pr(\overline{X} > 70) = \Pr(Z > 1.96)$$

$$= .0250$$

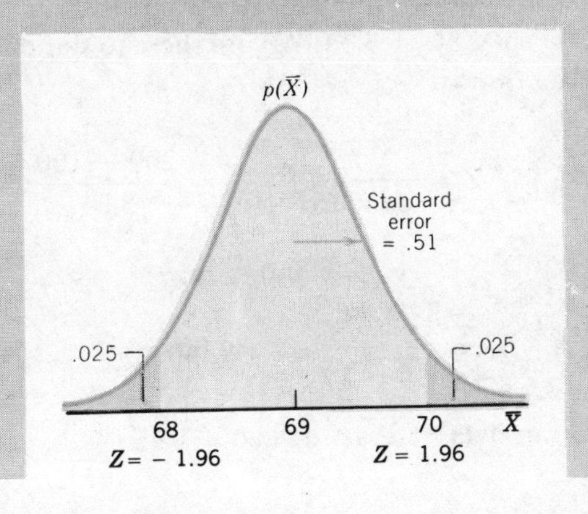

This is the right-hand tail shown in the figure. Because the normal distribution is symmetric, the left-hand tail has the same probability .025. Thus we can find the probability we want in the central chunk:

$$\text{Probability} = 1.00 - .025 - .025 = .95$$

We can therefore be highly confident (95%) that the sample mean will be within 1 inch of the population mean.

The next example will show how a problem involving a total can be solved by simply rephrasing it in terms of a mean.

Example 4-6

A ski lift is designed with a load limit of 10,000 pounds. It claims a capacity of 50 persons. Suppose the weights of all the people using the lift have a mean of 190 pounds and a standard deviation of 25 pounds. What is the probability that a random group of 50 persons will total more than the load limit of 10,000 pounds?

Solution

First, we rephrase the question: "A random sample of 50 persons will total more than 10,000 pounds" is exactly the same as "A random sample of 50 persons will *average* more than 10,000/50 = 200 pounds each." Thus we wish to make a statement about a sample mean. First, what is its distribution? From the Normal Approximation Theorem, $\overline{X}$ has an approximately normal distribution with expected value = μ = 190 and standard error = $\sigma/\sqrt{n} = 25/\sqrt{50} = 3.54$. We use these to standardize the average load limit of 200 pounds:

$$Z = \frac{\overline{X} - \mu}{\text{standard error}} = \frac{200 - 190}{3.54} = 2.83$$

Thus,

$$\Pr(\overline{X} > 200) = \Pr(Z > 2.83) = .0023$$

So the chance of an overload is only about 1/4%.

These examples have shown how easily the normal distribution can handle random sampling—either the sample mean, or the sample total via the mean.

*(b) Random Samples, Keeping Out the Chips

Recall that random sampling (VSRS), as defined in (4-1), does not hold if the population size N is small, and the chips are kept out as they are drawn. Keeping out the chips is more efficient because we do not risk drawing the same chip over again and repeating information already known. If N and n denote the population and sample size, the gain in efficiency may be expressed as

> When chips are kept out as they are sampled, the sampling fluctuation in $\overline{X}$ is reduced by the factor:
>
> $$\text{reduction factor} = \sqrt{\frac{N-n}{N-1}} \tag{4-9}$$

This means a change in the Normal Approximation Theorem (4-7). The standard error of $\overline{X}$ changes from $\sigma/\sqrt{n}$ to

$$\text{Standard error of } \overline{X} = \frac{\sigma}{\sqrt{n}}\sqrt{\frac{N-n}{N-1}} \tag{4-10}$$

Note that if the population is very much larger than the sample, the reduction factor (4-9) is nearly 1, and therefore can be ignored. For example, in the typical political poll where $N \simeq 100{,}000{,}000$ and $n \simeq 1000$,

$$\sqrt{\frac{N-n}{N-1}} = \sqrt{\frac{100{,}000{,}000 - 1000}{100{,}000{,}000 - 1}} = .999995 \simeq 1.00 \tag{4-11}$$

PROBLEMS

4-4 Suppose that 10 men were sampled randomly from the population of Table 4-1 and that their average height $\overline{X}$ was calculated. Then imagine that the experiment was repeated many, many times. Answer True or False; if False, correct it.

(a) The long and short men in the sample tend to average out, making $\overline{X}$ fluctuate less than a single observation.

(b) To be specific, the sample mean $\overline{X}$ fluctuates around its target μ with a standard error of only σ/n.

4-5 (a) A sociologist plans to randomly sample 25 incomes from a certain population of men under 30, a skewed population with mean μ = \$20,000 and standard deviation σ = \$6000. Her sample mean $\overline{X}$ will be a random variable that will only imperfectly reflect the population mean μ. In fact, the possible values of $\overline{X}$ will fluctuate around an expected value of ______ with a standard error of ______, and with a distribution shape that is ______.

(b) The sociologist is worried that her sample mean will be misleadingly high. A statistician assures her that it is unlikely that $\overline{X}$ will exceed μ by more than 10%. Calculate just how unlikely this is.

4-6 The weights of packages filled by a machine are normally distributed about a mean of 25 ounces, with a standard deviation of 2 ounces. What is the probability that a random sample of n packages from the machine will have an average weight of less than 24 ounces if:

(a) $n = 1$?
(b) $n = 4$?
(c) $n = 16$?
(d) $n = 64$?

4-7 Suppose that the education level among adults in a certain country has a mean of 11.1 years and a standard deviation of 3 years. What is the probability that in a random survey of 25 adults you will find an average level of schooling between 10 and 12 years?

4-8 A ski lift is designed with a load limit of 18,000 pounds. It claims a capacity of 100 persons. If the weights of all the people using the lift have a mean of 175 pounds and a standard deviation of 30 pounds, what is the probability that a group of 100 persons will exceed the load limit?

4-9 In order to be usable, suppose a bicycle chain must be within 1/4 inch of 54 inches. Because of slight variations in the manufacturing process, the individual links are not exactly uniform. Suppose they fluctuate slightly around a mean value of 1/2 inch, with a standard deviation of 1/200 inch.

(a) How many links should be strung together to form a chain?

(b) What proportion of the chains would then meet the given standard?

4-10 The 1200 tenants of a large apartment building have weights distributed as follows:

Weight (pounds)	*Proportion of Tenants*
50	.20
100	.30
150	.40
200	.10

Each elevator in the building has a load limit of 2800 pounds. If 20 tenants crowd into one elevator, what is the probability it will be overloaded?

4-11 In each of Problems 4-7 to 4-10, what crucial assumption did you implicitly make? Suggest some circumstances where it would be seriously violated. Then how would the correct answer differ?

4-12 In a young couple's club of 50 families, family size varied as follows:

X = Number of Children	*Frequency*
0	25
1	15
2	10
Total	50

Five couples were chosen as delegates to a conference on family relations. What is the chance that this group of 5 couples would have more than an average of 1 child per family:

(a) If they were the first 5 couples to volunteer?
(b) If they were chosen by lot?

4-4 PROPORTIONS

(a) Proportions as Sample Means

The first bowl of chips we sampled (men's heights in Figure 4-1) had all sorts of numbers written on them. In our next example we will be sampling from a bowl where there are only two numbers—0 and 1. This will provide a way to deal with a problem first stated in Chapter 1: How can we make statistical forecasts of an election?

Example 4-7

An election may be interpreted as asking every voter in the population, "How many votes do you cast for the Republican candidate?" If this is an honest election, the voter has to reply either 0 or 1. In the 1972 presidential election (the Republican Nixon versus the Democrat McGovern, ignoring third parties as usual), 60% voted Republican ($X = 1$). The population distribution was therefor as follows:

X = *Number of Republican Votes by an Individual*	$p(X)$ = *Relative Frequency*
0	.40
1	.60

(a) What is the population mean? The population variance? The population proportion of Republicans?
(b) When 10 voters were randomly polled, they gave the following answers:

1, 0, 1, 1, 1, 0, 1, 0, 1, 1

What is the sample mean? The sample proportion of Republicans?

Solution

(a) We calculate the population mean and variance in Table 4-4—a short and peculiar table. In fact, it's as short as a table can possibly be. This makes the arithmetic very simple, however, and we easily find the mean is .60—which exactly equals the proportion of Republicans.

We also find the variance is .24—which equals .60 × .40 (the proportion of Republicans and Democrats, respectively). It is remarkable that a table can be so simple to calculate, and at the same time yield such interesting answers.

(b) The sample mean is $\bar{X} = \Sigma X/n = 7/10 = .70$. The sample proportion is $P = 7/10$, which exactly equals the sample mean. Once again, just as in part (a), we have found *the proportion coincides with the mean.*

TABLE 4-4

Population Mean and Variance of a 0-1 Variable, When the Proportion of Republicans is $\pi = 60\% = .60$

X = *Number of Republican Votes by an Individual*	$p(X)$ = *Relative Frequency* = *Population Proportion*	$X\,p(X)$	$(X - \mu)$	$(X - \mu)^2$	$(X - \mu)^2 p(X)$
0	.40	0	−.60	.36	.144
1	.60	.60	.40	.16	.096
		$\mu = .60$			$\sigma^2 = .24$

From this example, we can easily generalize. To set things up, for a single voter drawn at random, let

$$X = \text{the number of Republican votes he casts} \qquad (4\text{-}12)$$

If this seems a strange way to define such a simple random variable, we could explicitly define it with a formula:

$$\begin{aligned} X &= 1 \text{ if he votes Republican} \\ &= 0 \text{ otherwise} \end{aligned} \qquad (4\text{-}13)$$

Thus the population of voters may be thought of as the bowl of chips, marked 0 or 1, shown in Figure 4-6. As usual, we shall let π denote the population proportion that is Republican (proportion of chips marked 1). Then we could find the moments of the population in terms of this proportion, just as in Example 4-7:

$$\text{Population mean } \mu = \text{population proportion } \pi \tag{4-14}$$

$$\text{Population standard deviation } \sigma = \sqrt{\pi(1-\pi)} \tag{4-15}$$

FIGURE 4-6 A 0-1 population (population of voters).

When we takc a sample, the sample mean is calculated by adding up the 1's (counting up the Republicans), and dividing by n. This of course yields the sample proportion of Republicans, so that

$$\boxed{\text{Sample mean } \overline{X} = \text{sample proportion } P} \tag{4-16}$$

like (4-14)

Thus we have found an ingenious way to handle a proportion: It is simply a disguised mean—of a 0-1 variable. (Since it provided such a simple way of counting the number of Republicans, the 0-1 variable is also called a *counting variable.* Other names are *on-off variable, binary variable,* or, most commonly, *dummy variable.*)

As means, proportions can be easily handled with the general theory of sampling already developed. How, for example, does the sample proportion P fluctuate around the population proportion π? Since P is just the sample mean $\overline{X}$ in disguise, we can find its expected value by substituting (4-16) and (4-14) into (4-4):

$$\boxed{\text{Expected value of } P = \pi} \tag{4-17}$$

The standard error of P is obtained by substituting (4-16) and (4-15) into (4-5):

$$\text{Standard error of } P = \sqrt{\frac{\pi(1-\pi)}{n}} \tag{4-18}$$

Finally, the Normal Approximation Theorem assures us that, for large samples, the distribution is approximately normal. We may therefore summarize:

> **The Normal Approximation Theorem for Proportions.** In random samples (VSRS) of size n, the sample proportion P fluctuates around the population proportion π with a standard error of $\sqrt{\pi(1-\pi)/n}$. As n increases, the distribution of P therefore fluctuates less and less around its target π. It also gets closer and closer to normal (bell-shaped).

(4-19)
like (4-7)

This theorem allows us to answer all kinds of interesting questions about proportions, as the following example illustrates.

Example 4-8

Consider again the 1972 American presidential election, in which 60% voted for the Republican candidate, Nixon. Just before the election, suppose a random sample of 30 voters was polled in order to forecast the election winner. If a minority such as 13/30 were Republicans in the sample, a Republican defeat would be forecast for the election. Since the majority (60%) of the population was Republican, this sample would clearly be misleading—an unlucky draw.

What would be the probability of drawing an unlucky sample with a Republican minority; that is, the chance of an erroneous forecast?

Solution

If P denotes the proportion of Republicans in the sample, we want the probability that P is less than 50% (.50). By the Normal Approximation Theorem (4-19), P fluctuates normally around $\pi = .60$ with a standard error $= \sqrt{\pi(1-\pi)/n} = \sqrt{.60(.40)/30} = .0894$. We use these to standardize the critical value $P = .50$,

$$Z = \frac{P - \pi}{\text{standard error}} = \frac{.50 - .60}{.0894} = -1.12 \qquad \text{like (4-8)}$$

Thus,

$$\begin{aligned} \Pr(P < .50) &= \Pr(Z < -1.12) \\ &= .1314 \end{aligned} \tag{4-20}$$

So the chance of an incorrect forecast is about 13%.

The next example will show how a problem involving a total count can be solved by simply rephrasing it in terms of a proportion.

Example 4-9

Of your first 15 grandchildren, what is the chance that there will be more than 10 boys?

Solution

First we rephrase the question: "More than 10 boys" is exactly the same as "The *proportion* of boys is more than 10/15." In this form, we can use the Normal Approximation theorem (4-19): P fluctuates normally around $\pi = .50$ with a standard error $= \sqrt{\pi(1-\pi)/n} = \sqrt{.50(.50)/15} = .129$. We use these to standardize the critical proportion $P = 10/15$,

$$Z = \frac{P - \pi}{\text{standard error}} = \frac{\frac{10}{15} - .50}{.129} = 1.29 \tag{4-21}$$

Thus,

$$\Pr\left(P > \frac{10}{15}\right) = \Pr(Z > 1.29) = .0985 \tag{4-22}$$

So the probability of more than 10 boys is about 10%.

Example 4-9 is recognized as a binomial problem. It could have been answered more precisely, but with a lot more work, by adding up the probability of getting 11 boys, plus the probability of getting 12 boys, and so on—with each of these probabilities calculated from the binomial formula (3-10). Since the normal approximation is so much easier to apply, we shall use it henceforth, and refer to it as the *normal approximation to the binomial.*

*(b) Continuity Correction

The Normal Approximation theorem (4-19) is only an approximation, and therefore does involve some error. To illustrate this error, note that we could have answered the question in Example 4-9 *equally well* by rephrasing it differently at the beginning: "More than 10 boys" is exactly the same as "11 or more boys," which in turn is exactly the same as "the *proportion* of boys is 11/15 or more." If we pursued this, we would get an answer slightly different from (4-22), but nevertheless an equally good approximation. Rather than having to choose between these two approximations, we could get an even better approximation by striking a compromise right at the beginning: Instead of 10 boys or 11 boys, use 10½ boys. When this is used in (4-21), we obtain

$$Z = \frac{P - \pi}{\text{standard error}} = \frac{\frac{10\frac{1}{2}}{15} - .50}{.129} = 1.55 \tag{4-23}$$

Thus,

$$\begin{aligned} \Pr\left(P > \frac{10\frac{1}{2}}{15}\right) &= \Pr(Z > 1.55) \\ &= .061 \end{aligned} \tag{4-24}$$

This compromise agrees very well with the correct answer of .059, which we would have obtained if we had gone back and answered the question absolutely correctly by using the binomial distribution. Specifically, this compromise is much better than the answer of .098 in (4-22). Why? Because we arrived at answer (4-22) by approximating a discrete distribution (the binomial) with a continuous distribution (the normal). The compromise (4-24) allowed us to correct for this mismatch, and so is customarily called the *continuity correction.* This correction is so helpful that it is worth stating in general:

The continuity correction to the binomial is obtained by first phrasing the question in the two possible ways (for example, "more than 10" or "11 or more.") Then the halfway value ($10\frac{1}{2}$) is used in the subsequent normal approximation.	(4-25)

PROBLEMS

If you want high accuracy in your answers (optional), you should do them with continuity correction (wcc) as given in (4-25), especially if the sample size n is small.

4-13 In a certain city, 55% of the eligible jurors are women. If a jury of 12 is picked fairly (at random), what is the probability that there would be 3 or fewer women?

4-14 In the 1980 presidential election, 34.9 million voted Democratic and 43.2 million voted Republican (Carter versus Reagan, ignoring third parties as usual). The typical political poll randomly samples 1500 voters. What is the probability that such a poll would correctly forecast the election winner, i.e., that a majority of the sample would be Republican?

4-15 In the New Year, what is the probability that of the first 50 babies born, 40% or more will be boys?

4-16 (a) If a fair coin is tossed $n = 15$ times, what is the chance of getting about 50% heads—specifically, between 40% and 60%?
(b) Repeat, for $n = 30$ tosses.
(c) Repeat, for $n = 100$ tosses.
(d) Repeat, for $n = 1000$ tosses.
Note how the chance of getting about 50% heads approaches certainty, as n increases.

4-17 Recall that probability was defined as limiting relative frequency. That means, for example, that if a fair die is thrown a million times, the relative frequency of aces (or proportion P) will likely be very close to 1/6. To be specific, calculate the probability that P will be within .001 of 1/6.

4-18 What is the chance that, of the first 10 babies born in the New Year, 7 or more will be boys? Answer in three ways:

(a) Exactly, using the binomial distribution.

(b) Approximately, using the normal distribution.

*(c) Approximately, using the normal approximation with continuity correction.

*(d) Answer True or False; if False, correct it.

Part (c) illustrates that the normal approximation with continuity correction can be an excellent approximation, even for n as small as 10.

CHAPTER 4 SUMMARY

4-1 When all individuals in a population are equally likely to be drawn in a sample, we can obtain unbiased estimates of the population. If the n observations in the sample are also independently drawn, the sample is called a VSRS (very simple random sample or, more briefly, a random sample). One way to get a random sample is to draw n chips from a very large bowlful. Random sampling is easily simulated, and is the fundamental sampling technique of this book.

4-2 From a simulated sample, we can calculate $\bar{X}$. We can try our luck again and again, drawing many many samples, and each time calculate the sample mean $\bar{X}$. If we then assemble all of these $\bar{X}$ values, they form the *Monte Carlo* sampling distribution of the mean.

4-3 The graph of the sampling distribution shows that $\bar{X}$ fluctuates normally around its target μ, with a standard error $= \sigma/\sqrt{n}$. This formula tells us that the larger the sample size n, the less $\bar{X}$ will fluctuate around μ.

4-4 Proportions are easily handled with general formulas already developed. We simply recognize a proportion as a disguised mean, using a 0-1 variable. We thus find that the sample proportion P fluctuates normally about its target π, with a standard error $= \sqrt{\pi(1 - \pi)/n}$.

REVIEW PROBLEMS

4-19 Match the symbol on the left with the phrase on the right:

μ	sample mean
$\overline{X}$	sample variance
σ^2	population proportion
s^2	population variance
π	sample proportion
P	population mean

4-20 Fill in the blank:

Suppose that in a certain election, the United States and California are alike in their proportion of Democrats, π, the only difference being that the United States is about 10 times as large a population. In order to get an equally reliable sampling estimate of π, the U.S. sample should be ______ as large as the California sample.

4-21 American women have heights that are approximately normally distributed around a mean of 64 inches, with a standard deviation of 3 inches. Calculate and illustrate graphically each of the following:

(a) If an American woman is drawn at random, what is the chance her height will exceed 66 inches (5 ft, 6 in.)?

(b) If a sample of 25 are randomly drawn, what is the chance that the sample average will exceed 66 inches?

4-22 When polarized light passes through α-lactose sugar, it is rotated by an angle of exactly 90°. The *observed* angle, however, is somewhat in error; suppose, it is normally distributed around 90° with a standard deviation of 1.2°. A sample of 4 independent observations is taken.

(a) What is the chance that the first observation exceeds 91°? (Since another sugar, D-xylose, rotates polarized light by 92°, the observer might then mistakenly think the α-lactose was D-xylose.)

(b) What is the chance that all 4 measurements will exceed 91°?

(c) What is the chance that the average of the 4 measurements will exceed 91°? [Before you calculate the chance, can you say how it will compare to (a) and (b)?]

4-23 Suppose the breaking strength of a rope is the sum of its 9 component strands of hemp; these strands are drawn at random from a large supply, whose mean breaking strength is 50 pounds and standard deviation is 12 pounds. What is the chance that a rope will be broken by a 400-pound load?

4-24 A large company uses many thousands of lightbulbs that burn with an average life of 80 days and a standard deviation of 20 days. If 1000 are installed on January 1, about how many will need to be replaced by April 1 (94 days later)?

4-25 (a) Suppose that the population of weights of airline passengers has mean 150 pounds and standard deviation 25 pounds. A certain plane has a capacity of 7800 pounds. What is the chance that a flight of 50 passengers will overload it?

(b) What assumptions are you making implicitly? In what way do you think they are questionable?

*(c) To reduce the chance of overload to just 1%, what must the capacity be?

*4-26 In a certain town, all traffic lights alternate red and green, 50 seconds each. To reach your destination requires passing 6 such lights, which are so far apart that there is no attempt to synchronize them. Thus the 6 times that you have to wait may be regarded as independent random variables. Each waiting time has half a chance to be 0 (if you are lucky and hit a green light) and half a chance to be approximately 5, 15, 25, 35, or 45 seconds (if you are unlucky and hit a red light).[2]

(a) Simulate such a 6-light trip, and calculate $\overline{X}$, the average waiting time per light.

(b) Have the instructor graph the frequency distribution of the answers to (a) obtained by all the class.

(c) Work out the theoretical sampling distribution of $\overline{X}$, and graph it. Compare it to (b).

(d) What is the chance that you will have to wait 2 minutes or more for the 6 lights, that is, an average of 20 seconds or more per light?

If you travelled this route 250 times next year on your way to work, about how many times would such a long wait occur?

---

[2]Strictly speaking, if there is a red light, the waiting time has a *continuous* distribution: *All* values from 0 to 50 are equally likely. We have grouped these values into cells of width 10, whose midpoints are 5, 15, 25, 35, 45, so that we can work out the problem with sums instead of calculus.

*4-27 At Las Vegas, roulette is played with a wheel that has 38 slots—20 losing slots and 18 winning slots. Your chances of losing your dollar are therefore 20/38, and of winning a dollar are 18/38.

(a) In the very long run, what is the average loss per play?

(b) What are a player's chances of ending up a net loser, if he plays:

(i) 5 times?

(ii) 25 times?

(iii) 125 times?

CHAPTER 5

CONFIDENCE INTERVALS

Education is a man's going forth from cocksure ignorance to thoughtful uncertainty.

Don Clark

5-1 INTRODUCTION: DEDUCTION AND INDUCTION

In the last chapter we started with a known population, and asked how a sample behaves. How close does the sample mean $\overline{X}$ come to the known population mean μ? This is called *deduction*—arguing from the general (population) to the specific (sample) as shown in Figure 5-1*a*.

In this chapter we will turn this argument around, to deal with the research problems that usually appear in practice: From a known sample we have drawn, what can we conclude about the unknown population from which it was taken? This is called *induction,* or statistical inference—arguing from the specific (sample) to the general (population) as shown in Figure 5-1*b*.

Before proceeding, however, in Table 5-1 we review the important differences between the sample and population. In the population, for example, it is essential to remember that the mean μ and variance σ^2 are fixed constants (though generally unknown). These are called *population parameters*.

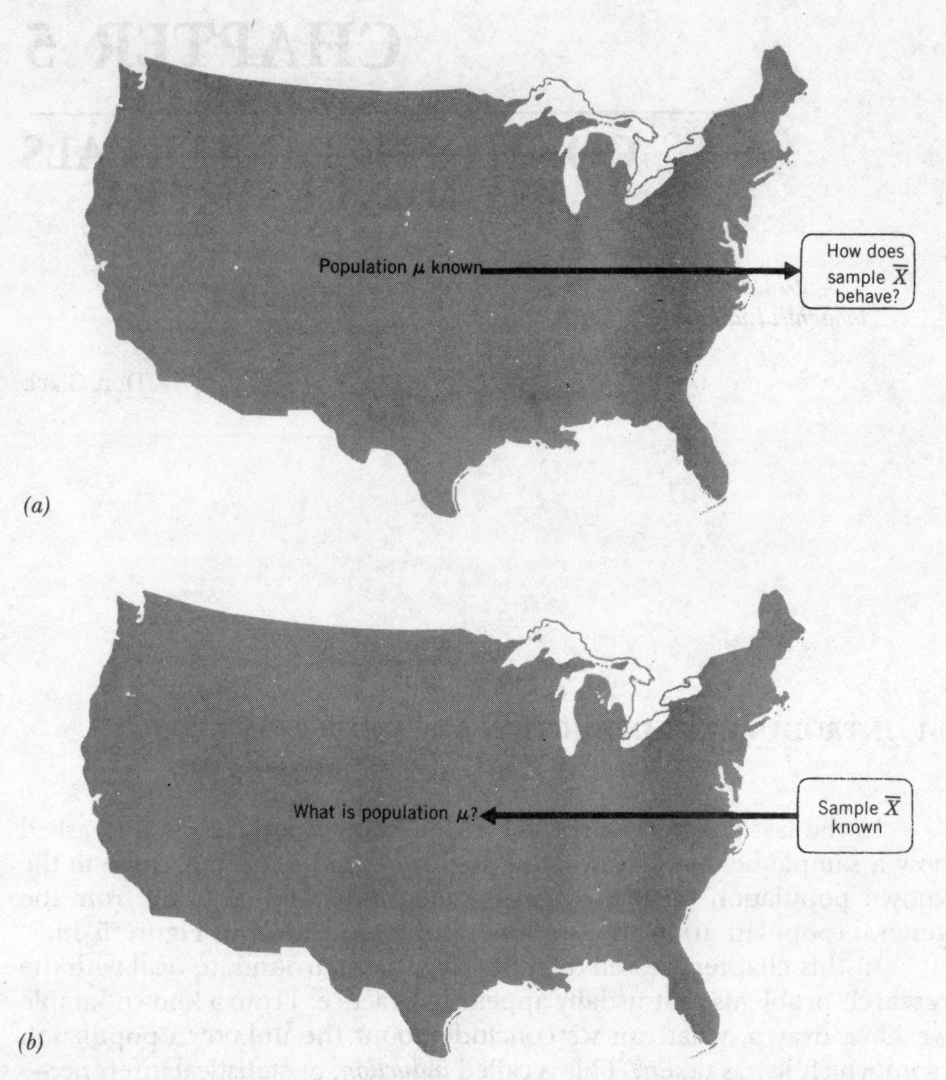

FIGURE 5-1 Deduction and induction contrasted. (*a*) Deduction (behavior of $\overline{X}$ in Chapter 4). (*b*) Induction (inference about μ, in Chapter 5).

In contrast, the sample mean $\overline{X}$ is a random variable, varying from sample to sample. Specifically, its distribution was found to be approximately normal in (4-7). A random variable such as $\overline{X}$ that is calculated from the observations in a sample is given the technical name *sample statistic.*

To keep clear the distinction between the sample and population we remind you of the two important conventions we use:

1. Sample statistics are denoted by ordinary English letters, whereas population parameters are denoted by Greek letters.
2. Samples are colored blue, and populations gray.

TABLE 5-1
Review of Sample Versus Population

Random Sample	*Population*
Relative frequencies f/n are used to compute $\bar{X}$ and s^2,	Probabilities $p(X)$ are used to compute μ and σ^2,
which are examples of random statistics or estimates.	which are examples of fixed parameters or targets.
ENGLISH LETTERS COLORED BLUE	GREEK LETTERS COLORED GRAY

5-2 95% CONFIDENCE INTERVAL FOR A MEAN μ

(a) Introduction

Statistical induction is based on deduction of the kind we developed in the last chapter. There we saw how closely we could expect a sample mean $\bar{X}$ to come to its target μ, the population mean. For instance, Example 4-5 showed that in sampling $n = 100$ observations from a population of men's heights, the sample mean $\bar{X}$ was almost certain (95% certain) to be less than an inch away from its target μ. In this chapter we don't know the population mean μ; instead, we know the sample mean $\bar{X}$ that we have observed. But we can still be almost certain that they are less than an inch apart. In other words, the population mean μ will be within an inch of the observed sample mean. We write this as

$$\mu = \bar{X} \pm 1 \text{ inch} \tag{5-1}$$

and call it a 95% confidence interval. That's statistical inference in a nutshell.

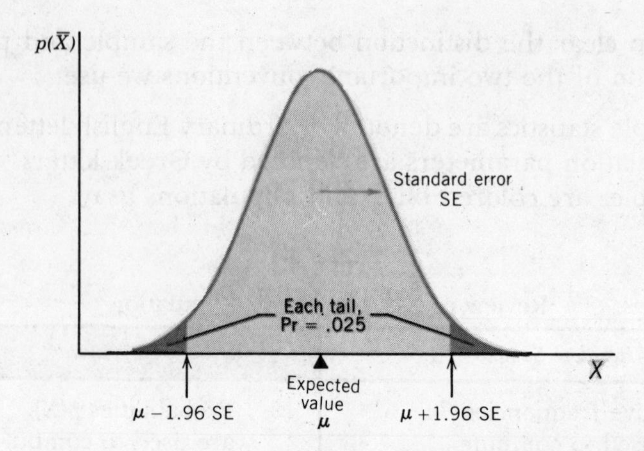

FIGURE 5-2 Normal distribution of the sample mean around the fixed but unknown parameter μ. 95% of the probability is contained within $\pm$ 1.96 standard errors.

(b) Confidence Intervals in General

Recall that the sample mean $\overline{X}$ is normally distributed around its target μ, as shown in Figure 5-2. For 95% confidence, therefore, we just leave 2½% probability in each tail of the normal distribution of $\overline{X}$. In Appendix Table IV, we find this requires going 1.96 standard deviations from μ. That is, there is a 95% chance that $\overline{X}$ and μ are within 1.96 standard errors (SE) of each other. The 95% confidence interval may therefore be written:

95% confidence interval for μ

$$\mu = \overline{X} \pm 1.96\,\text{SE} \tag{5-2}$$

where the standard error of $\overline{X}$ is

$$\text{SE} = \frac{\sigma}{\sqrt{n}} \tag{5-3}$$

(4-5) repeated

We must be exceedingly careful not to misinterpret this. The target μ has not become a variable, but remains a population constant. Equation

(5-2) is a probability statement about the random variable $\bar{X}$ or, more precisely, the *random interval* $\bar{X} \pm 1.96$ SE. It is this interval that varies, not μ.

(c) Illustration

To emphasize that it is the confidence interval that fluctuates while μ remains constant, let us return to our example: Suppose we have a random sample of $n = 25$ men to construct an interval estimate for μ, the mean height of the whole population of men on a large midwestern campus. In addition, to clearly illustrate what is going on, suppose we have supernatural knowledge of the population μ and σ; suppose we know that

$$\left.\begin{aligned} \mu &= 69 \\ \sigma &= 5.1 \end{aligned}\right\} \tag{5-4}$$

Then from (5-3), the standard error of $\bar{X}$ is

$$SE = \frac{5.1}{\sqrt{25}} = 1.02 \tag{5-5}$$

Now let us observe what happens when the statistician (who of course does not have our supernatural knowledge) tries to bracket μ with the confidence interval (5-2). And to appreciate the random nature of his task, let us make him re-estimate the confidence interval, say 50 times, each time using a different random sample of 25 men.

Figure 5-3 illustrates the statistician's typical experience: $\bar{X}$ is distributed around $\mu = 69$, with a 95% chance that it lies as close as

$$\pm 1.96\,SE = \pm 1.96(1.02) = \pm 2.0 \tag{5-6}$$

That is, there is a 95% chance that any $\bar{X}$ will fall in the range *AB* from 67 to 71 inches.

But the statistician does not know this. He can only blindly take his random sample, and use it to calculate $\overline{X}$; suppose this turns out to be 70. From (5-2) he calculates the appropriate 95% confidence interval for μ:

$$\mu = \overline{X} \pm 1.96\,\text{SE} \tag{5-7}$$
$$= 70 \pm 1.96\,(1.02) \tag{5-8}$$
$$= 70 \pm 2 \tag{5-9}$$
$$= 68 \text{ to } 72 \tag{5-10}$$

This interval estimate for μ is the first one shown in Figure 5-3. We note that in his first effort the statistician is right; μ is indeed enclosed in this interval.

In his second sample, suppose he happens to draw a shorter group of individuals, and duly computes $\overline{X}$ to be 68.2 inches. From a similar evaluation of (5-2) he comes up with his second interval shown in Figure 5-3. If he continues in this way to construct 50 interval estimates, about 95% of them will bracket the constant μ. Only about 2 or 3 will miss the mark.

We can easily see why the statistician is right so often. For each interval estimate, he is simply adding and subtracting 2 inches from the sample mean; and this is the same $\pm$ 2 inches that was calculated in (5-6) and defines the 95% range *AB* around μ. Thus, if and only if he observes a sample mean within this range, will his interval estimate bracket μ. And this happens 95% of the time, in the long run.

In practice, of course, a statistician would not take many samples—he only takes one. And once this interval estimate is made, he is either right or wrong; this interval brackets μ or it does not. But the important point to recognize is that the statistician is using a method with a 95% chance of success; in the long run, 95% of the intervals he constructs in this way will bracket μ.

(d) Analogy: Pitching Horse Shoes

Constructing 95% confidence intervals is like pitching horseshoes. In each case there is a fixed target, either the population μ or the stake. We are trying to bracket it with some chancy device, either the random interval or the horseshoe. This analogy is illustrated in Figure 5-4.

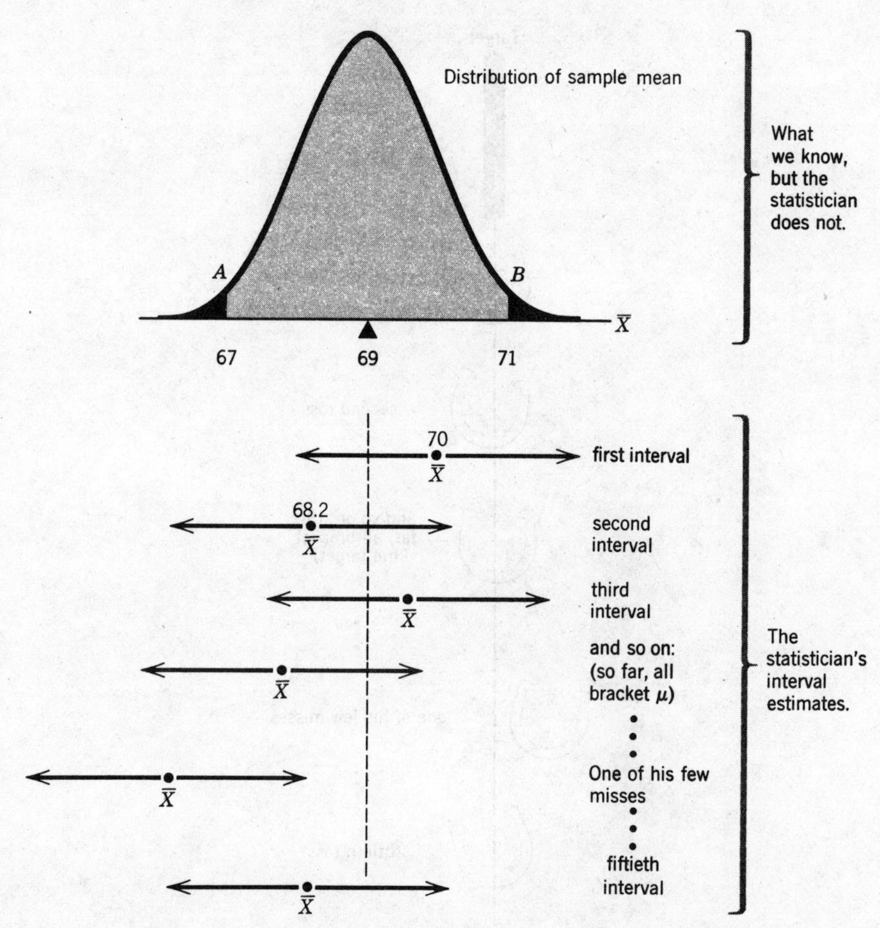

FIGURE 5-3 Constructing 50 interval estimates: About 95% correctly bracket the target μ.

There are two important ways, however, that confidence intervals differ from pitching horseshoes. First, only *one* confidence interval is customarily constructed. Second, the target μ is *not* visible like a horseshoe stake. Thus, whereas the horseshoe player always knows the score (and specifically, whether or not the last toss bracketed the stake), the statistician does not. He continues to "throw in the dark," without knowing whether or not a specific interval estimate has bracketed μ. All he has to go on is the statistical theory that assures him that, in the long run, he will succeed 95% of the time.

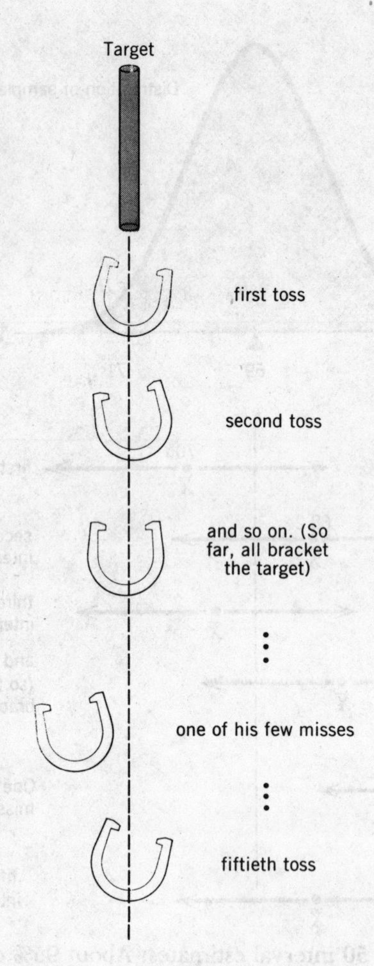

FIGURE 5-4 How an expert pitching 50 horsehoes is like the statistician in Figure 5-3: About 95% of his attempts correctly bracket the target.

(e) Review

To review, we briefly emphasize the main points:

1. The population parameter μ is constant, and remains constant. It is the interval estimate that is a random variable, because its center $\overline{X}$ is a random variable.

2. To appreciate where it came from, the confidence interval (5-2) may be written as

$$\boxed{\mu = \bar{X} \pm z_{.025}\,\text{SE}} \tag{5-11}$$

   where $z_{.025}$ is the value 1.96 obtained from Appendix Table IV, that is, the Z value that cuts off $2\frac{1}{2}\%$ from the upper tail (and by symmetry, also $2\frac{1}{2}\%$ from the lower tail). Equation (5-11) is extremely useful as the prototype for all confidence intervals that we will study. When we substitute (5-6) into (5-11), we obtain another very useful form:

   95% confidence interval

$$\mu = \bar{X} \pm z_{.025}\frac{\sigma}{\sqrt{n}} \tag{5-12}$$

3. As sample size is increased, $\bar{X}$ has a smaller standard error $\sigma/\sqrt{n}$, and therefore the confidence interval becomes more narrow and precise. This is the value of increasing sample size.
4. Suppose we wish to be more confident, for example, 99% confident. Then the range must be large enough to encompass 99% probability. Since this leaves .005 in each tail, the formula (5-12) would use $z_{.005} = 2.57$ (found in Table IV). Thus the confidence interval becomes more vague. This is exactly as we would expect: The more certain we want to be about a statement, the more vague we must make it.

An example will illustrate how easily confidence intervals can be calculated.

Example 5-1

Sixteen marks were randomly sampled from a very large class having a standard deviation of 12. If these 16 marks had a mean of 58, find a 95% confidence interval for the mean mark of the whole class.

Solution

Substitute $n = 16$, $\sigma = 12$, and $\overline{X} = 58$ into (5-12):

$$\mu = 58 \pm 1.96 \frac{12}{\sqrt{16}}$$

$$= 58 \pm 6$$

That is, the 95% confidence interval for μ is

$$52 < \mu < 64$$

The confidence intervals we develop in this chapter work well if the parent population is normal, or nearly normal. But in some cases where the parent population is a quite different shape, it may be better to construct a confidence interval around the sample median, rather than around the sample mean $\overline{X}$. (So-called *nonparametric* techniques of this kind are discussed in Chapter 13.)

PROBLEMS

5-1 Make the correct choice in each square bracket:

(a) The sample mean [$\overline{X}$, μ] is an estimate of the population mean [$\overline{X}$, μ].

(b) $\overline{X}$ fluctuates from sample to sample with a standard deviation equal to [σ/n, $\sigma/\sqrt{n}$], which is also called the [standard error or SE, population standard deviation].

(c) If we make an allowance of about [$\sqrt{n}$, 2] standard errors on either side of $\overline{X}$, we obtain an interval wide enough to contain the target μ, 95% of the time. This is called the [population parameter, 95% confidence interval].

(d) For greater confidence such as 99%, the confidence interval must be made [narrower, wider].

5-2 (a) Suppose that you took a random sample of 10 accounts in a large department-store chain, and found that the mean balance due was \$27.60. If you know that the standard deviation of all balances due is \$12.00, find the 95% confidence interval for the mean of all balances due.

(b) Explain to the vice president the meaning of your answer to (a) as simply as you can.

(c) Suppose that the skeptical vice president undertook a complete accounting of the whole population of balances due, and that the mean balance due turned out to be $29.10. What would you say?

5-3 A random sample of 4 infants from a population (hospital nursery) gave the following birth weights (kilograms):

3.1, 2.8, 3.6, 3.7

(a) If the population standard deviation is .4, find the 95% confidence interval for the population mean birth weight.
(b) Find the 99% confidence interval.

5-4 An anthropologist measured the heights (in inches) of a random sample of 100 men from a certain island, and found the sample mean to be 71.3. If the population standard deviation is 3 inches,
(a) Find a 95% confidence interval for the mean height μ of the whole population.
(b) Suppose the anthropologist wanted the 95% confidence interval to be narrower—to be specific, ± .50 inch. How large a sample should she gather?
(c) In part (b) we found out what sample size n was required to meet a specified sampling error allowance e = .50 inch. Show that the general formula for n in terms of e and σ is

$$\boxed{n = \left[1.96 \frac{\sigma}{e}\right]^2}$$

(d) If we want 25 times the accuracy (i.e., e reduced by a factor of 25), how much larger must the sample be?

5-3 USING t, WHEN σ IS ESTIMATED BY s

In the previous section it was assumed, somewhat unrealistically, that the statistician knows the true population standard deviation σ. In this section, we consider the more typical case where she does not know σ.

With σ unknown, the statistician wishing to evaluate the confidence interval (5-12) must use some estimate of σ—with the most obvious candidate being the *sample* standard deviation s. (Note that s, along with $\overline{X}$, can always be calculated from the sample data.) But the use of s introduces an additional source of unreliability, especially if the sample is small. To retain 95% confidence, we must therefore widen the interval, by replacing $z_{.025}$ with a larger number customarily denoted $t_{.025}$. When we substitute s and the compensating $t_{.025}$ into (5-12), we obtain

95% confidence interval for the population mean

$$\mu = \overline{X} \pm t_{.025} \frac{s}{\sqrt{n}} \tag{5-13}$$

The value $t_{.025}$ is listed in the shaded column of Appendix Table V, and is tabulated according to the degrees of freedom (d.f.):

$$\begin{aligned} \text{d.f.} &= \text{amount of information used in calculating } s^2 \\ &= \text{divisor in } s^2 \end{aligned} \tag{5-14}$$

The concept of d.f. was explained already in equation (2-14). Recall that altogether in the sample, there are n observations (pieces of information, or d.f.). One d.f. is used up in calculating the mean, leaving $n - 1$ d.f. for the variance:

$$\text{d.f.} = n - 1 \tag{5-15}$$

For example, if the sample size is $n = 6$, we read down Table V to d.f. $= 5$ (in the left-hand column), which gives $t_{.025} = 2.571$. Note that this is indeed larger than $z_{.025} = 1.96$.

In practice, when do we use the normal z table, and when the t table? (Incidentally, these are the two most useful tables in statistics. So useful, in fact, that we repeat them inside the front cover, for easy reference.) If σ is known, the normal z value in (5-12) is appropriate. If σ has to be estimated with s, the t value in (5-13) is appropriate—regardless of sample size. However, if the sample size is large, the normal z is an accurate enough approximation[1] to the t. So in practice, t is used only for *small samples* when *σ is unknown*. (You may think of t as standing for *tiny* sample adjustment.) The next example illustrates this.

Example 5-2

From a large class, a sample of 5 grades were drawn: 58, 60, 53, 81, and 73. Calculate a 95% confidence interval for the whole class mean μ.

Solution

Since $n = 5$, d.f. $= 4$; therefore, from Table V inside the front cover, $t_{.025} = 2.776$. In Table 5-2 we calculate $\bar{X} = 65$ and $s^2 = 134.5$. When all these are substituted into (5-13), we obtain

$$\begin{aligned} \mu &= 65 \pm 2.776 \frac{\sqrt{134.5}}{\sqrt{5}} \\ &= 65 \pm 14 \end{aligned} \tag{5-16}$$

That is, the mean grade of the whole class is between 51 and 79. This is a pretty vague interval—because the sample is so small.

TABLE 5-2
Analysis of One Sample

Observed Grade X	$(X - \bar{X})$	$(X - \bar{X})^2$
58	−7	49
60	−5	25
53	−12	144
81	+16	256
73	+8	64
$\bar{X} = \frac{325}{5} = 65$	0✓	$s^2 = \frac{538}{4} = 134.5$

[1]This may be verified from Table V. For example, a 95% confidence interval with d.f. $= 60$ should use $t_{.025} = 2.00$; but using $z_{.025} = 1.96$ is a very good approximation.

To sum up, we note that (5-13) is of the form

$$\mu = \bar{X} \pm t_{.025} \text{ (estimated standard error)} \tag{5-17}$$

This is quite analogous to (5-11), the only difference being that the *estimated* standard error in (5-17) requires using the larger t instead of z.

PROBLEMS

5-5 Answer True or False; if False, correct it.

If σ is unknown, then we must use (5-13) instead of (5-12). This involves replacing σ with its estimator s, an additional source of unreliability; and to allow for this, $z_{.025}$ is replaced by the larger $t_{.025}$ value in order to keep the confidence level at 95%.

5-6 A random sample of 5 women had the following cholesterol levels in their blood (grams per liter):

3.0, 1.8, 2.1, 2.7, 1.4

(a) Calculate a 95% confidence interval for the mean cholesterol level for the whole population of women.
(b) Construct a 99% confidence interval.

5-7 The reaction times of 30 randomly selected drivers were found to have a mean of .83 second and standard deviation of .20 second.

(a) Find a 95% confidence interval for the mean reaction time of the whole population of drivers.
(b) One more driver was randomly selected, and found to have a mean reaction time of .98 second. Is this surprising? Explain.

5-8 From a very large class in statistics, the following 40 marks were selected randomly:

71	74	65	72	64	42	62	62	58	82
49	83	58	65	68	60	76	86	74	53
78	64	55	87	56	50	71	58	57	75
58	86	64	56	45	73	54	86	70	73

Construct a 95% confidence interval for the average mark of the whole class. (*Hint:* You may find it reduces your work to group the observations into cells of width 5.)

5-4 DIFFERENCE IN TWO MEANS ($\mu_1 - \mu_2$)

Two population means are commonly compared by forming their difference:

$$\mu_1 - \mu_2 \tag{5-18}$$

A reasonable estimate of this is the corresponding difference in *sample* means:

$$\bar{X}_1 - \bar{X}_2$$

Our interest again is in constructing an interval estimate around this.

(a) If Population Variances Known

We can use an argument similar to the one set out in Section 5-2 to develop the appropriate confidence interval:

$$(\mu_1 - \mu_2) = (\bar{X}_1 - \bar{X}_2) \pm z_{.025}\,\text{SE} \tag{5-19}$$

like (5-11)

But what is the standard error of $(\bar{X}_1 - \bar{X}_2)$? It may be proved that the variance of $(\bar{X}_1 - \bar{X}_2)$ is just the sum of the variances of $\bar{X}_1$ and $\bar{X}_2$, that is,

$$\text{var}\,(\bar{X}_1 - \bar{X}_2) = \frac{\sigma_1^2}{n_1} + \frac{\sigma_2^2}{n_2} \tag{5-20}$$

Therefore,

$$\text{SE} = \sqrt{\frac{\sigma_1^2}{n_1} + \frac{\sigma_2^2}{n_2}} \tag{5-21}$$

Substituting (5-21) into (5-19) yields

95% confidence interval for the difference in means, in independent samples

$$(\mu_1 - \mu_2) = (\bar{X}_1 - \bar{X}_2) \pm z_{.025} \sqrt{\frac{\sigma_1^2}{n_1} + \frac{\sigma_2^2}{n_2}} \tag{5-22}$$

When σ_1 and σ_2 are known to have a common value, say σ, the 95% confidence interval reduces to

$$(\mu_1 - \mu_2) = (\bar{X}_1 - \bar{X}_2) \pm z_{.025}\, \sigma \sqrt{\frac{1}{n_1} + \frac{1}{n_2}} \tag{5-23}$$

(b) If Population Variances Unknown

In practice, the population σ is not known, and has to be estimated from the sample information. This requires $z_{.025}$ to be replaced with the broader value $t_{.025}$ in (5-23), of course, and so we obtain:

95% confidence interval for independent samples, when the population variances are equal and unknown

$$(\mu_1 - \mu_2) = (\bar{X}_1 - \bar{X}_2) \pm t_{.025}\, s_p \sqrt{\frac{1}{n_1} + \frac{1}{n_2}} \tag{5-24}$$

where s_p is an estimate of σ. How do we derive it? Since both populations have the same variance σ^2, it is appropriate to pool the information from *both samples* to estimate it. So our estimate is called the *pooled* variance s_p^2: We add up all the squared deviations from both samples, and then divide by the total d.f. in both samples, $(n_1 - 1) + (n_2 - 1)$. That is,

$$s_p^2 = \frac{\sum(X_1 - \bar{X}_1)^2 + \sum(X_2 - \bar{X}_2)^2}{(n_1 - 1) + (n_2 - 1)} \tag{5-25}$$

where X_1 (or X_2) represents the typical observation in the first (or second) sample. To complete (5-24), we need the d.f. for t. According to (5-14), this is just the divisor used in calculating s_p^2 above:

$$\text{d.f.} = (n_1 - 1) + (n_2 - 1) \tag{5-26}$$

To show that this confidence interval is not too complicated after all, we illustrate with an example.

Example 5-3

From a large class, a sample of 4 grades were drawn: 64, 66, 89, and 77. From a second large class, an independent sample of 3 grades were drawn: 56, 71, and 53. If it is reasonable to assume that the variances in the two classes are roughly the same, calculate a 95% confidence interval for the difference between the two class means, $\mu_1 - \mu_2$.

Solution

In Table 5-3 we calculate the sample means and the squared deviations. Thus, from (5-25),

$$s_p^2 = \frac{398 + 186}{3 + 2} = \frac{584}{5} = 117$$

Since d.f. = 5, from Table V, $t_{.025} = 2.571$. Substituting into (5-24), we obtain

$$\begin{aligned}(\mu_1 - \mu_2) &= (74.0 - 60.0) \pm 2.571\sqrt{117}\sqrt{\frac{1}{4} + \frac{1}{3}} \\ &= 14 \pm 21 \qquad (5\text{-}27) \\ &= -7 \text{ to } +35\end{aligned}$$

Thus with 95% confidence we conclude that the average of the first class (μ_1) may be 7 marks below the average of the second class (μ_2). Or μ_1 may be 35 marks above μ_2—or anywhere in between. There is a very large error allowance (± 21) in this example because the samples were so small.

TABLE 5-3
Analysis of Two Independent Samples

Class 1			*Class 2*		
Observed X_1	$(X_1 - \overline{X}_1)$	$(X_1 - \overline{X}_1)^2$	*Observed X_2*	$(X_2 - \overline{X}_2)$	$(X_2 - \overline{X}_2)^2$
64	−10	100	56	−4	16
66	−8	64	71	11	121
89	15	225	53	−7	49
77	3	9			
$\overline{X}_1 = \frac{296}{4} = 74.0$	0✓	398	$\overline{X}_2 = \frac{180}{3} = 60.0$	0✓	186

Equation (5-24) requires that the two samples be *independent,* as in Example 5-3. As another example, consider just one class of students, examined at two different times—say fall and spring terms. The fall population and spring population of grades could then be sampled and compared using (5-24)—provided, of course, that the two samples are independently drawn. (They would not be independently drawn, for example, if we made a point of canvassing the same students twice.)

(c) Paired Samples

Suppose in our comparison of fall and spring grades that we *do* wish to use the same individuals in both samples; then (5-24), which requires independent samples, is no longer applicable. Instead, we proceed as follows: The paired grades (spring X_1 and fall X_2) for each of the 4 students in our sample are set out in Table 5-4. The natural first step is to see how each student changed, that is, calculate the difference, $D = X_1 - X_2$, for each student. Once these differences are calculated, then the original data, having served their purpose, can be discarded. We proceed to treat the differences D now as a *single sample,* and analyze them just as we analyze any other single sample (e.g., the sample in Table 5-2). First, we calculate the average difference $\overline{D}$. Then we use this *sample* $\overline{D}$ appropriately in (5-13) to construct a confidence interval for the average *population* difference Δ, obtaining

95% confidence interval for paired samples

$$\Delta = \overline{D} \pm t_{.025} \frac{s_D}{\sqrt{n}} \tag{5-28}$$

TABLE 5-4

Analysis of two Paired Samples

Observed Grades			*Difference*		
Student	X_1 *(Spring)*	X_2 *(Fall)*	$D = X_1 - X_2$	$(D - \overline{D})$	$(D - \overline{D})^2$
A	64	54	10	−4	16
B	66	54	12	−2	4
C	89	70	19	5	25
D	77	62	15	1	1
			$\overline{D} = \frac{56}{4} = 14.0$	0✓	$s_D^2 = \frac{46}{3} = 15.3$

For our example in Table 5-4, we calculate $\overline{D} = 14$, $s_D = \sqrt{15.3}$. Also d.f. $= n - 1 = 3$ so that $t_{.025} = 3.182$. Substituting into (5-28), we obtain

$$\begin{aligned} \Delta &= 14 \pm 3.182 \frac{\sqrt{15.3}}{\sqrt{4}} \\ &= 14 \pm 6 \\ &= 8 \text{ to } 20 \end{aligned} \tag{5-29}$$

(d) Why Paired Samples?

It is interesting to compare the case of paired samples (5-29) with the case of independent samples (5-27). The sampling error for the paired data is much reduced (± 6 vs. ± 21). The reason for this is intuitively clear. Pairing achieves a match that keeps many of the extraneous variables constant. In using the same four students, we kept sex, IQ, and many other factors exactly the same in both samples. We therefore had more leverage on the problem at hand—the difference in fall versus spring grades.

To summarize the chapter so far, we finish with an example that illustrates most of the important formulas for means.

Example 5-4

To measure the effect of a fitness campaign, a university randomly sampled five employees before the campaign, and another 5 employees after. The weights were as follows (along with the person's initial):

Before J.H. 168, K.L. 195, M.M. 155, T.R. 183, M.T. 169
After L.W. 183, V.G. 177, E.P. 148, J.C. 162, M.W. 180

(a) Calculate a 95% confidence interval for:
   (i) The mean weight before the campaign
   (ii) The mean weight after the campaign
   (iii) The mean weight loss during the campaign

(b) It was decided that a better sampling design would be to measure the same people after, as before. Their figures were:

After K.L. 197, M.T. 163, T.R. 180, M.M. 150, J.H. 160

On the basis of these people, calculate a 95% confidence interval for the mean weight loss during the campaign.

Solution

Before			*After*		
X_1	$(X_1 - \overline{X}_1)$	$(X_1 - \overline{X}_1)^2$	X_2	$(X_2 - \overline{X}_2)$	$(X_2 - \overline{X}_2)^2$
168	−6	36	183	13	169
195	21	441	177	7	49
155	−19	361	148	−22	484
183	9	81	162	−8	64
169	−5	25	180	+10	100
$\overline{X}_1 = \frac{870}{5} = 174$	0√	944	$\overline{X}_2 = \frac{850}{5} = 170$	0√	866

(a) (i) By (5-13)

$$\mu_1 = 174 \pm 2.776 \frac{\sqrt{944/4}}{\sqrt{5}}$$

$$= 174 \pm 19$$

(ii) $$\mu_2 = 170 \pm 2.776 \frac{\sqrt{886/4}}{\sqrt{5}}$$

$$= 170 \pm 18$$

(iii) By (5-24),

$$\mu_1 - \mu_2 = (174 - 170) \pm 2.306 \sqrt{\frac{944 + 866}{4 + 4}} \sqrt{\frac{1}{5} + \frac{1}{5}}$$

$$= 4 \pm 22$$

(b) We must be sure to list the people in the same matched order so that it is meaningful to calculate the individual weight losses:

Person	X_1	X_2	$D = X_1 - X_2$	$(D - \overline{D})$	$(D - \overline{D})^2$
J.H.	168	160	+ 8	4	16
K.L.	195	197	− 2	−6	36
M.M.	155	150	+ 5	1	1
T.R.	183	180	+ 3	−1	1
M.T.	169	163	+ 6	2	4
			$\overline{D} = \frac{20}{5}$, $\overline{D} = 4$	0✓	$s_D^2 = \frac{58}{4} = 14.5$

By (5-28),

$$\Delta = 4 \pm 2.776 \frac{\sqrt{14.5}}{\sqrt{5}}$$

$$= 4 \pm 5$$

Remarks. The paired samples gave a much more precise interval for the weight loss (± 5 vs. ± 22). This is because, in using exactly the same 5 individuals, we kept many things such as sex, age, race, etc. equal.

Pairing is obviously a desirable feature to design into any experiment, where feasible. If pairing cannot be achieved by using the same individual twice, we should look for other ways. For example, we might use pairs of twins—ideally, identical twins This would keep genetic and many environmental factors constant. Of course, to decide which person within the pair is to be given the treatment, and which is to be left as the control, we would have to be fair and unbiased—that is, do it at random (with the flip of a coin, for example).

PROBLEMS

5-9 Two samples of 10 seedlings each, were grown with different fertilizers. The first sample had an average height $\overline{X} = 12.8$ inches and

$\Sigma(X - \overline{X})^2 = 54$. The second sample had an average height $\overline{X} = 10.5$ inches and $\Sigma(X - \overline{X})^2 = 72$. Construct a confidence interval for the difference between the average population heights $(\mu_1 - \mu_2)$:

(a) At the 95% level of confidence.
(b) At the 90% level of confidence.

5-10 Independent random samples of adult whites and blacks were taken in 1972. They gave the following years of school completed:

Whites	8, 18, 10, 10, 14
Blacks	9, 12, 5, 10, 14

Construct a 95% confidence interval for:

(a) The white population mean.
(b) The black population mean.
(c) The mean difference between whites and blacks.

5-11 Five people selected at random had their breathing capacity measured before and after a certain treatment, obtaining the data below.

Before (X)	*After (Y)*
2750	2850
2360	2380
2950	2930
2830	2860
2260	2330

Construct a 95% confidence interval for the mean increase in breathing capacity in the whole population.

5-12 How much does an interesting environment affect the actual physical development of the brain? To answer this, for rats at least, Rosenzweig and others (1964) took 10 litters of purebred rats. From each litter, one rat was selected at random for the treatment group, and one for the control group. The two groups were treated the same, except that the treated rats lived altogether in a cage with interesting playthings, while the control rats lived in bare isolation. After a month, every rat was killed and its cortex (highly developed part of the brain) was weighed, with the following results (in centigrams) for the 10 pairs of littermates:

Treatment	*Control*
68	65
65	62
66	64
66	65
67	65
66	64
66	59
64	63
69	65
63	58

(a) Construct an appropriate 95% confidence interval.
(b) State your answer to (a) in a sentence or two that would be intelligible to a layman.

5-13 In a large American university in 1969, the men and women professors were sampled independently, yielding the annual salaries given below (in thousands of dollars. From Katz, 1973).

Women	*Men*
9	16
12	19
8	12
10	11
16	22

(a) Calculate a 95% confidence interval for the mean salary difference between men and women.
(b) How well does this show the university's discrimination against women?

5-14 Repeat Problem 5-13, using the complete data given in Problems 2-1 and 2-2—independent random samples with the following characteristics:

Women	*Men*
$n = 5$	$n = 25$
$\overline{X} = 11.0$	$\overline{X} = 16.0$
$\Sigma(X - \overline{X})^2 = 40$	$\Sigma(X - \overline{X})^2 = 590$

5-15 To determine which of two seeds was better, a state agricultural station chose 7 two-acre plots of land randomly within the state. Each plot was split in half, and a coin was tossed to determine in an unbiased way which half would be sown with seed *A*, and which with seed *B*. The yields, in bushels, were as follows:

County	*Seed A*	*Seed B*
B	82	88
R	68	66
T	109	121
S	95	106
A	112	116
M	76	79
C	81	89

Which seed do you think is better? To back up your answer, construct a 95% confidence interval.

5-16 How seriously does alcohol affect prenatal brain development? To study this issue (Jones and others, 1974), six women were found who had been chronic alcoholics during pregnancy. The resulting children were tested at age seven, producing a sample of six IQ scores with a mean of 78, and $\Sigma(X - \overline{X})^2 = 1805$. A control group of 46 women were found who on the whole were similar in many respects (same average age, education, marital status, etc.)—but without alcoholism in their pregnancy. Their 46 children had IQ scores that averaged 99, and $\Sigma(X - \overline{X})^2 = 11{,}520$.

(a) If this had been a randomized controlled study, what would be the 95% confidence interval for the difference in IQ that prenatal alcoholism makes?

(b) Since this in fact was an observational study, how should you modify your claim in (a)?

5-5 PROPORTIONS

(a) Large Sample Formula

In Example 4-7 we saw that a sample proportion P is just a disguised sample mean $\overline{X}$ drawn from a 0-1 population. For example, if we observe 7 Republicans in a sample of 10, then the sample proportion of Republicans is:

$$P = \overline{X} = \tfrac{1}{10}(1 + 0 + 1 + 1 + 0 + 1 + 1 + 0 + 1 + 1) = \tfrac{7}{10}$$

Similarly the population proportion π is just the disguised mean μ in a 0-1 population. Therefore, the simplest method to derive an interval estimate for a proportion is to modify the interval estimate for a mean. We substitute P for $\overline{X}$, and $\sqrt{\pi(1 - \pi)}$ for σ according to (4-15). Then (5-12) becomes

$$\pi = P \pm 1.96\sqrt{\frac{\pi(1-\pi)}{n}} \tag{5-30}$$

What can we do about the unknown π that appears in the right-hand side of (5-30)? Fortunately, we can substitute the sample P for π. (This is a strategy we have used before, when we substituted s for σ in the confidence interval for μ.) This approximation introduces another source of error; but with a large sample size, this is no problem. Thus

95% confidence interval for the proportion, for large n

$$\pi = P \pm 1.96\sqrt{\frac{P(1-P)}{n}} \tag{5-31}$$

For this to be a good approximation, the sample size n ought to be large enough so that at least 5 successes and 5 failures turn up. As an example, the voter poll in Chapter 1 used this formula.

(b) Graphical Method, Large or Small Samples

There is a graphical way to find an interval estimate for π, which works for both large and small sample sizes. This method, based on Figure 5-5, is extremely easy to use. For example, suppose we observe 16 Republicans in a sample of 20 individuals in a Kansas City suburb.

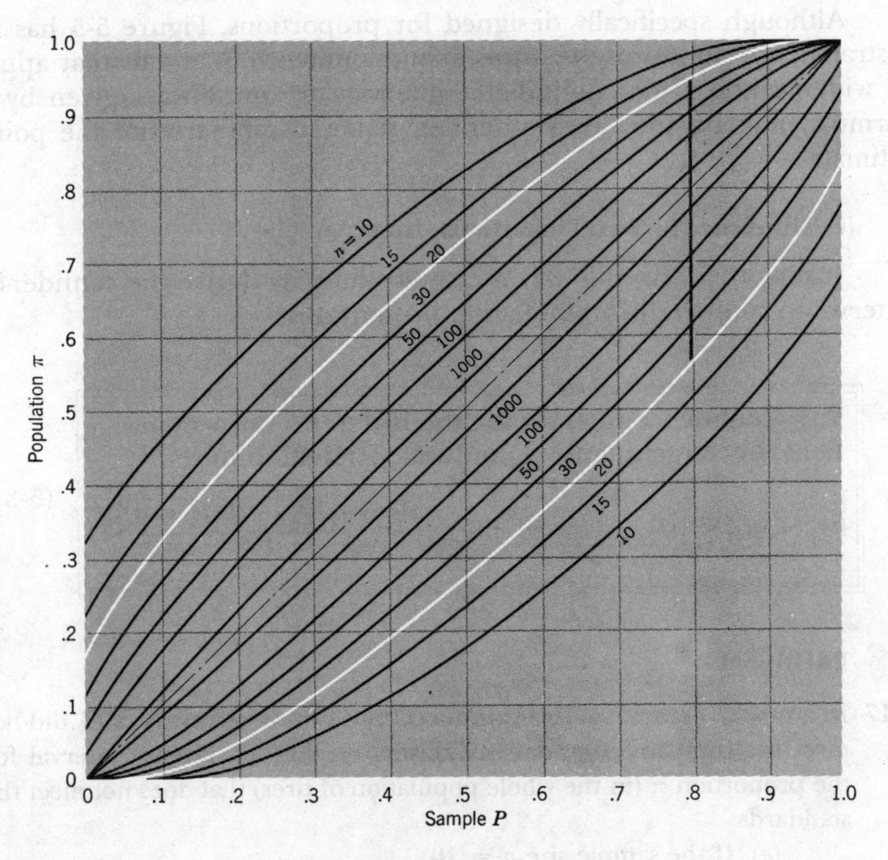

FIGURE 5-5 95% confidence intervals for the population proportion π. (Reproduced with the kind permission of Clopper and Pearson, 1934.)

We first calculate $P = 16/20 = .80$. We then read up the vertical line passing through $P = .8$, noting where it cuts the curves labelled $n = 20$ (highlighted here in white). These two points of intersection define our confidence interval for π, shown in color:

$$.56 < \pi < .96 \tag{5-32}$$

This confidence interval is not symmetric about the estimate $P = .80$; the sampling allowance below is $.80 - .56 = .24$, while the allowance above is only $.96 - .80 = .16$. Why is it longer below? The answer may be seen by examining a more extreme case, where $P = 1.00$. Then the sampling allowance would have to be *entirely* below P, since π cannot exceed 1.00. (From Figure 5-5, we find $.83 < \pi < 1.00$.)

Although specifically designed for proportions, Figure 5-5 has illustrated two interesting features about confidence intervals that apply in wider contexts too: Confidence intervals are not always given by a formula, nor are they always defined symmetrically around the point estimate.

(c) Difference in Two Proportions, Large Samples

Just as we derived (5-31), we could similarly derive the confidence interval to compare two population proportions:

95% confidence interval for the difference in proportions, for large n_1 and n_2, and independent samples

$$(\pi_1 - \pi_2) = (P_1 - P_2) \pm 1.96\sqrt{\frac{P_1(1 - P_1)}{n_1} + \frac{P_2(1 - P_2)}{n_2}} \tag{5-33}$$

PROBLEMS

5-17 In a random sample of tires produced by a certain company, 20% did not meet the company's standards. Construct a 95% confidence interval for the proportion π (in the whole population of tires) that does not meet the standards:

(a) If the sample size $n = 10$.
(b) If $n = 25$.
(c) If $n = 2500$.

5-18 In a poll of 1063 college students in 1970, the Gallup poll found that 49% of those interviewed believed that change in America is likely to occur in the next 25 years through relatively peaceful means (rather than through a revolution). Although the poll was not exactly a random sample, treat it as such in order to get an approximate 95% confidence interval for the population proportion π.

5-19 In response to the question in 5-18, the same affirmative reply was given by 50% of students 18 years old and under, and 69% of those 24 years old and over.

(a) Construct a 95% confidence interval for the difference in these two subpopulation proportions, if $n_1 = 300$ and $n_2 = 300$.

(b) If $n_1 = 500$ and $n_2 = 100$, would you guess the 95% confidence interval will be the same width? Why, or why not? Then verify your guess by calculating the confidence interval.

5-20 In a 1974 poll (*New York Times,* June 2, 1974, p. E6), 1650 Americans were asked for their opinions on the following issue:

The U.S. Supreme Court has ruled that a woman may go to a doctor to end pregnancy at any time during the first three months of pregnancy. Do you favor or oppose this ruling?

A week later, an independent sample of 1650 Americans were asked the same question, except that the words "to end pregnancy" were changed to "for an abortion." The responses were as follows:

Response → / ↓ *Wording*	*In Favor*	*Opposed*	*No Opinion*
"To end pregnancy"	46%	39%	15%
"For an abortion"	41%	49%	10%

(a) Let the proportion of voters in favor be denoted by π_1 for the first wording ("to end pregnancy") and by π_2 for the second wording ("for an abortion"). Construct a 95% confidence interval for the difference $\pi_2 - \pi_1$.

(b) Repeat (a) for the proportion of voters *who had an opinion.* Now, for example, $P_1 = .46/(.46 + .39) = .54$.

(c) Would you agree with the conclusion, "If you want to know what Americans really think about abortion, *you have to ask them about abortion.*"

5-21 Ten days before the 1980 presidential election, a Gallup poll of 600 men showed 49% in favor of Carter; and of 600 women, 58% in favor of Carter (ignoring third-party candidates, as usual). Construct an appropriate 95% confidence interval to show:

(a) What was the difference between men and women in the whole population.

(b) What proportion of all the voters (men *and* women) were in favor of Carter.

5-6 ONE-SIDED CONFIDENCE INTERVALS

(a) Simplest Case

There are occasions when, in order to establish a point, we wish to make a statement that a population value is at *least as large* as a certain value. The appropriate technique is then a one-sided confidence interval, which puts the 5% error allowance all in one tail:

95% confidence interval (one-sided)

$$\mu > \bar{X} - z_{.05} \frac{\sigma}{\sqrt{n}} \qquad (5\text{-}34)$$

like (5-12)

To illustrate this, we rework the data of Example 5-1.

Example 5-5

Sixteen marks were randomly sampled from a very large class that had a standard deviation of 12. If the sample mean was 58, construct the appropriate one-sided confidence interval to show how good the whole class mean is.

Solution

Substitute $n = 16$, $\sigma = 12$, and $\bar{X} = 58$ into (5-34). And noting from Table IV (or Table V, last row) that $z_{.05} = 1.64$,

$$\mu > 58 - 1.64 \frac{12}{\sqrt{16}} = 58 - 5$$

$$\mu > 53 \qquad (5\text{-}35)$$

Thus the instructor can conclude, with 95% confidence, that the class mean is at least 53.

Remarks. The confidence interval covers all values above the sample mean $\overline{X} = 58$, and also a sufficient range of values below it to ensure 95% confidence of being correct.

Although the one-sided confidence interval gives a better lower bound than the two-sided confidence interval, we must pay a pretty high price: The one-sided confidence interval has no upper bound at all.

(b) Other cases

Any two-sided confidence interval may be similarly adjusted to give a one-sided confidence interval. For example, when σ is unknown and must be replaced with s, (5-34) becomes

$$\mu > \overline{X} - t_{.05} \frac{s}{\sqrt{n}} \qquad \text{(5-36) like (5-13)}$$

Similarly, for two samples,

$$(\mu_1 - \mu_2) > (\overline{X}_1 - \overline{X}_2) - t_{.05}\, s_p \sqrt{\frac{1}{n_1} + \frac{1}{n_2}} \qquad \text{(5-37) like (5-24)}$$

Of course, if we want to state that a population value is *below* a certain figure, we would use a one-sided confidence interval of the form

$$\mu < \overline{X} + t_{.05} \frac{s}{\sqrt{n}} \qquad \text{(5-38)}$$

Example 5-6

In planning a dam, suppose the government wishes to estimate μ, the mean annual irrigation benefit per acre. They therefore take a random sample of 25 one-acre plots, and find that the average benefit is \$8.10 and the standard deviation is \$2.40.

The government wishes to make a claim about how large the mean benefit μ is, and they want 99% confidence in their claim. Construct the appropriate confidence interval.

Solution

We use a confidence interval like (5-36), except that 99% confidence requires use of $t_{.01} = 2.492$ (from Table V, with d.f. = 24). We therefore obtain

$$\mu > 8.10 - 2.492 \frac{2.40}{\sqrt{25}}$$

$$\mu > 6.90$$

Thus, with 99% confidence, the government can claim that the mean benefit is at least $6.90 per acre.

PROBLEMS

5-22 A manufacturing process has produced millions of TV tubes, with a mean life $\mu = 1200$ hours, and $\sigma = 300$ hours. A new process produced a sample of 100 tubes with $\overline{X} = 1265$.

(a) To state how good the new process is, calculate a one-sided 95% confidence interval.

(b) Does this indicate that the new process is better than the old?

5-23 A random sample of 5 students' grades were drawn, yielding $\overline{X} = 65$ and $s = 11.6$. Calculate a one-sided 95% confidence interval to show how high the population mean grade is.

5-24 A random sample of 10 men professors' salaries gave a mean of 16 (thousand dollars, annually); a random sample of 5 women professors' salaries gave a mean of only 11. The pooled variance s_p^2 was 11.7. Calculate a one-sided 95% confidence interval to show how much men's mean salary is higher than women's.

5-25 Construct a one-sided 95% confidence interval for Problems 5-15 and 5-16.

CHAPTER 5 SUMMARY

5-1 To distinguish between the population (target) and the sample (estimate), we use Greek letters and gray for the population, and ordinary English letters and blue for the sample.

5-2 From the normal tables, we find that 95% of the time, $\bar{X}$ will be within 1.96 standard errors of μ. This yields the 95% confidence interval (5-12):

$$\mu = \bar{X} \pm 1.96 \frac{\sigma}{\sqrt{n}}$$

5-3 In practice, σ is unknown in the formula above, and has to be estimated with s. To allow for the additional uncertainty, we replace the value $z_{.025} = 1.96$ with the wider value $t_{.025}$.

5-4 To estimate the difference in two population means $(\mu_1 - \mu_2)$, matched samples should be used whenever feasible. They are more efficient than independent samples.

5-5 Proportions are easily handled with the confidence-interval formulas already developed. We simply treat a proportion as a disguised mean.

5-6 To make a claim as strong as possible, a one-sided 95% confidence interval is often useful. It puts all of the 5% error into one tail.

We have developed many confidence intervals in this chapter, and will develop more throughout the rest of the book, since they are the ideal way to allow for statistical fluctuation. To make it easy for you to refer to them, we have listed all these confidence intervals two pages inside the back cover.

In addition, just inside the back cover is a guide called WHERE TO FIND IT. Here we list the commonest statistical problems that occur in practice, and the solution to each, including the appropriate confidence interval and a worked example or two. This table provides an excellent reference to prepare for exams—or real life.

REVIEW PROBLEMS

5-26 To determine the difference in gasoline *A* and *B*, four cars chosen at random were driven over the same route twice, once with gasoline *A* and once with gasoline *B*. The mileages (in miles per gallon) were as follows:

Car	*gas A*	*gas B*
1	23	20
2	17	16
3	16	14
4	20	18

Find a 95% confidence interval for the mileage difference in the two gasolines.

5-27 Soon after he took office in 1963, President Johnson was approved by 160 out of a sample of 200 Americans. With growing controversy over his Vietnam policy, by 1968 he was approved by only 70 out of a sample of 200 Americans.

Using a 95% confidence interval, calculate the drop in approval from 1963 to 1968.

5-28 In 1977, a sample of five men and five women was drawn from the population of American college graduates, and their incomes recorded (thousands of dollars, annually):

Men	*Women*
14	7
36	15
22	22
23	15
20	11

Calculate a 95% confidence interval to show how much men's mean income is higher than women's.

5-29 Suppose that a sample of 100 men at a midwestern university had the following distribution of IQ scores:

Midpoint	*IQ Range*	*Frequency*
100	93–107	29
115	108–122	38
130	123–137	20
145	138–152	10
160	153–167	3

(a) Graph the frequency distribution.
(b) Calculate a 95% confidence interval for the mean IQ of all men at this university.
(c) Construct a 95% confidence interval for the proportion of men who have an IQ over 137.5.

5-30 A "Union Shop" clause in a contract requires every worker to join the union soon after starting to work for the company. In 1973 there were 31 states that permitted the Union Shop, and 19 states (mostly southern) that had earlier passed "Right-to-Work" laws that outlawed the Union Shop and certain other practises. A random sample of 5 states from each group showed the following average hourly wage within the state:

States with Union Shop, etc.	*States with Right-to-Work*
$4.00	$3.50
3.10	3.60
3.60	3.20
4.20	3.90
$4.60	$2.80

On the basis of these figures, a friend claims that the Right-to-Work laws are costing the average worker 50¢ per hour. Do you think this claim should be modified? If so, how?

5-31 An analysis was carried out (Gilbert and others, 1977) on 44 research papers that used randomized clinical trials to compare an innovative treatment (*I*) with a standard treatment (*S*), in surgery and anaesthesia. In 23 of the papers, I was preferred to *S* (and in the other 21 papers, *S* was preferred to *I*).

(a) Assuming the 44 papers constitute a random sample from the population of all research papers in this field, construct a 95% confidence interval for the population proportion where *I* is preferred to *S*.

(b) Do you agree with their interpretation?

"... When assessed by randomized clinical trials, innovations are successful only about half the time. Since innovations brought to the stage of randomized trials are usually expected by the innovators to be sure winners, we see that ... the value of the innovation needs empirical checking."

5-32 Lack of experimental control (e.g., failure to randomly assign the treatment and control) may affect the degree of enthusiasm with which a new medical treatment is reported. To test this hypothesis, 38 studies of a certain operation (portacaval shunt) were classified as follows (Gilbert and others, 1977):

Degree of Control \ Reported Effectiveness of Operation	Moderate or Marked	None
Well Controlled	3	3
Uncontrolled	31	1

(a) Assuming these 38 studies constitute a random sample, construct an appropriate 95% confidence interval. [Although (5-33) is only a rough approximation for small samples, it's the best you have, so use it.]

(b) Do you agree with the following interpretations reported by the authors?

"... Nothing improves the performance of an innovation as much as the lack of controls."

"... weakly controlled trials ... may make proper studies more difficult to mount, as physicians become less and less inclined, for ethical reasons, to subject the issue to a carefully controlled trial lest the 'benefits' of a seemingly proven useful therapy be withheld from some patients in the study."

5-33 In 1954 a large-scale experiment was carried out to test the effectiveness of a new polio vaccine. Among 740,000 children selected from grade 2 classes throughout the United States, 400,000 volunteered. Half of the volunteers were randomly selected for the vaccine shot; the remaining half were given a placebo shot of salt water. The results were as follows (taken, with rounding, from Meier, 1977):

Group	*Number of Children*	*Number of Cases of Polio*
Vaccinated	200,000	57
Placebo (control)	200,000	142
Refused to volunteer	340,000	157

(a) For each of the three groups, calculate the polio rate (cases per 100,000).

(b) Estimate the reduction in the polio rate that vaccination produces, including a 95% confidence interval.

*(c) Suppose *all* the volunteers had been vaccinated, leaving the refusals as the control group:

(i) Before analyzing the data, criticize this procedure.

(ii) What kind of data would you have obtained? Would it have given the correct answer to question (b)?

*5-34 To test the effect of a drug, 3 subjects were randomly selected and tested with and without the drug. The test consisted of measuring the pulse rate just before and after a 200-yard run. The following pulse rates were obtained:

Subject	*With drug*		*Without Drug*	
	Before run	*After run*	*Before run*	*After Run*
MacNeil	108	123	104	113
Bellhouse	94	104	99	107
Koval	80	93	81	87

To what extent does the drug increase pulse rate after running? Answer with an appropriate 95% confidence interval.

CHAPTER 6

Hypothesis Testing

There are no whole truths: all truths are half-truths.

A. N. Whitehead

6-1 HYPOTHESIS TESTING USING CONFIDENCE INTERVALS

(a) A Modern Approach

A statistical *hypothesis* is simply a claim about a population that can be put to a test by drawing a random sample. A typical hypothesis, for example, is that the Republicans are a minority in the voter population in the Kansas City suburb analyzed in (5-32). In fact, this hypothesis can be rejected, because the sample of 25 voters showed (at a 95% level of confidence) that between 56% and 96% of all voters were Republicans—and that is not a minority.

This voter example illustrates how we can use a confidence interval to test an hypothesis. Another example will show this in more detail. (Recall that in the format of this book, a numbered Example is a problem for you to work on first; the solution in the text should only be consulted later as a check.)

Example 6-1

In a large American university, 10 men and 5 women professors were independently sampled in 1969, yielding the annual salaries given below (in thousands of dollars. Same source as Problems 2-1 and 5-13.)

Men (X_1)	
12	20
11	14
19	17
16	14
22	15
$\overline{X}_1 = 16$	

Women (X_2)
9
12
8
10
16
$\overline{X}_2 = 11$

These sample means give a rough estimate of the underlying population means μ_1 and μ_2. Perhaps they can be used to settle an argument: A husband claims that there is no difference between men's salaries (μ_1) and women's salaries (μ_2). In other words, if we denote the difference as $\Delta = \mu_1 - \mu_2$, he claims that

$$\Delta = 0 \tag{6-1}$$

His wife, however, claims that the difference is as large as 7 thousand dollars:

$$\Delta = 7 \tag{6-2}$$

Settle this argument by constructing a 95% confidence interval.

Solution

The 95% confidence interval is, from (5-24),

$$\Delta = (\overline{X}_1 - \overline{X}_2) \pm t_{.025}\, s_p \sqrt{\frac{1}{n_1} + \frac{1}{n_2}}$$

$$= (16 - 11) \pm 2.16 \sqrt{\frac{152}{13}} \sqrt{\frac{1}{10} + \frac{1}{5}}$$

$$= 5.0 \pm 2.16\,(1.87) \tag{6-3}$$

$$= 5.0 \pm 4.0 \tag{6-4}$$

Thus, with 95% confidence, Δ is estimated to be between 1.0 and 9.0. Thus the claim $\Delta = 0$ (the husband's hypothesis) seems implausible, because it falls outside the confidence interval. On the other hand the claim $\Delta = 7$ (the wife's hypothesis) seems more plausible, because it falls within the confidence interval.

In general, any hypothesis that lies outside the confidence interval may be judged implausible, that is, may be *rejected*. On the other hand, any hypothesis that lies within the confidence interval may be judged plausible or *acceptable*. Thus

A confidence interval may be regarded as just the set of acceptable hypotheses.	(6-5)

If a 95% confidence interval is being used, it would be natural to speak of the hypothesis as being tested at a 95% *confidence level.* In conforming to tradition, however, we usually speak of testing at an *error level* of 5% (the complement of the confidence level of 95%).

Thus, to return to our example, we formally conclude from (6-4) that the hypothesis $\Delta = 0$ is rejected at the 5% error level. In other words, we have collected sufficient sample evidence (and, consequently, have a small enough sampling allowance) so that we can discern a difference between men's and women's salaries. We therefore call the difference *statistically discernible* at the 5% error level.

Our formal statistical language must not obscure the important common-sense aspects of this problem, of course. Although we have shown (at the 5% error level) that men's and women's salaries are different, we have *not* shown that discrimination necessarily exists. There are many alternative explanations. For example, men may have a longer period of education than women, on average. What we really should do then is compare men and women of the *same qualifications*. (This will in fact be done later, using *multiple regression* analysis in Problem 9-13.)

Example 6-2

Suppose the confidence interval (6-4) had been based on a smaller sample and, consequently, had been vaguer. (Note how smaller sample sizes n_1 and n_2 increase the size of the sampling allowance.) Specifically, suppose we calculated the confidence interval to be

$$\Delta = 5 \pm 8 \tag{6-6}$$

$$-3 < \Delta < 13 \tag{6-7}$$

Since the hypothesis $\Delta = 0$ falls within this interval, it cannot be rejected. In other words, these results are no longer statistically discernible, so we may call them *statistically indiscernible* at the 5% error level. Are the following interpretations true or false?

(a) The true (population) difference may well be 0. That is, the population of men's salaries may be the same as the women's on average. The difference in *sample* means ($\overline{X}_1 - \overline{X}_2 = 5$) may represent only random fluctuation, and therefore cannot be used to show that a real difference exists in the population means.

(b) In (6-7), we see that the plausible population differences include both negative and positive values, that is, we cannot even decide whether men's salaries on the whole are better or worse than women's.

(c) In (6-6), we see that the sampling allowance ($\pm$ 8) overwhelms the estimated difference (5). Whenever there is this much sampling error, we call the result statistically indiscernible.

Solution

Each of these statements is correct, and a reasonable interpretation.

In summary, if a confidence interval has already been calculated, then it can be used immediately, without any further calculations, to test any hypothesis.

(b) The Traditional Approach

The hypothesis $\Delta = 0$ in (6-1) is of particular interest. Since it represents no difference whatsoever, it is called the *null hypothesis* H_0. In rejecting it because it lies outside the confidence interval (6-4), we established the important claim that there was indeed a difference between men and women's income. Such a result has traditionally been called *statistically significant* at the 5% *significance level.*

There is a problem with the term "statistical significance". It is a technical phrase that simply means that enough data has been collected to establish that a difference does exist. It does *not* mean that the difference is necessarily important. For example, if we had taken huge samples from nearly identical populations the 95% confidence interval, instead of (6-4), might be

$$\Delta = .005 \pm .004 \tag{6-8}$$

This difference is so miniscule that we could dismiss it as being of no *real* significance, even though it is just as *statistically* significant as (6-4). In other words, *statistical* significance is a technical term, with a far different meaning than *ordinary* significance.

There is also a problem with the term "5% significance level". It sounds like the higher this value (say 10% rather than 5%) the better the test. But precisely the reverse is true. (Our level of confidence would only be 90%, rather than 95%).

Unfortunately, but understandably, many people tend to confuse statistical significance with ordinary significance. To reduce the confusion, we prefer the word discernible to the word significant. In conclusion, therefore, the traditional phrase *statistically significant at the 5% significance level* technically means exactly the same thing as the more modern phrase *statistically discernible at the 5% error level.* We prefer the modern phrase, because it is less likely to be misinterpreted.

PROBLEMS

6-1 For each of Problems 5-26 and 5-27, state and test the null hypothesis, indicating whether the difference is statistically discernible at the 5% level.

6-2 (This problem will be analyzed two ways. Here we will use the familiar technique of confidence intervals. Later, in Example 6-3, we will develop a new technique.)

A manufacturing process has produced millions of TV tubes with a mean life $\mu = 1200$ hours, and standard deviation $\sigma = 300$ hours. A new process is tried on a sample of 100 tubes, producing a sample average $\bar{X} = 1265$ hours. Will this new process produce a *long-run* average that is better than the null hypothesis $\mu = 1200$?

(a) Specifically, is the sample mean $\bar{X} = 1265$ statistically discernible (i.e., discernibly different from the H_0 value of 1200) at the 5% error level? Answer by seeing whether the one-sided 95% confidence interval excludes $\mu = 1200$.

(b) Repeat, for the 1% error level.

(c) Repeat, for the .1% error level.

6-2 PROB-VALUE (ONE-SIDED)

(a) What is Prob-Value?

In Section 6-1 we developed a simple way to test *any* hypothesis, by examining whether or not it falls within the confidence interval. Now we take a new perspective by concentrating on just one hypothesis, the null hypothesis H_0. We will calculate just how much (or how little) it is supported by the data.

Example 6-3

A traditional manufacturing process has produced millions of TV tubes, with a mean life $\mu = 1200$ hours and a standard deviation $\sigma = 300$ hours. A new process, recommended by the engineering department as better, produces a sample of 100 tubes, with an average $\bar{X} = 1265$. Although this sample makes the new process look better, is this just a sampling fluke? Is it possible that the new process is really no better than the old, and we have just turned up a misleading sample?

To formulate this problem more specifically, we state the *null hypothesis:* the new process would produce a population that is no different from the old, that is, H_0: $\mu = 1200$. This is sometimes abbreviated to

$$\mu_0 = 1200 \tag{6-9}$$

The claim of the engineering department is called the *alternative hypothesis,* H_1: $\mu > 1200$. This is sometimes abbreviated to

$$\mu_1 > 1200$$

How consistent is the sample $\overline{X} = 1265$ with the null hypothesis $\mu_0 = 1200$? Specifically, if the null hypothesis were true, what is the probability that $\overline{X}$ would be as high as 1265?

Solution

In Figure 6-1 we show the hypothetical distribution of $\overline{X}$, if H_0 is true. (Here, for the first time, we show the convention of drawing hypothetical distributions in ghostly white.) By the Normal Approximation Theorem, this distribution is normal, with mean $\mu_0 = 1200$, and standard error (SE) $= \sigma/\sqrt{n} = 300/\sqrt{100} = 30$. We use these to standardize the observed value $\overline{X} = 1265$:

$$Z = \frac{\overline{X} - \mu_0}{SE} \tag{6-10}$$

$$= \frac{1265 - 1200}{30} = 2.17 \tag{6-11}$$

Thus,

$$\Pr(\overline{X} \geq 1265) = \Pr(Z \geq 2.17) \simeq .015 \tag{6-12}$$

Remarks. This means that, if in fact the new process is no better (that is, if H_0 is true), there would be only a $1\frac{1}{2}\%$ chance of observing $\overline{X}$ as large as 1265. We call $1\frac{1}{2}\%$ the *prob-value for* H_0. (Although it is customarily called the *p-value,* or simply *p,* we chose the name *prob-value* to distinguish it from the other *p*'s appearing in this text.)

The prob-value summarizes very clearly how much agreement there is between the data and H_0: In this example the data provides very little

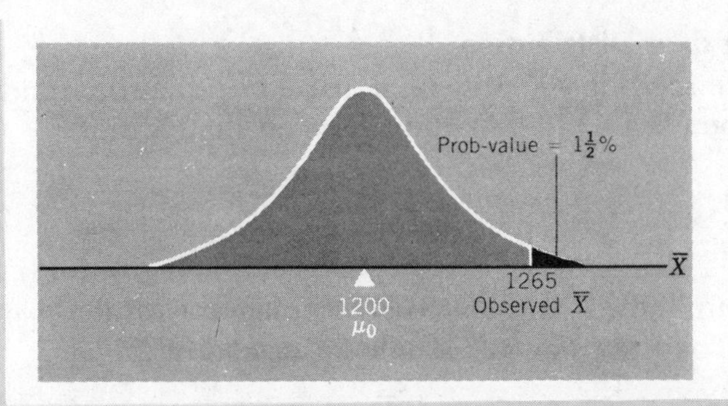

FIGURE 6-1 Prob-value ≡ Pr($\overline{X}$ would be large as the value actually observed, if H_0 is true.)

support for H_0; but if $\overline{X}$ had been observed closer to H_0 in Figure 6-1, the prob-value would have been larger.

In general, for any hypothesis being tested, we define the prob-value for H_0 as

$$\text{Prob-value} \equiv \Pr\left(\begin{array}{c}\text{the sample statistic would be as large}\\ \text{as the value actually observed,}\\ \text{if } H_0 \text{ is true}\end{array}\right) \tag{6-13}$$

The prob-value in Figure 6-1 is calculated in the right-hand tail, because the alternative hypothesis is on the right side ($\mu > 1200$). On the other hand, if the alternative hypothesis were on the left ($\mu < 1200$), then the prob-value would be calculated in the left-hand tail; that is,

$$\text{Prob-value} = \Pr\left(\begin{array}{c}\text{the sample statistic would be } \textit{as small}\\ \textit{as} \text{ the value actually observed,}\\ \text{if } H_0 \text{ is true}\end{array}\right)$$

Whether right-sided or left-sided, the prob-value is an excellent way to *summarize what the data says about the credibility of* H_0.

(b) Using the t Distribution

We have seen how $\overline{X}$ was standardized so that the standard normal table could be used. The key statistic we evaluated was

$$Z = \frac{\overline{X} - \mu_0}{\sigma/\sqrt{n}} \qquad \text{(6-14) like (6-10)}$$

Usually σ is unknown, and has to be estimated with the *sample* standard deviation s. Then the statistic is called t instead of Z:

$$t = \frac{\overline{X} - \mu_0}{s/\sqrt{n}} \qquad \text{(6-15)}$$

Since $\overline{X}$ fluctuates around μ_0, Z fluctuates around 0. Similarly t fluctuates around 0—but with wider variability, as already noted in Chapter 5. (We no longer know the exact value σ, but instead have to use an estimate s, with its inevitable uncertainty.) The resulting wider distribution of t is shown in Figure 6-2.

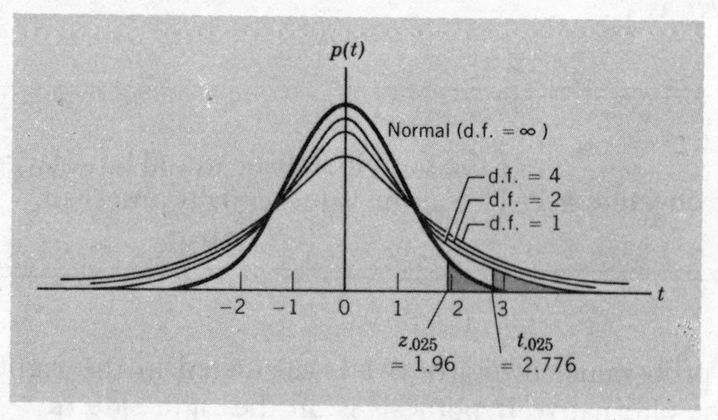

FIGURE 6-2 The standard normal distribution and the t distributions compared.

There are many t distributions, one for each sample size; hence, one for each d.f. (degrees of freedom). In Figure 6-2 we see that the larger the sample size, the less spread out is the t distribution. (The larger the sample, the more reliable is the estimate s, and consequently the less variable is t.) Eventually, as sample size approaches infinity, s^2 estimates σ^2 dead on, and then the t distribution coincides with the normal z. This is reflected in the t table. Each of the distributions in Figure 6-2 corresponds to a row in Appendix Table V. In the last row, where d.f. = ∞, the t and z values become identical.

But in all other cases, t values are larger than z (as first noted in Chapter 5). For example, in Figure 6-2, $z_{.025}$ is 1.96 while the corresponding $t_{.025}$ with 4 d.f. is 2.776. The use of t, when the sample size is small and σ has to be estimated with s, is easily illustrated with an example.

Example 6-4

A sample of $n = 5$ grades gave $\overline{X} = 65$, and $s = 11.6$ (as in Problem 5-23). Suppose a claim is made that the population mean is only 50. What is the prob-value for this hypothesis?

Solution

From (6-15) we calculate

$$t = \frac{65 - 50}{11.6/\sqrt{5}} = 2.89 \tag{6-16}$$

In calculating s, we used d.f. $= n - 1 = 4$. We therefor scan along the fourth row of Table V, and find that the observed t value of 2.89 lies beyond $t_{.025} = 2.776$. As Figure 6-3 shows, this means the tail probability is smaller than .025. That is,

$$\boxed{\text{Prob-value} < .025} \tag{6-17}$$

Since the prob-value is a measure of credibility for H_0, such a low value leads us to conclude that H_0 is an implausible hypothesis. In other words: If H_0 were true (population mean = 50) there would be less than $2\frac{1}{2}$ chances in a hundred of getting a sample mean as high as the 65 actually observed.

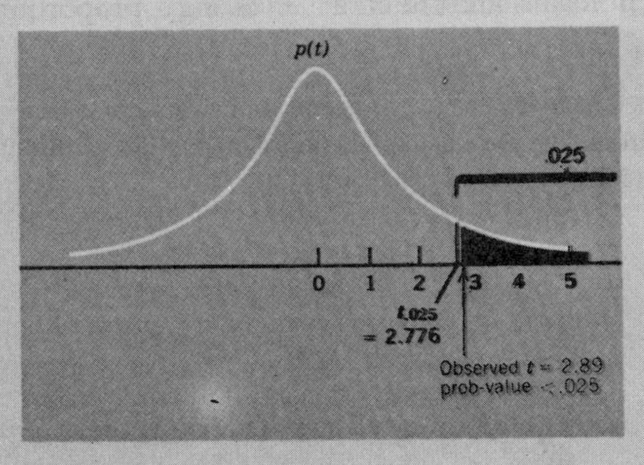

FIGURE 6-3 Prob-value found using t in Table V.

We can easily generalize the use of t to cover other hypothesis tests. In (6-15), the numerator is the difference between the estimated value and the hypothetical value being tested (null hypothesis). The denominator is the estimated standard error. The generalization of (6-15) is therefore

$$t = \frac{\text{estimate} - \text{null hypothesis}}{\text{standard error}} \tag{6-18}$$

Often the null hypothesis is 0, in which case (6-18) takes on a very simple form:

$$t = \frac{\text{estimate}}{\text{standard error}} \tag{6-19}$$

This makes good intuitive sense: The t ratio simply measures how large the estimate is, relative to its standard error.

Our earlier advice on when to use t is worth repeating. Use t when σ is unknown, and unreliably estimated in a small sample (n less than 30, 60, or 120, depending on the accuracy required). Otherwise, use z as a reasonable approximation, so that you can use the more detailed z table (Table IV) rather than the t table (Table V).

Finally, to illustrate the diversity of distributions for which the prob-value can be calculated, we finish with two more examples—one involving a difference in means and the other involving a proportion.

Example 6-5

For the mean difference in men's and women's salaries (thousands of dollars, annually), we calculated the following 95% confidence interval (in Problem 5-24):

$$(\mu_1 - \mu_2) > (\overline{X}_1 - \overline{X}_2) - t_{.05}\,\text{SE}$$
$$> 5.0 - 1.771\,(1.87) \tag{6-20}$$

$$(\mu_1 - \mu_2) > 1.7 \tag{6-21}$$

where there were 13 d.f. in calculating SE, and therefor 13 d.f. for the t distribution.

The null hypothesis ($H_0 : \mu_1 - \mu_2 = 0$) is implausible, because it is excluded from the confidence interval (6-21). To express just how little credibility the data attributes to H_0, let us calculate its prob-value.

Solution

The null hypothesis is $\mu_1 - \mu_2 = 0$, so that (6-19) is appropriate:

$$t = \frac{\text{estimate}}{\text{SE}} = \frac{(\bar{X}_1 - \bar{X}_2)}{\text{SE}}$$

Of course, the values we need here are the ones already calculated for the confidence interval (6-20). Thus

$$t = \frac{5.0}{1.87} = 2.67$$

Since d.f. = 13, we scan along the thirteenth row of Table V, and find that the observed t value of 2.67 lies beyond $t_{.010} = 2.650$. Thus

$$\text{Prob-value} < .010 \tag{6-22}$$

Accordingly, we conclude that H_0 has very little credibility indeed.

Example 6-6

To investigate whether black children a generation ago showed racial awareness and antiblack prejudice, a group of 252 black children was studied (Clark and Clark, 1958). Each child was told to choose a doll to play with from among a group of four dolls, two white and two nonwhite. A white doll was chosen by 169 of the 252 children.

What is the prob-value for the null hypothesis that the children ignore color? (The alternative hypothesis is that the children are prejudiced against black, that is, in favor of white.)

Solution

First we formulate the problem mathematically. Suppose the 252 children can be regarded as a random sample from a large population of black children. (This is quite an assumption, so the final prob-value we calculate should be interpreted cautiously.) In any case, the null hypothesis is that the population proportion choosing a white doll is 50-50; that is,

$$\pi_0 = .50 \tag{6-23}$$

The observed sample proportion is $P = 169/252 = .67$. Its standard error is given by (4-18) as $\sqrt{\pi(1 - \pi)/n}$ and we use the null value $\pi = .50$, since prob-value is always based on the null hypothesis. (Remember that prob-value was defined as the probability of . . . , assuming H_0 is true.) Thus

$$t = \frac{\text{estimate} - \text{null hypothesis}}{\text{standard error}} \qquad \text{(6-18)repeated}$$

$$= \frac{.67 - .50}{\sqrt{.50(.50)/252}} = 5.40 \qquad \text{(6-24)}$$

Since the normal z distribution (Table IV) can be used in this large sample case instead of t,

$$\text{Prob-value} = \Pr(Z \geq 5.40)$$
$$< .000000287$$

Since the prob-value is miniscule, there is almost no credibility for the null hypothesis. Thus, to the extent that this sample reflects the properties of a random sample, it can be concluded that a generation ago, even black children were prejudiced in favor of white.

PROBLEMS

6-3 In Boston in 1968, Dr. Benjamin Spock, a famous pediatrician and activist against the Vietnamese war, was tried for conspiracy to violate the Selective Service Act. The judge who tried Dr. Spock had an interesting record: Of the 700 people the judge had selected for jury duty in his past few trials, only 15% were women (a simplified version of Zeisel and Kalven, 1972). Yet in the city as a whole, about 29% of the eligible jurors were women.

To evaluate the judge's fairness in selecting women,

(a) State the null hypothesis, in words and symbols.

(b) Calculate the prob-value for H_0.

6-4 A doctor tested the effectiveness of a certain drug by giving it to a group of rats, while a second group of rats was kept under identical conditions as a control. The difference (in mean weight increases) between the treated group T and the control group C yielded a t value of 2.23, and hence a prob-value of about .013, for the null hypothesis ($\mu_T = \mu_C$) against the alternative hypothesis ($\mu_T > \mu_C$). Answer True or False; if False, state why.

(a) The initial assignment of the rats to the treatment or control group should be done at random, to avoid bias.

(b) The probability of H_0 is .013.

(c) If we repeated the experiment, and if H_0 were true, the probability is .013 of getting a t value at least as large as the one we observed.

6-5 A random sample of 6 students in a phys-ed class had their pulse rate (beats per minute) measured before and after the 50-yard dash, with the following results:

Before	*After*
74	83
87	96
74	97
96	110
103	130
82	96

(a) Calculate the one-sided 95% confidence interval for the mean increase in pulse rate.

(b) State the null hypothesis, in words and symbols. Then calculate its prob-value.

6-6 A random sample of two thousand drivers killed in auto crashes were classified as to whether or not they had alcohol in their blood, and whether or not they were responsible for the accident.

Alcohol? \ *Responsible?*	*Yes*	*No*
Yes	650	150
No	700	500

In the whole population, is there a difference between the drivers with and without alcohol? To answer this,

(a) Calculate the appropriate prob-value.

(b) Calculate the appropriate 95% confidence interval (one-sided) to show how much difference we can claim.

(c) To what extent does this data show that alcohol increases accidents?

6-3 CLASSICAL HYPOTHESIS TESTS

(a) What is a Classical Test?

Suppose we have the same data as in Example 6-3. Recall that the traditional manufacturing process had produced a population of millions of TV tubes, with a mean life $\mu = 1200$ hours and a standard deviation $\sigma = 300$ hours. To apply a classical hypothesis test of whether a new process is better, we will proceed in three steps—the first two before any data is collected:

1. The null hypothesis (H_0:$\mu = 1200$) is formally stated. At the same time, we set the sample size (such as $n = 100$), and the error level of the test (such as 5%) hereafter referred to as α.
2. We now assume temporarily that the null hypothesis is true—just as we did in calculating the prob-value. And we ask, what can we expect of a sample mean drawn from this sort of world? Its specific distribution is again shown in Figure 6-4, just as it was in Figure 6-1. But there is one important difference in these two diagrams: whereas in Figure 6-1 the shaded prob-value was calculated from the observed $\overline{X}$, in Figure 6-4 the shaded area is arbitrarily set at $\alpha = 5\%$. This defines the critical range for rejecting the null hypothesis (shown as the big arrow). All this is done before any data is observed.

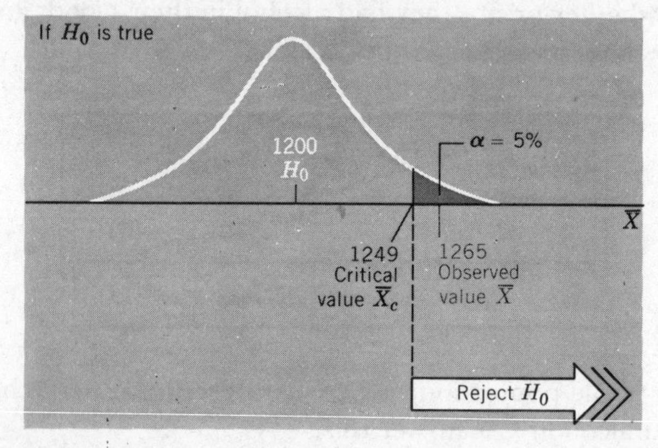

FIGURE 6-4 A classical test at level $\alpha = 5\%$.

3. The sample is now taken. If the observed $\overline{X}$ falls in the rejection region in Figure 6-4, then it is judged sufficiently in conflict with the null hypothesis H_0 to reject H_0. Otherwise H_0 is acceptable.

The critical value $\overline{X}_c = 1249$ for this test was calculated by noting from Appendix Table IV that a 5% tail is cut off the normal distribution by a critical Z value of $z_{.05} = 1.64$; that is,

$$\text{Critical } Z = \frac{\overline{X}_c - \mu_0}{\sigma/\sqrt{n}} = 1.64 \qquad \text{(6-25) like (6-14)}$$

$$\frac{\overline{X}_c - 1200}{300/\sqrt{100}} = 1.64$$

$$\overline{X}_c = 1249 \qquad \text{(6-26)}$$

In our example, the observed $\overline{X} = 1265$ is beyond this critical value, thus leading us to reject H_0 at the 5% error level.

There is another way of looking at this testing procedure. If we get an observed $\overline{X}$ exceeding 1249, there are two explanations:

1. H_0 is true, but we have been exceedingly unlucky and got a very improbable sample $\overline{X}$. (We're born to be losers; even when we bet with odds of 19 to 1 in our favor, we still lose).
2. H_0 is not true after all. Thus it is no surprise that the observed $\overline{X}$ was so high.

We opt for the more plausible second explanation. But we are left in some doubt; it is just possible that the first explanation is the correct one. For this reason we qualify our conclusion to be "at the 5% error level."

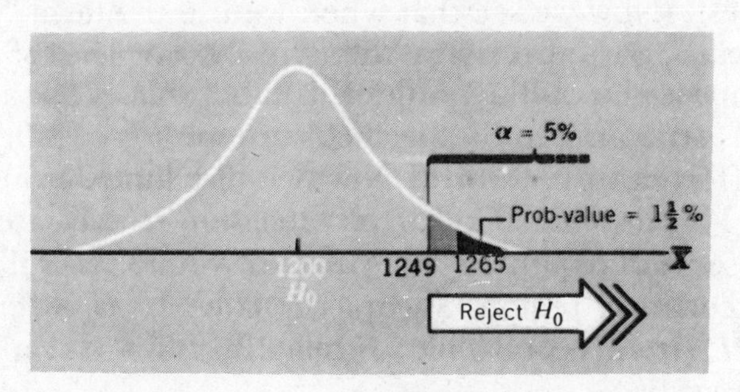

FIGURE 6-5 Classical hypothesis testing and prob-value.

(b) Classical Hypothesis Testing and Prob-Value

For the example above, a comparison of classical testing and prob-value is set out in Figure 6-5. Since the prob-value [$1\frac{1}{2}\%$ from (6-12)] is less than $\alpha = 5\%$, the observed $\bar{X}$ is correspondingly in the rejection region. That is,

$$\boxed{\text{Reject } H_0 \text{ if prob-value} \leq \alpha} \tag{6-27}$$

To restate this, we recall that the prob-value is a measure of the credibility of H_0. If this credibility sinks below α, then H_0 is rejected.

Applied statisticians increasingly prefer prob-values to classical testing, because classical tests involve setting α arbitrarily (usually at 5%). Rather than introduce such an arbitrary element, it is often preferable just to quote the prob-value, leaving the reader to pass her own judgement on H_0. [Formally, by determining whatever level of α she deems appropriate for her purpose, the reader may reach her own decision using (6-27).]

*(c) Type I and Type II Errors

In the decision-making process we run the risk of committing two distinct kinds of error. The first is shown in Figure 6-6*a* (a reproduction of Figure 6-4), which shows what the world looks like if H_0 is true. In this event, there is a 5% chance that we will observe $\bar{X}$ in the shaded region, and thus erroneously reject the true H_0. Rejecting H_0 when it is true is called a *type I error,* with its probability of course being α, the error level of the test. (Now we see that when we use the term "error level of a test" we mean, more precisely, "the type I error level of a test.")

But suppose the null hypothesis is false, that is, the alternative hypothesis H_1 is true and, to be specific, suppose $\mu = 1240$. Then we are living in a different sort of world. Now $\bar{X}$ is distributed around $\mu = 1240$, as shown in Figure 6-6*b*. The correct decision in this case would be to reject the false null hypothesis H_0. An error would occur if $\bar{X}$ were to fall in the H_0 acceptance region. Such acceptance of H_0 when it is false is called a *type II error.* Its probability is called β, and is shown as the shaded area in Figure 6-6*b*.

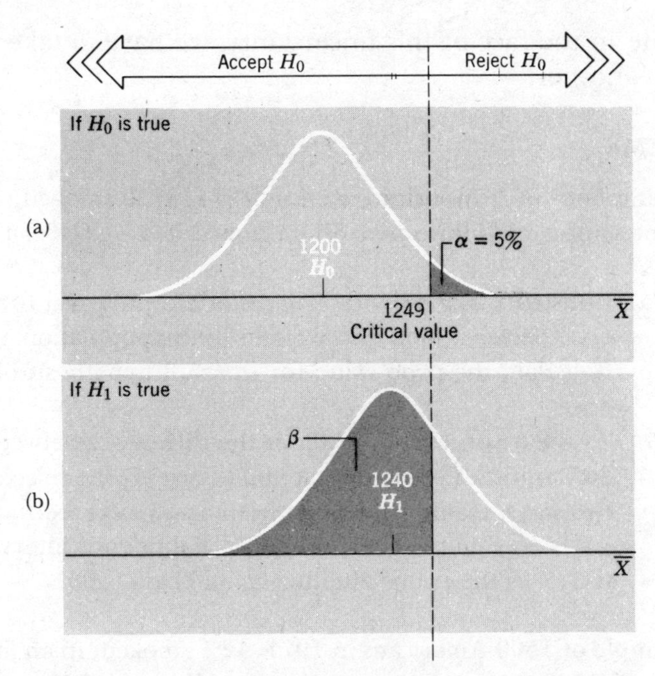

FIGURE 6-6 The two kinds of error that can occur in a classical test. (*a*) If H_0 is true, then α = probability of erring (by calling the true hypothesis H_0 "rejected"). (*b*) If H_1 is true then β = probability of erring (by calling the false hypothesis H_0 "acceptable").

TABLE 6-1

Possible Results of an Hypothesis Test (Derived from Figure 6-6)

State of the World \ *Decision*	*Accept H_0*	*Reject H_0*
If H_0 is true	Correct decision. Probability = $1 - \alpha$ = confidence level	Type I error. Probability = α = level of the test
If H_0 is false	Type II error. Probability = β	Correct decision. Probability = $1 - \beta$

Table 6-1 summarizes the dilemma of hypothesis testing: The state of the real world, whether H_0 is true or false, is unknown. If a decision must be made in the face of this uncertainty, we have to take the risk of one error or another.

PROBLEMS

6-7 In a sample of 750 American men in 1974, 45% smoked; in an independent sample of 750 women, 36% smoked (*Gallup Opinion Index*, June 1974).

(a) Construct a one-sided 95% confidence interval for the difference between men and women in the population.

(b) Calculate the prob-value for the null hypothesis of no difference.

(c) At the error level $\alpha = .05$, is the difference between 45% and 36% statistically discernible; that is, can H_0 be rejected? Answer two ways, making sure that both answers agree:

(i) Is H_0 excluded from the 95% confidence interval?

(ii) Does the prob-value for H_0 fall below .05?

6-8 In a sample of 1500 Americans in 1971, 42% smoked; in an independent sample of 1500 Americans a year later, 43% smoked. For the increase over the year, repeat Problem 6-7.

6-9 (Acceptance Sampling.) The prospective purchaser of several new shipments of waterproof gloves hopes they are as good as the old shipments, which had a 10% rate of defective pairs. But he fears that they may be worse. So for each shipment, he takes a random sample of 100 pairs and counts the proportion P of defective pairs so that he can run a classical test. The level of this test was determined by such relevant factors as the cost of a bad shipment, the cost of an alternative supplier, and the new shipper's reputation. When all these factors were taken into account, suppose the appropriate level of this test was $\alpha = .09$.

(a) State the null and alternative hypotheses, in words and symbols.

(b) What is the critical value of P? That is, how large must P be in order to reject the null hypothesis (i.e., reject the shipment)?

(c) Suppose that for 6 shipments, the values of P turned out to be 12%, 25%, 8%, 16%, 24% , and 21%. Which of these shipments should be rejected?

*6-10 In Problem 6-9, suppose that the purchaser tries to get away with a small sample of only 10 pairs. Suppose that instead of setting $\alpha = .09$, he sets the rejection region to be $P \geq 20\%$ (i.e., he will reject the shipment if there are 2 or more defective pairs among the 10 pairs in the sample).

For this test, what is α? (*Hint:* Rather than using the normal approximation, the binomial distribution is easier and more accurate.)

*6-11 Consider the problem facing an air traffic controller at Chicago's O'Hare Airport (the world's largest, in traffic). If a small irregular dot appears on the screen, approaching the flight path of a large jet, she must decide between:

H_0: all is well. It's only a bit of interference on the screen.
H_1: a collision with a small private plane is imminent.
Fill in the blanks:

A "false alarm" is a type __ error, and its probability is denoted by __. A "missed alarm" is a type __ error, and its probability is denoted by __. By making the equipment more sensitive and reliable, it is possible to reduce both __ and __.

*6-12 In manufacturing machine bolts, the quality control engineer finds that it takes a sample of $n = 100$ to detect an accidental change of .5 millimeter in the mean length of the manufactured bolt. Suppose he wants more precision, to detect a change of only .1 millimeter, with the same α and β. How much larger must his sample be? (*Hint:* It is easy, if you rephrase it in terms of confidence intervals. To make a confidence interval 5 times as precise, how much larger must the sample be?)

*6-13 Consider the classical test shown in Figure 6-6.

(a) If the observed $\bar{X}$ is 1245, do you reject H_0?

(b) Suppose that you had designed this new process, and you were convinced that it was better than the old; specifically, on the basis of sound engineering principles, you really believed that H_1:$\mu = 1240$. Common sense suggests that H_0 should be rejected in favor of H_1. In this conflict of common sense with classical testing, which path would you follow? Why?

(c) Suppose that sample size is doubled to 200, while you keep $\alpha = 5\%$ and you continue to observe $\bar{X}$ to be 1245. What would be the classical test decision now? Would it, therefore, be true to say that your problem in part (a) may have been inadequate sample size?

(d) Suppose now that your sample size n was increased to a million, and in that huge sample you observed $\overline{X} = 1201$. An improvement of only one unit over the old process is of no economic significance (i.e., does not justify retooling, etc.); but is it statistically significant (discernible)? Is it therefore true to say that a sufficiently large sample size may provide the grounds for rejecting any specific H_0 – no matter how nearly true it may be—simply because it is not *exactly* true?

*6-4 CLASSICAL TESTS RECONSIDERED

(a) Reducing α and β

In panel *(a)* of Figure 6-7 we show again the two error probabilities of Figure 6-6: α, if H_0 is true; and β, if H_1 is true. In panel *(b)*, we illustrate how decreasing α (by moving the critical point to the right to, say, 1270) will at the same time increase β. In statistics, as in economics, the problem is trading off conflicting objectives. There is an interesting legal analogy. In a murder trial, the jury is being asked to decide between H_0, the hypothesis that the accused is innocent, and the alternative H_1, that he is guilty. A type I error is committed if an innocent man is condemned, while a type II error occurs if a guilty man is set free. The judge's admonition to the jury that "guilt must be proved beyond a reasonable doubt" means that α should be kept very small. There have been many legal reforms (e.g., in limiting the power of the police to obtain a confession) that have been designed to reduce α, the probability that an innocent man will be condemned. But these same reforms have increased β, the probability that a guilty man will evade punishment. There is no way of pushing α down to 0 (ensuring absolutely against convicting an innocent man) without raising β to 1 (letting every defendent go free and making the trial meaningless). The one way that α and β can both be reduced is by increased evidence—or, to return to our statistical example, by increasing sample size as in Figure 6-7*c*.

(b) Some Difficulties with Classical Tests

Figure 6-7 can be used to illustrate some of the difficulties that may be encountered in applying a classical reject-or-accept hypothesis test at a prespecified level α. Suppose in Figure 6-7*a* that we have observed a sample $\overline{X} = 1245$. This is not quite extreme enough to allow rejection of H_0 at level $\alpha = 5\%$, so $\mu_0 = 1200$ is accepted. But if we had set

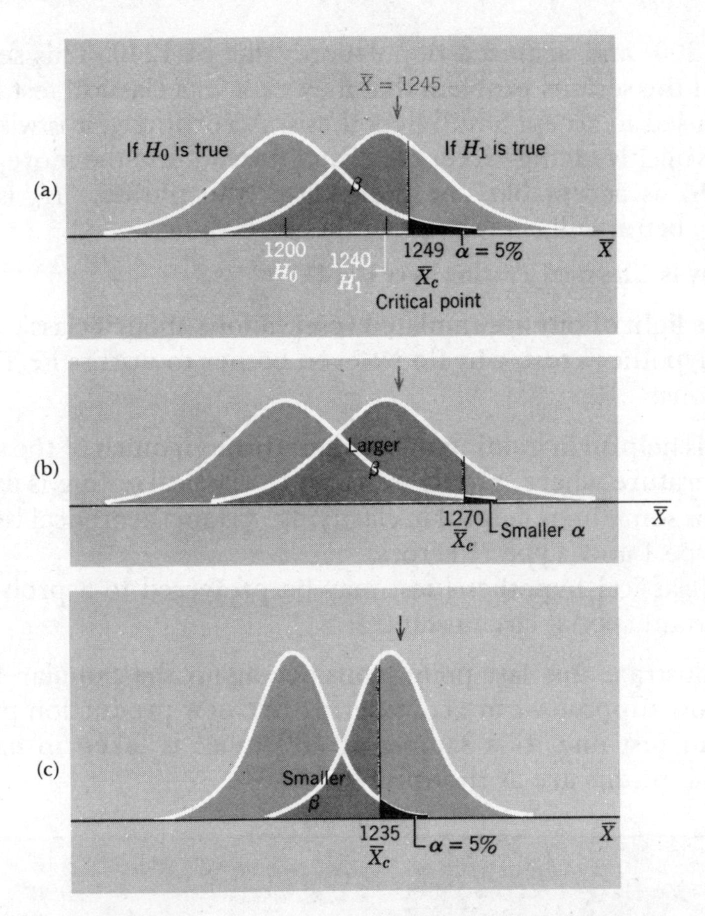

FIGURE 6-7 (*a*) Hypothesis test of Figure 6-6 showing α and β. (*b*) How a reduction in α increases β, other things being equal. (*c*) How an increase in sample size allows one error probability (β) to be reduced, while holding constant the other (α).

$\alpha = 10\%$, then H_0 would have been rejected. This illustrates once again how an arbitrary specification of α leads to an arbitrary decision.

There is an even deeper problem. Accepting $\mu_0 = 1200$ (at level $\alpha = 5\%$) is a disaster, if we had prior grounds for believing H_1 is true; that is, for expecting that the new process would yield a mean $\mu_1 = 1240$. In this case our prior belief in the new process is strongly supported by the sample observation of 1245. Yet we have used this *confirming* sample result (in this classical hypothesis test) to *reverse* our original view. That is, we have used an observation of 1245 to judge in favor of a population value of 1200, and against a population value of 1240. This serves as a warning of the serious problem that may exist in a classical test if a small sample is used to accept a null hypothesis. Accordingly, it is wise to stop short of explicitly saying "accept H_0," and instead, use the more reserved phrase, "H_0 is acceptable," or even better the phrase, "H_0 is not rejected."Or, better still, simply quote the prob-value.

(c) Why Is Classical Testing Ever Used?

In the light of our accumulated reservations about a classical accept-or-reject hypothesis test, why do we even bother to discuss it? There are three reasons:

1. It is helpful in guiding the student through much of the statistical literature where, rightly or wrongly, classical testing is used a lot.
2. It is sometimes helpful in clarifying certain theoretical issues, like Type I and Type II errors.
3. A classical hypothesis test may be preferred to a prob-value in certain special circumstances.

To illustrate this last point, consider again the familiar Example 6-3; but now suppose we are considering five new production processes, rather than just one. If a sample of 100 tubes is taken in each case, suppose the results are as shown in Table 6-2.

TABLE 6-2

New Process	$\bar{X}$	*Is Process Really Better Than the Old Process (where $\mu = 1200$)?*
1	1265	?
2	1240	?
3	1280	?
4	1150	?
5	1210	?

We now have two options. We can calculate five prob-values for these five processes, just as we calculated the prob-value for the first process in Example 6-3. But this will involve a lot more work than using a classical testing approach, which requires only that we specify α (at say 5%) and then calculate one single figure, the critical value $\bar{X}_c = 1249$ derived in (6-26). Then all five of the sample values can immediately be evaluated without any further calculation. (Note that H_0 is rejected only for processes 1 and 3; in other words, these are the only two that may be judged superior to the old method.)

But if a classical hypothesis test is to be used in this way, the level α should not be arbitrarily set. Instead, α should be determined rationally, on the basis of two considerations:

1. *Prior belief.* To again use our example, the less confidence we have in the engineering department that assured us these new processes are better, the smaller we will set α (i.e., the stronger the evidence we will require to judge in their favor). So we want to answer questions such as: Are the engineers' votes divided? Have they ever been wrong before?
2. *Losses involved in making a wrong decision.* The greater the costs of a type I error (i.e., needlessly retooling for a new process that is actually no better), the smaller we will set α, the probability of making that sort of error. Similarly, the greater the cost of a type II error (i.e., failing to switch to a new process that is actually better), the smaller we will set β. The problem is that with limited resources, we can't make both α and β smaller (recall Figure 6-7*b*). But in setting their level, we must obviously take into account the relative costs of the two errors.

It is possible, in an advanced course, to take account of prior beliefs and losses in a more formal rational way. Called *Bayesian decision analysis,* this theory is developed, for example, in Raiffa, 1968.

*PROBLEMS

6-14 Answer True or False. If False, correct it.

(a) Comparing hypothesis testing to a jury trial, we may say that the type I error is like the error of condemning a guilty man.

(b) Suppose, in a certain test, that the prob-value turns out to be .013. Then H_0 would be acceptable at the 5% level and also at the 1% level.

6-15 In the jury trial analogy, what would be the effect on α and β of reintroducing capital punishment on a large scale? (If the issue is not clear for murder, suppose it were reintroduced even for rape.)

6-16 Consider a very simple example, in order to keep the philosophical issues clear. Suppose that you are gambling with a single die, and lose whenever the die shows 1 dot (ace). After 100 throws, you notice that you have suffered a few too many losses—20 aces. This makes you suspect that your opponent is using a loaded die; specifically, you begin to wonder whether this is one of the crooked dice recently advertised as giving aces one-quarter of the time.

(a) Find the critical proportion of aces beyond which you would reject H_0 at the 5% level.

(b) Illustrate this test with a diagram similar to Figure 6-6. From this diagram, roughly estimate α and β.

(c) With your observation of 20 aces, what is your decision? Suppose that you are playing against a strange character you have just met on a Mississippi steamboat. A friend passes you a note indicating that this stranger cheated him at poker last night; and you are playing for a great deal of money. Are you happy with your decision?

(d) If you double α, use the diagram in (b) to roughly estimate what happens to β.

*6-17 In Problem 6-10, suppose the alternative hypothesis is that the shipment is 30% defective.

(a) State this in symbols.

(b) Calculate β.

*6-18 Repeat the problem above, for Problem 6-9 in place of 6-10. Then, is it fair to say that an increased sample size reduced both α and β?

*6-5 PROB-VALUE (TWO-SIDED)

In the previous three sections, we discussed only the one-sided test, in which the alternative hypothesis and, consequently, the rejection region and prob-value are just on one side. The one-sided test, like the one-sided confidence interval, is appropriate when there is a claim to be made, such as, "more than," "less than," "better than," "worse than," "at least," etc.

However, there are occasions when it is more appropriate to use a two-sided test or a two-sided confidence interval. These occasions often may be recognized by key symmetrical phrases such as, "different from," "changed for better or worse," "unequal," etc. This section discusses the minor modifications required for such two-sided prob-values.

Example 6-7

Consider again the testing of TV tubes in Example 6-3. Suppose that the null hypothesis remains as

$$H_0: \mu = 1200$$

But now change the alternative hypothesis by supposing that our engineers cannot advocate the new process as better but must concede that it may be worse. Then the alternative hypothesis would be

$$H_1: \mu > 1200 \text{ or } \mu < 1200$$

that is:

$$H_1: \mu \neq 1200 \tag{6-28}$$

In other words, we now are testing whether the new process is *different* (whereas in Example 6-3, we were testing whether it was *better*). Thus, even before we collect any data, we can agree that a value of $\overline{X}$ well below 1200 would be just as strong evidence against H_0 as a value of $\overline{X}$ well above 1200; that is, what counts is how far away $\overline{X}$ is, on *either side*.

If the sample mean $\overline{X} = 1265$, what is the two-sided prob-value? That is, what is the probability that $\overline{X}$ will be at least as distant (in either direction) from the null hypothesis as 1265 is? (For your solution, recall that $\overline{X}$ was based on a sample of $n = 100$ observations, and the population standard deviation was given as $\sigma = 300$.)

Solution

To measure how far away $\overline{X}$ is from the null hypothesis, we start with $\overline{X} - \mu_0 = 1265 - 1200$. We then calculate the standardized Z value by dividing by the standard error $\sigma/\sqrt{n}$. Thus

$$Z = \frac{1265 - 1200}{300/\sqrt{100}} = 2.17 \tag{6-29}$$

The prob-value is the probability that we will observe a Z value this extreme, that is, a Z value above 2.17 or below -2.17 (see Figure 6-8). From Table IV we find

$$\Pr(Z \geq 2.17) = .015$$

By symmetry, the probability of a Z value below -2.17 is the same, .015. The probability of being extreme, one way or another, is therefor .030. This is the two-sided prob-value.

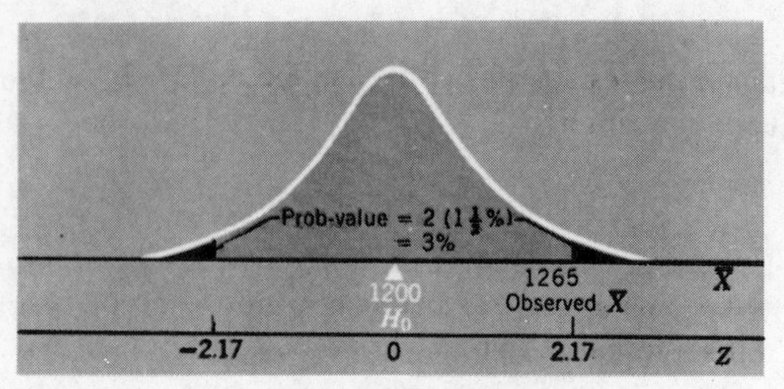

FIGURE 6-8 Two-sided prob-value ≡ Pr ($\bar{X}$ would be as extreme as the value actually observed, if H_0 is true). Compare with Figure 6-1.

In general, whenever the alternative hypothesis is two-sided, it is appropriate to calculate the two-sided prob-value for H_0. As we have just seen in Figure 6-8, whenever the distribution is symmetric, the two-sided prob-value is just double the one-sided prob-value.

6-19 Answer True or False; if False, correct it.

(a) If the alternative hypothesis is two-sided, then the prob-value, classical test, and confidence interval should be two-sided too.

(b) To decide whether the probability π of a die coming up ace is fair, suppose we are testing the hypothesis

$$H_0: \pi = 1/6 \quad \text{against} \quad H_1: \pi < 1/6$$

Then we should use a two-sided test, rejecting H_0 when π turns out to be large.

6-20 In a Gallup poll of 1500 Americans in 1975, 45% answered "yes" to the question, "Is there any area right around here—that is, within a mile—where you would be afraid to walk alone at night?" In an earlier poll of 1500 Americans in 1972, only 42% had answered "yes."

(a) Construct a two-sided confidence interval for the change in proportion who are afraid.

(b) Calculate the two-sided prob-value for the null hypothesis of no change.

(c) At the error level $\alpha = 5\%$, is the increase from 42% in 1972 to 45% in 1975 statistically discernible? That is, can H_0 be rejected? Answer in two ways, making sure that both answers agree:
(i) Is H_0 excluded from the 95% confidence interval?
(ii) Does the prob-value for H_0 fall below 5%?

6-21 Repeat Problem 6-20, with the appropriate minor changes in wording, for the following income data sampled from U.S. physicians in 1970:

	General Practitioners	*Pediatricians*
Sample size n	200	200
Mean income $\overline{X}$	\$38,000	\$36,000
Standard deviation s	\$18,000	\$20,000

pooled standard deviation s_p = \$19,000

CHAPTER 6 SUMMARY

6-1 The values inside a 95% confidence interval are called acceptable hypotheses (at the 5% error level), while the values outside are called rejected hypotheses. The no-difference hypothesis is called the null hypothesis H_0; when it is rejected, a difference is established and so the result is called statistically discernible (significant).

6-2 Prob-value is defined as the chance of getting a value of $\overline{X}$ as large as the one actually observed, if H_0 were true. That is, it is the tail area (of the distribution centered on μ_0) beyond the observed value of $\overline{X}$. Prob-value therefore measures the credibility of H_0.

6-3 For a classical test, we reject H_0 if the prob-value sinks below a specified error level α (usually 5%). A classical test is mainly of theoretical interest—for example, to clarify the two kinds of possible error (type I and type II). In practice, a classical test is less informative than a prob-value.

*6-4 Classical tests can be misleading, especially if H_0 is accepted on the basis of a small sample.

*6-5 The two-sided prob-value is double the one-sided prob-value, and is appropriate whenever the alternative hypothesis is two-sided.

REVIEW PROBLEMS

6-22 A professor was evaluated on a five-point scale by 15 students randomly selected from one class, and 40 students from another, with the following results:

Scale Value	*Frequencies*	
	10:30 AM Class	*3:30 PM Class*
1 (terrible)	0	4
2	0	7
3	4	13
4	7	9
5 (excellent)	4	7
	$n = 15$	$n = 40$

(a) Graph the frequency distributions of the two samples.

(b) Calculate the two sample means and standard deviations, and show them on the graph.

6-23 In Problem 6-22, the registrar wanted to know whether the difference in the two samples was just a sampling fluke, or reflected a real difference in the classes. In fact she suspected it showed a real difference, because it was customary for late classes to be restless and give their professors rather low ratings.

Answer her as thoroughly as you can. (*Hint:* We suggest you calculate the one-sided confidence interval, and prob-value for H_0.)

6-24 Suppose that a scientist concludes that a difference in sample means is "statistically significant (discernible) at the 1% level." Answer True or False; if False, correct it.

(a) There is at least a 99% chance that there is a real difference in the population means.

(b) The prob-value for H_0 (population means are exactly equal) is 1% or less.

(c) If there were no difference in the population means, the chance of getting such a difference (or more) in the sample means is 1% or less.

(d) The scientist's conclusion is sound evidence that a difference in population means exists. Yet in itself it gives no evidence whatever that this difference is large enough to be of practical importance. This illustrates that statistical significance and practical significance are two entirely different concepts.

6-25 When 287 mothers were observed on how they held their newborn babies, most held them on the left side. This was true of both left- and right-handed mothers, as the following breakdown shows. (From Salk, 1973. Another interesting account of this is given in Nemenyi and others, 1977, pp. 3–9.)

Mother Is \ *She Holds Baby on*	*Right Side*	*Left Side*	*Totals*
Right-handed	43	212	255
Left-handed	7	25	32
Totals	50	237	287

Assuming these 287 are a random sample:

(a) Calculate a 95% confidence interval for the proportion of mothers in the population who carry their babies on the left side. Is this proportion statistically discernible from the 50-50 proportion you would expect from sheer chance?

(b) Calculate a 95% confidence interval for the difference between the right- and left-handed mothers. Is the difference statistically discernible?

(c) Would you care to guess why there is a strong preference for the left side?

6-26 In Problem 6-25, Dr. Salk conjectured that the mother holds the baby on the left instinctively, so that it can better hear the reassuring beat of its mother's heart. To test this hypothesis, he randomly divided newborns into two groups: The treated group of 45 babies had a tape of an adult heartbeat played into the nursery; the control group of 45 babies had no heartbeat tape, but otherwise were treated the same. For each baby, the weight increase from the first to the fourth day was recorded, and graphed as follows (again, from Salk, 1973; and we report only the group of babies of middling birth weight):

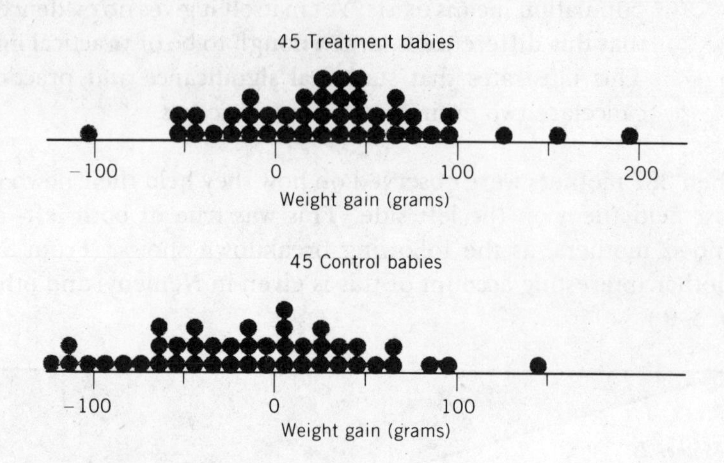

Note that the treatment babies on the whole gain weight, while the control babies lose slightly. Specifically, here are the calculated sample statistics:

	Sample Mean $\overline{X}$	*Squared Deviations* $\Sigma(X - \overline{X})^2$
Treatment	37	158,000
Control	−17	186,000

Assuming that these babies represent a random sample from a population:

(a) Calculate the one-sided 95% confidence interval for the mean difference that the treatment makes.

(b) Calculate the prob-value for the null hypothesis that the heartbeat tape does not improve weight gain.

(c) At the 5% level, can the null hypothesis be rejected, that is, is the difference in means statistically discernible?

*6-27 Let us look more closely at the assumption made in Problem 6-26, that the 90 babies were "a random sample from a population." In fact, they were not. Like most researchers, Dr. Salk found it completely impractical to take a random sample from the whole U.S. population. Instead he merely studied whatever babies were convenient, in a New York City hospital nursery.

Although the samples were not randomly drawn, nevertheless the really important randomization was done: The division of the babies into treatment and control groups was randomized, to ensure that it was fair and unbiased.

(a) Does it make any sense at all to talk about an underlying population?

(b) If there is no underlying population, does a confidence interval make any sense? Does a prob-value?

*6-28 With current treatment, 40% of all patients suffering from a rare tropical disease improved (within a year), and 60% did not. A new treatment was given to 8 such patients, in a pilot study to see whether it was effective. It was agreed that if 6 or more patients improved, the new treatment would be judged more effective than the old.

(a) State the null and alternative hypotheses, in words and symbols.

(b) State the region where H_0 is to be rejected, in symbols.

(c) What is α? Also state your answer in words.

(d) Suppose in the long run (unknown to the researchers, of course) that 70% of the patients given the new treatment would improve. What is the chance that the pilot study will miss finding this treatment more effective than the old? What is this number called, technically?

PART III

RELATING TWO OR MORE VARIABLES

CHAPTER 7

Simple Regression

Whatever moves is moved by another.

St. Thomas Aquinas

7-1 INTRODUCTION

In practice, we often want to study much more than an isolated characteristic of a single variable, such as its mean. We usually want to look at how the variable is related to other variables—what statisticians call *regression.*

To illustrate, consider how wheat yield depends on the amount of fertilizer applied. If we graph the yield Y that follows from various amounts of fertilizer X, a scatter similar to Figure 7-1 might be observed. From this scatter it seems clear that the amount of fertilizer does affect yield. In addition, it should be possible to describe *how,* by fitting a curve through this scatter. In this chapter, we will stick to the simplest case, where Y is related to one variable X by a straight line, called the *simple regression* of Y on X.

Since yield depends on fertilizer, it is called the *dependent variable* or *response* Y. Since fertilizer application is not dependent on yield, but instead is determined independently by the experimenter, it is referred to as the *independent variable* or *factor* or *regressor* X. An example will illustrate.

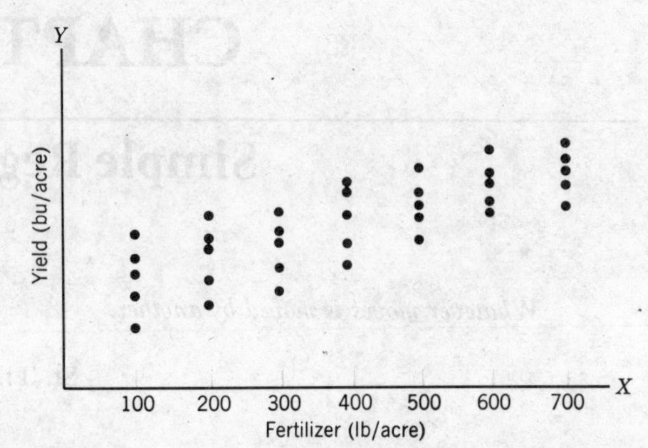

FIGURE 7-1 Observed relation of wheat yield to fertilizer application on 35 experimental plots.

Example 7-1

In a study of how wheat yield depends on fertilizer, suppose that funds are available for only seven experimental observations. So the experimenter sets X at seven different levels, taking only one observation Y in each case, as shown in Table 7-1.

(a) Graph these points, and roughly fit a line by eye.

(b) Use this line to predict yield, if fertilizer application is 400 pounds.

TABLE 7-1

Observations of Fertilizer and Yield

X *Fertilizer* *(lb/acre*	Y *Yield* *(bu/acre)*
100	40
200	50
300	50
400	70
500	65
600	65
700	80

Solution

(a) The scatter of observations is graphed in Figure 7-2. The line you fit by eye to this scatter should look something like the one we have drawn.

(b) With a fertilizer application of $X = 400$ pounds, the predicted yield is the height $\hat{Y} \simeq 60$ bushels shown on the fitted line in Figure 7-2. The deviation of the actual value Y from the predicted

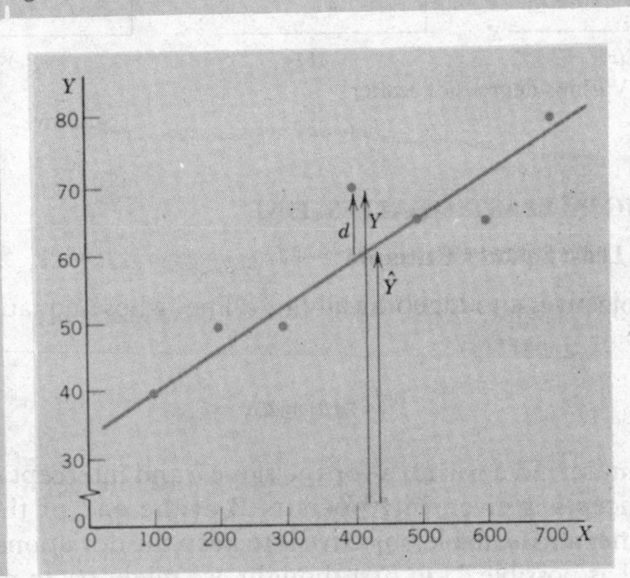

FIGURE 7-2 Line fitted by eye to the data of Table 7-1.

value $\hat{Y}$ is of particular interest. Denoted by d, we can write it algebraically as

$$d = Y - \hat{Y}$$

Roughly speaking, we have tried to keep all such deviations (errors) as small as we can in selecting our line by eye.

How good is a rough fit by eye, such as that used in Example 7-1? In Figure 7-3*a*, we note that if all the points were exactly in a line, then the fitted line could be drawn in with a ruler perfectly accurately. But as we progress to the highly scattered case shown in Figure 7-3*c*, we need to find another method—a method that is more objective, and is easily computerized. The following section, therefore, sets forth algebraic formulas for fitting a line.

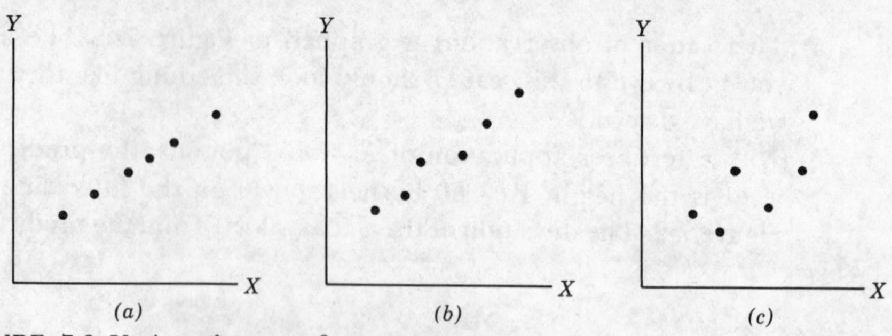

FIGURE 7-3 Various degrees of scatter.

7-2 FITTING A LEAST SQUARES LINE

(a) The Least Squares Criterion

Our objective is to algebraically fit a line, whose equation is of the form

$$\hat{Y} = a + bX \tag{7-1}$$

That is, we must find a formula for the slope b and intercept a. (A review of these concepts is given in Appendix B at the end of the book.) In fitting this line, a reasonable objective is to keep the deviations d in Figure 7-2 "as small as possible." On first thought, we might try to minimize the total amount of deviation Σd. But because some of the points are above the line and others below, some deviations d will be positive and others negative; to the extent they cancel, they will make the total (Σd) deceptively near zero. To overcome this "positive-negative" problem, we first square the deviations to make them all positive, and then sum them to obtain *the least squares criterion:*

$$\text{minimize } \Sigma d^2 = \Sigma(Y - \hat{Y})^2 \tag{7-2}$$

This criterion selects a single line of best fit, called the *least squares line.*

(b) The Least Squares Formulas

The formula for the least squares slope b turns out to be remarkably simple (derived, for example, in Wonnacott, 1977):

$$b = \frac{\Sigma(X - \overline{X})(Y - \overline{Y})}{\Sigma(X - \overline{X})^2} \tag{7-3}$$

The deviations $(X - \overline{X})$ and $(Y - \overline{Y})$ appear so often in this chapter that it is worthwhile abbreviating them. Let

$$\left.\begin{aligned} X - \overline{X} &\equiv x \\ Y - \overline{Y} &\equiv y \end{aligned}\right\} \tag{7-4}$$

This small x (or y) notation is chosen as a reminder that the deviations x are smaller numbers than the original observed values X, on the whole. (This is apparent in the first four columns of Table 7-2.) With this notation, the formula for b in (7-3) above can now be simplified:

TABLE 7-2

Fitting a Least Squares Line to the Data of Example 7-1

Data		*Deviation Form*		*Products*	
X	Y	$x = X - \overline{X}$ $= X - 400$	$y = Y - \overline{Y}$ $= Y - 60$	xy	x^2
100	40	−300	−20	6000	90,000
200	50	−200	−10	2000	40,000
300	50	−100	−10	1000	10,000
400	70	0	10	0	0
500	65	100	5	500	10,000
600	65	200	5	1000	40,000
700	80	300	20	6000	90,000
$\overline{X} = 400$	$\overline{Y} = 60$	$\sum x = 0\checkmark$	$\sum y = 0\checkmark$	$\sum xy = 16{,}500$	$\sum x^2 = 280{,}000$

$$b = \frac{\Sigma xy}{\Sigma x^2} \tag{7-5}$$

Once b is calculated, the intercept a can then be found from another simple formula:

$$a = \overline{Y} - b\overline{X} \tag{7-6}$$

For the data in Example 7-1, the calculations for a and b are laid out in Table 7-2. We calculate Σxy and Σx^2, and substitute them into (7-5):

$$b = \frac{\Sigma xy}{\Sigma x^2} = \frac{16{,}500}{280{,}000} = .059$$

Then we use this slope b (along with $\overline{X}$ and $\overline{Y}$ calculated in the first two columns of Table 7-2) to calculate the intercept a from (7-6):

$$a = \overline{Y} - b\overline{X} = 60 - .059(400) = 36.4$$

Plugging these estimated values a and b into (7-1) yields the equation of the least squares line:

$$\hat{Y} = 36.4 + .059X \tag{7-7}$$

This is graphed in Figure 7-4. It is so close to the fit by eye in Figure 7-2 that you have to look hard for the difference.

(c) Meaning of the Slope b

By definition, the slope of a line is the change in height Y, when we move to the right by one unit in the X direction. That is,

$$\text{Slope } b = \text{change in } Y \text{ that accompanies a unit change in } X \tag{7-8}$$

This is illustrated geometrically in Appendix B at the end of the book. It is so important that it is worthwhile giving an alternative algebraic

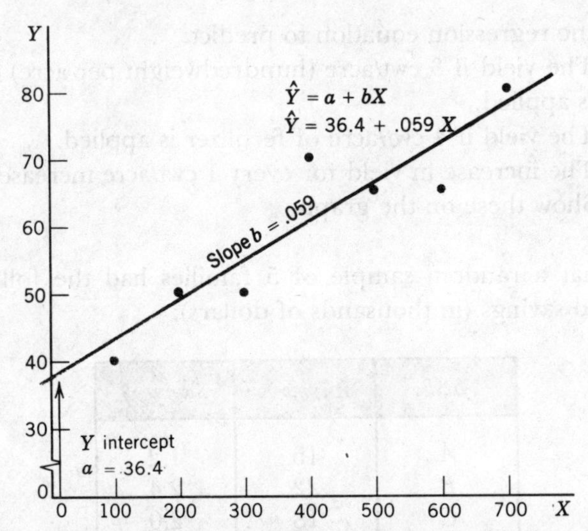

FIGURE 7-4 The least squares line fitted to the data of Example 7-1.

derivation here. Suppose in (7-7), for example, that X was increased by 1 unit, from 75 to 76 pounds, say. Then

$$\begin{aligned} \text{Initial } Y &= 36.4 + .059(75) \\ \text{New } Y &= 36.4 + .059(75 + 1) \\ &= 36.4 + .059(75) + .059 \\ &= \text{initial } Y + .059 \end{aligned}$$

That is, Y has increased by .059 as X has increased by 1—which is the slope b. And so (7-8) is established.

PROBLEMS

7-1 Suppose that 4 levels of fertilizer were applied to 4 randomly chosen plots, resulting in the following yields of corn:

Fertilizer X (cwt/acre)	*Yield Y (bus/acre)*
1	70
2	70
4	80
5	100

(a) Calculate the regression line of yield on fertilizer.
(b) Graph the 4 points and the regression line. Check that the line fits the data reasonably well.

(c) Use the regression equation to predict:
  (i) The yield if 3 cwt/acre (hundredweight per acre) of fertilizer is applied.
  (ii) The yield if 4 cwt/acre of fertilizer is applied.
  (iii) The increase in yield for every 1 cwt/acre increase in fertilizer. Show these on the graph.

7-2 Suppose that a random sample of 5 families had the following annual incomes and savings (in thousands of dollars):

Family	*Income X*	*Saving S*
A	16	1.2
B	22	2.4
C	18	2.0
D	12	1.4
E	12	.6

Calculate the regression line of S on X.

7-3 During the 1950s, radioactive waste leaked from a storage area near Hanford, Washington, into the Columbia River nearby. For nine counties downstream in Oregon, an index of exposure X was calculated (based on distance from Hanford, and distance of the average citizen from the river, etc.). Also, the cancer mortality Y was calculated (deaths per 100,000 person-years, 1959–1964). The data was as follows (from Fadeley, 1965; via Anderson and Sclove, 1978):

County	*Radioactive Exposure X*	*Cancer Mortality Y*
Clatsop	8.3	210
Columbia	6.4	180
Gilliam	3.4	130
Hood River	3.8	170
Morrow	2.6	130
Portland	11.6	210
Sherman	1.2	120
Umatilla	2.5	150
Wasco	1.6	140

From this data, summary statistics were computed:

$$\bar{X} = 4.6 \qquad \bar{Y} = 160$$

$$\Sigma x^2 = 97.0 \qquad \Sigma y^2 = 9400 \qquad \Sigma xy = 876$$

(a) Calculate the regression line for predicting Y from X.
(b) Estimate the cancer mortality if X were 5.0. And if X were 0.
(c) Graph the nine counties, and your answers in (a) and (b).
(d) To what extent does this data prove the harmfulness of radioactive exposure?

7-4 In the early 1900s, six provinces in Bavaria had recorded their infant mortality (deaths per 1000 live births) and breast-feeding (percentage of infants breast fed) as follows (Knodel, 1977):

Province	*Mortality (deaths per 1000)*	*Breast-feeding (%)*
Mittelfranken	250	60
Niederbayern	320	30
Oberfranken	170	90
Oberpfalz	300	60
Schwaben	270	40
Unterfranken	190	80
Mean	250	60%

(a) Calculate the appropriate regression line to predict the infant mortality of two other provinces, Oberbayern and Pfalz, which had breast-feeding rates of 37% and 85%.

If the actual infant mortalities were 290 and 168, calculate the prediction errors.

(b) To what extent does this data prove the benefits of breast-feeding?

7-3 THE REGRESSION MODEL

Our treatment of a sample of points so far has involved only fitting a line. Now we wish to make inferences about the parent population from which this sample was drawn. To do so, we must build a mathematical model that allows us to construct confidence intervals and test hypotheses.

(a) Simplifying Assumptions

Suppose in Figure 7-5*a* that we set fertilizer at level X_1 for many many plots. The resulting yields will not all be the same, of course; the weather might be better for some plots, the soil might be better for others, and so on. Thus we would get a distribution (or population) of Y values, appropriately called the probability distribution of Y_1 given X_1, or $p(Y_1/X_1)$. There will similarly be a distribution of Y_2 at X_2, and so forth. We can therefore visualize a whole set of Y populations such as those shown in Figure 7-5*a*. There would obviously be great problems in ana-

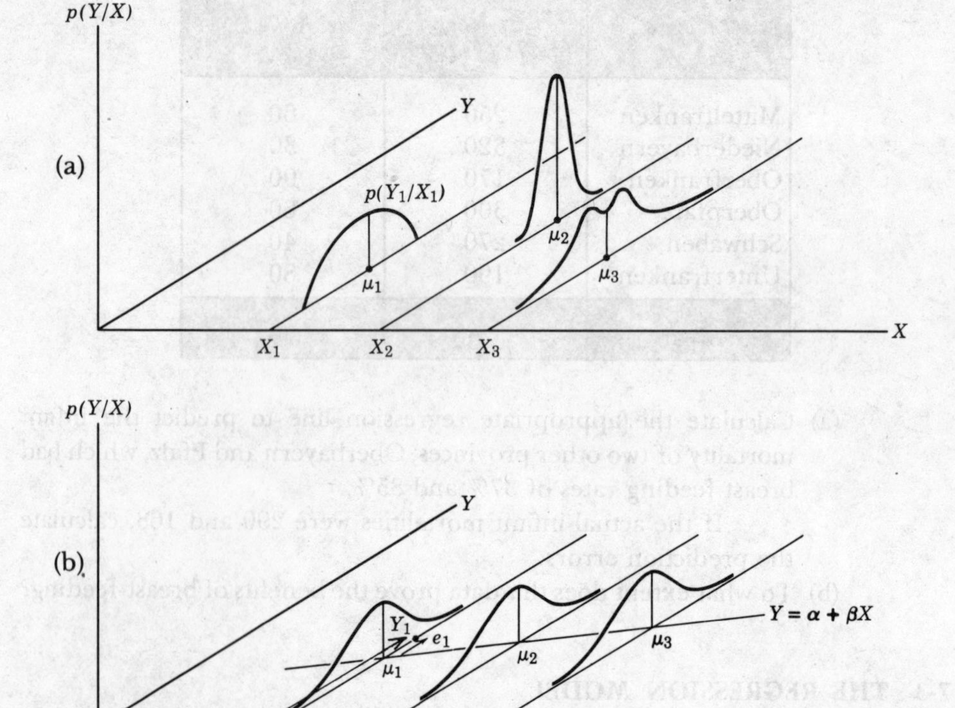

FIGURE 7-5 *(a)* General populations of Y, given X. *(b)* The special form of the populations of Y assumed in simple linear regression.

lyzing populations as peculiar as these. To keep the problem manageable, therefore, we make several assumptions about the regularity of these Y distributions; as shown in Figure 7-5*b,* we assume:

1. All the Y distributions have the same spread. Formally, this means the probability distributions $p(Y_i/X_i)$ have the same variance σ^2 for all X_i $(i = 1, 2, \ldots, n)$.
2. The means of all the distributions lie on a straight line, known as the true (population) regression line:

$$\text{Expected value of } Y_i = \mu_i = \alpha + \beta X_i \tag{7-9}$$

   The population parameters α and β specify the line, and are to be estimated from sample information.
3. The random variables Y_i are statistically independent. For example, if Y_1 happens to be large, there is no reason to expect Y_2 to be large (or small); that is, Y_2 is statistically unrelated to Y_1.

These three assumptions may be written more concisely as:

The random variables Y_i are statistically independent, with

$$\text{Mean} = \mu_i = \alpha + \beta X_i \qquad \text{Variance} = \sigma^2 \tag{7-10}$$

On occasion, it is useful to describe the deviation of Y_i from its expected value as the error or disturbance term e_i, so that the model may alternatively be written

$$Y_i = \alpha + \beta X_i + e_i$$

where the e_i are independent random variables, with

$$\text{Mean} = 0 \qquad \text{Variance} = \sigma^2 \tag{7-11}$$

For example, in Figure 7-5*b* the first observed value Y_1 is shown in color, along with the corresponding error term e_1.

As for the shape of the distribution of Y for a given X, if it is normal, then the least squares estimates in this chapter are absolutely the best. To the extent the shape is nonnormal, other techniques may be better.

(b) The Nature of the Error Term

Now let us consider in more detail the purely random part of Y, the error or disturbance term e. Why does it exist? Why doesn't a precise and exact value of Y follow, once the value of X is given? The random error may be regarded as the sum of two components:

1. *Measurement error.* There are various reasons why Y may be measured incorrectly. In measuring crop yield, there may be an error resulting from sloppy harvesting or inaccurate weighing. If the example is a study of the consumption of families at various income levels, the measurement error in consumption might consist of budget and reporting inaccuracies.
2. *Inherent variability* occurs inevitably in biological and social phenomena. Even if there were no measurement error, repetition of an experiment using exactly the same amount of fertilizer would result in somewhat different yields. The differences could be reduced by tighter experimental control—for example, by holding constant soil conditions, amount of water, and so on. But complete control is impossible—for example, seeds cannot be exactly duplicated.

(c) Estimating α and β

Suppose the true (population) regression, $Y = \alpha + \beta X$, is the black line shown in Figure 7-6. This is unknown to the statistician, who must estimate it as best he can by observing X and Y. If at the first level X_1, the random error e_1 happens to take on a negative value as shown in the diagram, he will observe Y_1 below the true line. Similarly, if the random errors e_2 and e_3 happen to take on positive values, he will observe Y_2 and Y_3 above the true line.

Now the statistician applies the least squares formulas to the only information he has—the sample points Y_1, Y_2, and Y_3. This produces an estimating line $\hat{Y} = a + bX$, which we color blue in Figure 7-6 (following our usual color convention for samples).

Figure 7-6 is a critical diagram. Before proceeding, you should be sure that you can clearly distinguish between: (1) the true regression and its surrounding e distribution; since these are population values and cannot be observed they are shown in black. (2) the Y observations and the resulting fitted regression line; since these are sample values, they are known to the statistician and colored blue.

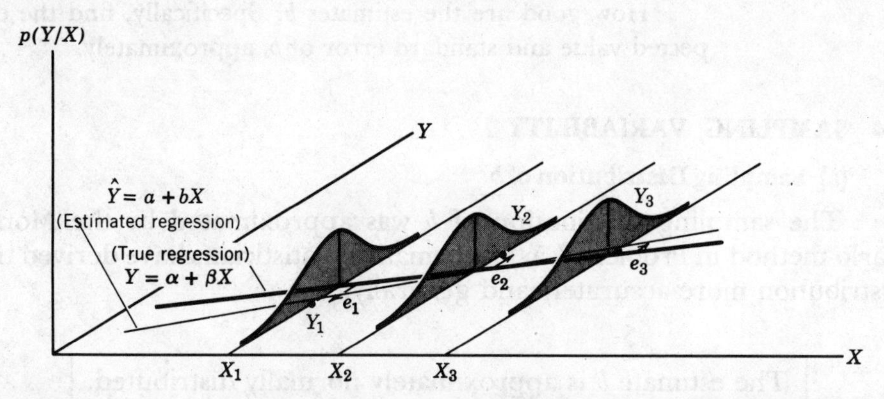

FIGURE 7-6 True (population) regression and estimated (sample) regression.

Unless the statistician is very lucky indeed, it is obvious that his estimated line will not exactly coincide with the true population line. The best he can hope for is that it will be reasonably close to the target. In the next section we develop this idea of "closeness" in more detail.

PROBLEMS

7-5 (Monte Carlo). Suppose the true (long-run average) relation of corn yield Y to fertilizer X is given by the line

$$Y = 2.40 + .30X$$

where Y is measured in tons of corn per acre, and X is measured in hundreds of pounds of fertilizer per acre.

(a) First, play the role of nature. Graph the line for $0 \leq X \leq 12$. Suppose the yield varies about its expected value on this line, with a standard deviation $\sigma = 1$, and a distribution that is normal. Simulate a sample of five such yields, one each for $X = 2, 4, 6, 8, 10$. (*Hint:* First calculate the five mean values from the given line. Then add to each a random normal error e from Appendix Table II.)

(b) Now play the role of the statistician. Calculate the least squares line that best fits the sample. Graph the sample and the fitted line.

(c) Finally, let the instructor be the evaluator. Have him graph several of the lines found in (b). Then have him tabulate all the values of b, and graph their relative frequency distribution.

How good are the estimates b? Specifically, find the expected value and standard error of b, approximately.

7-4 SAMPLING VARIABILITY

(a) Sampling Distribution of b

The sampling distribution of b was approximated by the Monte Carlo method in Problem 7-5. Mathematical statisticians have derived the distribution more accurately and generally:

> The estimate b is approximately normally distributed, with
>
> $$\text{Expected value of } b = \beta \tag{7-12}$$
>
> $$\text{Standard error of } b = \frac{\sigma}{\sqrt{\Sigma x^2}} \tag{7-13}$$

where $x = X - \overline{X}$ is the customary deviation from the mean. This distribution of b is graphed in Figure 7-7. Once again we see an estimate that is normally distributed.

(b) Experimental Design

The formula for the standard error in (7-13) has some interesting implications for experimental design. Suppose the experiment has been badly designed, with the X values close together. This makes the devia-

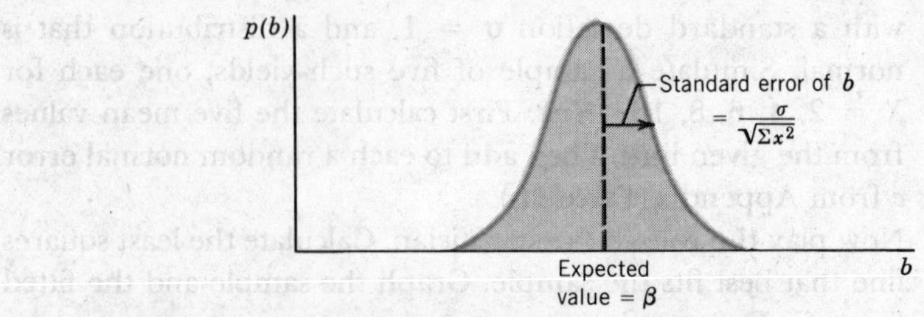

FIGURE 7-7 The sampling distribution of the estimate b.

tions x small; hence Σx^2 is small. Then the standard error of b in (7-13) is large, and b is a comparatively unreliable estimate. To check the intuitive validity of this, consider the scatter diagram in Figure 7-8*a*. The bunching of the X values means that the small part of the line being investigated is obscured by the error e, making the slope estimate b very unreliable. In this specific instance, our estimate has been pulled badly out of line by the errors—in particular by the one indicated by the arrow.

By contrast, in Figure 7-8*b*, we show the case where the X values are reasonably spread out. Even though the errors e remain the same, the estimate b is much more reliable, because errors no longer exert the same leverage.

As a concrete example, suppose we wish to examine how sensitive U.S. imports Y are to the international value of the dollar X. A much more reliable estimate should be possible using the recent periods when the dollar has been floating (and taking on a range of values) than in earlier periods when currencies were fixed (i.e., the dollar was fixed in value and only allowed to fluctuate within very narrow limits).

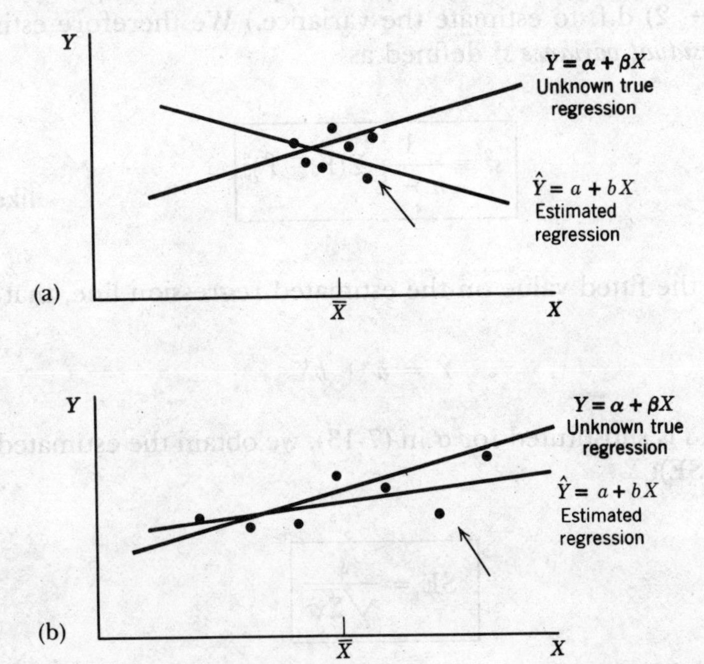

FIGURE 7-8 (*a*) Unreliable fit when the X values are very close. (*b*) More reliable fit when the X values are spread out.

7-5 CONFIDENCE INTERVALS AND TESTS FOR β

(a) Estimating the Standard Error of b

With the expected value, standard error, and normality of b established, statistical inferences about β are now in order. But first there is one remaining problem: From (7-13), the standard error of b is $\sqrt{\sigma^2/\Sigma x^2}$, where σ^2 is the variance of the Y observations about the population line. But σ^2 is generally unknown, and must be estimated. A natural way to estimate σ^2 is to use the deviations in Y about the *fitted* line. Specifically, consider the mean squared deviation about the fitted line:

$$\frac{1}{n}\Sigma d^2 = \frac{1}{n}\Sigma (Y - \hat{Y})^2 \qquad (7\text{-}14)$$

like (7-2)

Instead of n, however, we should use the divisor $n - 2$. (The reason is familiar: Two d.f. have already been used up in calculating a and b, which leaves $(n - 2)$ d.f. to estimate the variance.) We therefore estimate σ^2 with the *residual variance* s^2 defined as

$$\boxed{s^2 \equiv \frac{1}{n-2}\Sigma(Y - \hat{Y})^2} \qquad (7\text{-}15)$$

like (2-10)

where $\hat{Y}$ is the fitted value on the estimated regression line, that is,

$$\hat{Y} = a + bX$$

When s is substituted for σ in (7-13), we obtain the estimated standard error (SE):

$$\boxed{\text{SE} = \frac{s}{\sqrt{\Sigma x^2}}} \qquad (7\text{-}16)$$

With this in hand, statistical inferences can now be made.

(b) Confidence Intervals

Using the same argument as in Chapter 5 earlier, we could easily show that the 95% confidence interval for β is

$$\beta = b \pm t_{.025}\text{SE} \qquad (7\text{-}17) \text{ like } (5\text{-}17)$$

Substituting SE from (7-16) yields

95% confidence interval for the slope:

$$\beta = b \pm t_{.025}\frac{s}{\sqrt{\Sigma x^2}} \qquad (7\text{-}18)$$

where the degrees of freedom for t are the same as the divisor used in s^2:

$$\text{d.f.} = n - 2 \qquad (7\text{-}19)$$

Example 7-2

Find the 95% confidence interval for the slope that related wheat yield to fertilizer in (7-7).

Solution

TABLE 7-3

Calculations for the Residual Variance s^2

X	Y	$\hat{Y} = 36.4 + .059X$	$Y - \hat{Y}$	$(Y - \hat{Y})^2$
100	40	42.3	−2.3	5.29
200	50	48.2	1.8	3.24
300	50	54.1	−4.1	16.81
400	70	60.0	10.0	100.00
500	65	65.9	−0.9	.81
600	65	71.8	−6.8	46.24
700	80	77.7	2.3	5.29
				$s^2 = \frac{177.68}{7-2} = 35.54$

We first use (7-15) to calculate s^2 in Table 7-3. The critical t value then has d.f. $= n - 2 = 7 - 2 = 5$ (the same as the divisor in s^2). From Appendix Table V, $t_{.025}$ is found to be 2.571. Finally, note that b and Σx^2 were already calculated in Table 7-2. When these values are all substituted into (7-18),

$$\beta = .059 \pm 2.571 \frac{\sqrt{35.54}}{\sqrt{280{,}000}}$$

$$= .059 \pm 2.571\,(.0113) \tag{7-20}$$

$$= .059 \pm .029 \tag{7-21}$$

$$.030 < \beta < .088 \tag{7-22}$$

(c) Testing hypotheses

The hypothesis that X and Y are unrelated may be stated mathematically as $\beta = 0$. To test this hypothesis, we merely note whether the value 0 is contained in the confidence interval.

Example 7-3

In Example 7-2, test at the 5% level the hypothesis that yield is unrelated to fertilizer.

Solution

Since $\beta = 0$ is excluded from the confidence interval (7-22), we reject it, and conclude that yield is indeed related to fertilizer.

(d) Prob-Value

Rather than simply accept or reject, a more appropriate form for a test is the calculation of the prob-value. We first calculate the t statistic:

$$\boxed{t = \frac{b}{\text{SE}}} \qquad \text{(7-23) like (6-19)}$$

Then we look up the probability in the tail beyond this observed value of t; this is the prob-value.

Example 7-4

In Example 7-2, what is the prob-value for the null hypothesis that yield does not increase with fertilizer?

Solution

In the confidence interval (7-20) we have already calculated b and its standard errot SE, which we can now substitute into (7-23):

$$t = \frac{.059}{.0113} = 5.2 \qquad (7\text{-}24)$$

In Appendix Table V, we scan the row where d.f. = 5, and find the observed t value of 5.2 lies beyond $t_{.0025} = 4.773$. Thus

$$\text{Prob-value} < .0025 \qquad (7\text{-}25)$$

This is so little credibility for H_0 that we could reject it, and conclude that yield does indeed increase with fertilizer.

PROBLEMS

7-6 For Problems 7-1 and 7-2, construct a 95% confidence interval for the population slope β.

7-7 Suppose that a random sample of 4 families had the following annual incomes and savings:

Family	*Income X (thousands of $)*	*Saving S (thousands of $)*
A	22	2.0
B	18	2.0
C	17	1.6
D	27	3.2

(a) Estimate the population regression line $S = \alpha + \beta X$.
(b) Construct a 95% confidence interval for the slope β.
(c) Graph the 4 points and the fitted line, and then indicate as well as you can the acceptable slopes given by the confidence interval in (b).

7-8 Which of the following hypotheses is rejected by the data of Problem 7-7 at the 5% level? [*Hint:* See (6-5).]
(a) $\beta = 0$
(b) $\beta = .05$

(c) $\beta = .10$
(d) $\beta = .50$

7-9 In Problem 7-7, suppose the population marginal propensity to save (β) is known to be positive, if it is not 0.

(a) State the null and alternative hypotheses in symbols.
(b) Calculate the prob-value for H_0.
(c) Suppose that we are interested in making a statement of the form, "The population marginal propensity to save is at least as large as such and such." Construct a one-sided 95% confidence interval of this form.
(d) At the 5% level, can we reject the null hypothesis $\beta = 0$? Test in two ways:
   (i) Is the prob-value less than 5%?
   (ii) Is $\beta = 0$ excluded from the confidence interval?

7-10 Repeat Problem 7-9 using the data of Problem 7-2 instead of 7-7.

CHAPTER 7 SUMMARY

7-1 To show how a response Y is related to a factor (regressor) X, a line can be fitted called the regression line, $\hat{Y} = a + bX$.

7-2 The slope b and intercept a can be calculated from simple formulas, called *least squares*.

7-3 The actual observations (dots) must be recognized as just a sample from an underlying population. For this population, as usual we use a Greek letter for the slope β. This is the target being estimated by the sample slope b.

7-4 If sampling is random, then b fluctuates from sample to sample in a well understood way: b fluctuates around β with a specified standard error—just as $\overline{X}$ fluctuated around μ in Chapter 4.

7-5 We can therefore construct a confidence interval for β, or calculate the prob-value for $\beta = 0$. From either of these we can then test the hypothesis that $\beta = 0$.

REVIEW PROBLEMS

7-11 In 1970, a random sample of 50 American men aged 35 to 54 showed the following relation between annual income Y (in dollars) and education X (in years). (Reconstructed using the 1970 U.S. Census 1 in 10,000 sample):

$$\hat{Y} = 1200 + 800X$$

Average income was $\overline{Y} = \$10{,}000$ and average education was $\overline{X} = 11.0$ years, with $\Sigma x^2 = 900$. The residual standard deviation about the fitted line was $s = \$7300$.

(a) Predict the income of a man who completed 2 years of high school ($X = 10$).

(b) Calculate a 95% confidence interval for the population slope.

(c) Is the relation of income to education statistically discernible at the 5% level?

(d) Would it be fair to say that each year's education is worth $800?

7-12 A random sample of 5 boys had their heights (in inches) measured at age 4 and again at age 18, with the following results:

Boys' Initial	*Age 4*	*Age 18*
J.K.	40	68
A.M.	43	74
D.B.	40	70
I.S.	40	68
J.B.	42	70
Average	41	70

(a) Suppose another boy is randomly drawn from the same population. If he was 42 inches tall at age 4, would you predict he would be 70 inches tall at age 18 (like the last boy in the sample)? If not, why not? What *would* you predict?

(b) Graph the given 5 points, and your answer in (a).

(c) If the whole population of boys was graphed, and its least squares line was calculated, what would be its slope? Answer with an interval wide enough so that you would be willing to bet on it, 95 cents to a nickel.

7-13 A 95% confidence interval for a regression slope was calculated on the basis of 1000 observations: $\beta = .48 \pm .27$. Calculate the prob-value for the null hypothesis that Y does not increase with X.

CHAPTER 8

Multiple Regression

Truth is rarely pure and never simple.

Oscar Wilde

8-1 INTRODUCTION

Multiple regression is the extension of simple regression, to take account of more than one factor X. It is obviously the appropriate technique when we want to investigate the effect on Y of several factors simultaneously. Yet, even if we are interested in the effect of only one factor, it is still essential to do multiple rather than simple regression if the data comes from an observational study. As we stressed in Chapter 1, multiple regression will reduce the bias that would occur in a simple regression that did not control for extraneous factors. In this chapter, we will make this explicitly clear.

Example 8-1

Suppose the fertilizer and yield observations in Example 7-1 were taken at seven different agricultural experiment stations across the country. If soil conditions and temperature were essentially the same in all these areas, we still might ask whether part of the fluctuation in Y can be explained by varying levels of rainfall in different areas. A better prediction of yield may be possible if *both* fertilizer and rainfall are examined. The observed levels of rainfall are therefore given in Table 8-1, along with the original observations of yield and fertilizer.

TABLE 8-1
Observations of Yield, Fertilizer Application, and Rainfall

Y *Wheat Yield (bu/acre)*	X *Fertilizer (lb/acre)*	Z *Rainfall (inches)*
40	100	10
50	200	20
50	300	10
70	400	30
65	500	20
65	600	20
80	700	30

(a) On Figure 7-2, tag each point with its value of rainfall Z. Then, considering just those points with low rainfall ($Z = 10$), roughly fit a line by eye. Next, repeat for the points with moderate rainfall ($Z = 20$), and then for the points with high rainfall ($Z = 30$).
(b) Now, if rainfall were kept constant, roughly estimate what would be the slope of yield on fertilizer. That is, what would be the increase in yield per pound of additional fertilizer?
(c) If fertilizer were kept constant, roughly estimate what would be the increase in yield per inch of additional rainfall.
(d) Roughly estimate the yield if fertilizer were 400 pounds, and rainfall were 10 inches.

Solution

(a)

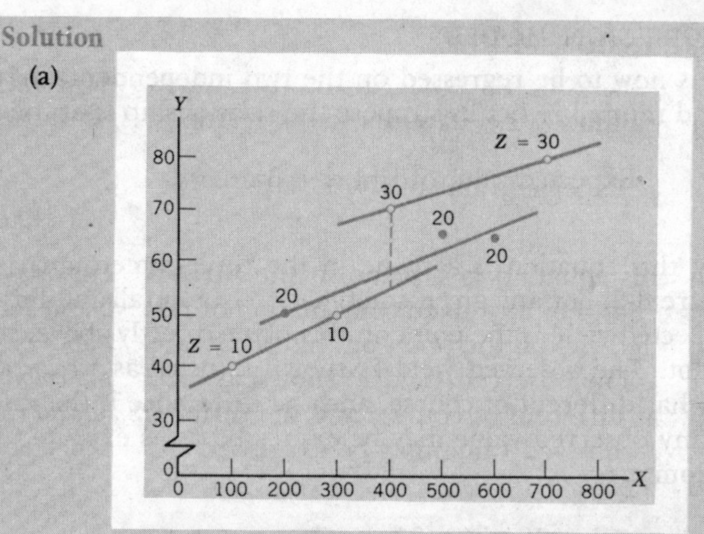

FIGURE 8-1 How yield depends on 2 variables (fertilizer X and rainfall Z).

(b) We note the largest slope in Figure 8-1 is 10/200 = .05 for the line $Z = 10$, while the smallest slope is 10/300 = .033 for the line $Z = 30$; on average these slopes are about .04 bushel per pound of fertilizer.

(c) Let us keep fertilizer constant, at the center of the data, for example, where $X = 400$. The dashed line shows the vertical distance between the line where rainfall $Z = 10$ and the line where $Z = 30$, which is about 15 bushels. Since this increase of 15 bushels comes from an increase of 20 inches of rain, this means that rain increases yield by about 15/20 = .75 bushel per inch of rainfall.

(d) On Figure 8-1 we use the line where $Z = 10$, at the point where $X = 400$, obtaining yield $\simeq$ 55 bushels.

Figure 8-1 shows clearly why adding an additional variable Z gives a better idea of the effect of fertilizer on the crop. If rainfall is not held constant and is ignored, we obtain the slope in Figure 7-2—which is larger because high rainfall tends to accompany high fertilizer. Thus the slope in Figure 7-2 is thrown off because we erroneously attribute to fertilizer the effects of both fertilizer *and* rainfall.

We must admit that the fitting by eye in Example 8-1 was vastly oversimplified. In the next two sections we will therefore develop a more objective and easily computerized method that will handle more complicated cases.

8-2 THE REGRESSION MODEL

Yield Y is now to be regressed on the two independent variables, fertilizer X and rainfall Z. Let us suppose the relationship is of the form

$$\text{Expected value of } Y = \alpha + \beta X + \gamma Z \tag{8-1}$$

like (7-9)

Geometrically this equation is a plane in the three-dimensional space shown in Figure 8-2. For any given combination of rainfall and fertilizer (X, Z), the expected yield is the point on this plane directly above, shown as a hollow dot. The observed yield Y, shown as usual as a *colored* dot, will be somewhat different of course, and the difference is the random error. Thus any observed value may be expressed as its expected value plus the random error e:

$$Y = \alpha + \beta X + \gamma Z + e \tag{8-2}$$

like (7-11)

with the same assumptions about e as in Chapter 7.

β is geometrically interpreted as the slope of the plane as we move in the X direction keeping Z constant, sometimes called the marginal effect of fertilizer X on yield Y. Similarly γ is the slope of the plane as we move in the Z direction keeping X constant, called the marginal effect of Z on Y.

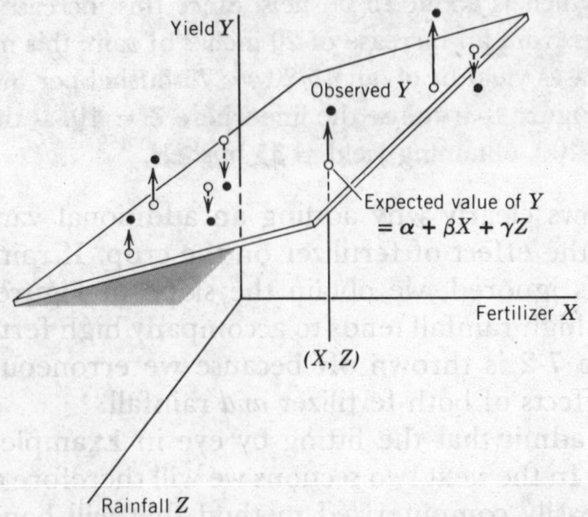

FIGURE 8-2 Scatter of observed points about the true regression plane.

8-3 THE LEAST SQUARES FITTED PLANE

As in simple regression, the problem is that the statistician does not know the *true* relationship [the true plane (8-1) shown in Figure 8-2]. Instead, she must fit an *estimated* plane of the form

$$\hat{Y} = a + bX + cZ \tag{8-3}$$

When we apply the usual least squares criterion for a good fit in multiple regression, we unfortunately do not get easy formulas for *a, b,* and *c.* Instead, we get 3 equations to solve, sometimes called *estimating equations* or *normal equations:*

$$\Sigma xy = b\Sigma x^2 + c\Sigma xz \tag{8-4}$$

$$\Sigma zy = b\Sigma xz + c\Sigma z^2 \tag{8-5}$$

$$a = \overline{Y} - b\overline{X} - c\overline{Z} \tag{8-6}$$

where once again we have used the convenient deviations:

$$\left.\begin{aligned} x &\equiv X - \overline{X} \\ z &\equiv Z - \overline{Z} \\ y &\equiv Y - \overline{Y} \end{aligned}\right\} \tag{8-7}$$

Equations (8-4) and (8-5) may be simultaneously solved for b and c. (The algebraic solution of a set of simultaneous linear equations is reviewed in Appendix C at the back of the book.) Then the last equation may be solved for a. For the data of Example 8-1, these calculations are shown in Table 8-2, and yield the fitted multiple regression equation

$$\hat{Y} = 28 + .038X + .83Z \tag{8-11}$$

As we mentioned in the preface, statistical calculations are usually done on a computer using a standard program. This is particularly true of multiple regression, where the calculations generally become much too complicated to do by hand. (Imagine what Table 8-2 would look like with 100 rows of data and 5 regressors—a fairly typical situation.) We therefore give, in Table 8-3, a typical computer solution. (This solution was computed on the MINITAB interactive computing system, as described in Ryan, Joiner, and Ryan, 1976). The given data in Table 8-1

TABLE 8-2

Calculations for the Multiple Regression of Y on X and Z

Data			*Deviation Form*			*Products*				
Y	X	Z	$y = Y - \overline{Y}$	$x = X - \overline{X}$	$z = Z - \overline{Z}$	xy	zy	x^2	z^2	xz
40	100	10	−20	−300	−10	6000	200	90,000	100	3000
50	200	20	−10	−200	0	2000	0	40,000	0	0
50	300	10	−10	−100	−10	1000	100	10,000	100	1000
70	400	30	10	0	10	0	100	0	100	0
65	500	20	5	100	0	500	0	10,000	0	0
65	600	20	5	200	0	1000	0	40,000	0	0
80	700	30	20	300	10	6000	200	90,000	100	3000
$\overline{Y} = 60$	$\overline{X} = 400$	$\overline{Z} = 20$	0✓	0✓	0✓	$\sum xy = 16{,}500$	$\sum zy = 600$	$\sum x^2 = 280{,}000$	$\sum z^2 = 400$	$\sum xz = 7000$

Estimating equations (8-4) and (8-5)
$$\begin{cases} 16{,}500 = 280{,}000\,b + 7000c \\ 600 = 7000b + 400c \end{cases} \quad (8\text{-}8)$$

Solution
$$\begin{cases} b = .0381 \\ c = .833 \end{cases} \quad (8\text{-}9)$$

From (8-6), $a = 60 - .0381(400) - .833(20)$

$$a = 28.1 \quad (8\text{-}10)$$

TABLE 8-3

Computer Solution for Wheat Yield Multiple Regression (Data in Table 8-1)

```
-- READ COLUMNS C1, C2, C3
--              40 100  10
--              50 200  20
--              50 300  10
--              70 400  30
--              65 500  20
--              65 600  20
--              80 700  30
--

-- REGRESS C1 ON 2 PREDICTORS IN C2,C3

THE REGRESSION EQUATION IS
Y =     28.1 +0.0381 X1 + 0.833 X2
```

was first read into three columns (C1, C2, and C3). Then the computer was instructed to regress the variable in C1 (yield) on the 2 regressors given in C2 and C3 (fertilizer and rainfall). The computer replied with the regression equation, which confirms the laborious calculations in Table 8-2. Now that you understand these calculations and how the computer carries them out, you may never have to go through them by hand again.

We recommend that you use a computer yourself to confirm the estimated regression (8-11), and to calculate the homework problems. It is a good opportunity to meet the machine, since a standard multiple regression program is available in practically every computer center (or even in some sophisticated pocket calculators).

If a computer is not feasible, you can still calculate the regression in a tableau like Table 8-2. Or you can skip the calculation entirely, and go on to the rest of the problem where the interpretation of the regression is discussed—this is the part that requires the human touch. For instance, the next example will show how the multiple regression equation (8-11) nicely answers the same questions as Example 8-1.

Example 8-2

(a) Graph the relation of Y to X given by (8-11), when rainfall has the constant value:

(i) $Z = 10$

(ii) $Z = 20$

(iii) $Z = 30$

(b) Compare to the figure in Example 8-1.

Solution

(a) Substitute $Z = 10$ into (8-11):

$$\hat{Y} = 28 + .038X + .83Z \qquad \text{(8-11) repeated}$$

$$\hat{Y} = 28 + .038X + .83(10)$$

$$= 36.3 + .038X \qquad \text{(8-12)}$$

This is a line with slope .038 and Y intercept 36.3. Similarly, when we substitute $Z = 20$ and then $Z = 30$ into (8-11), we obtain the lines

$$\hat{Y} = 44.6 + .038X \qquad \text{(8-13)}$$

$$\hat{Y} = 52.9 + .038X \qquad \text{(8-14)}$$

These are lines with the same slope .038, but higher and higher intercepts, as shown in Figure 8-3.

(b) The three lines given by the multiple regression model (8-11) are evenly spread because we had evenly spaced values of Z in part (a). The lines are also parallel, because they all have the same slope $b = .038$

In contrast, the three earlier lines fitted in Example 8-1 were not constrained to have the same slope.

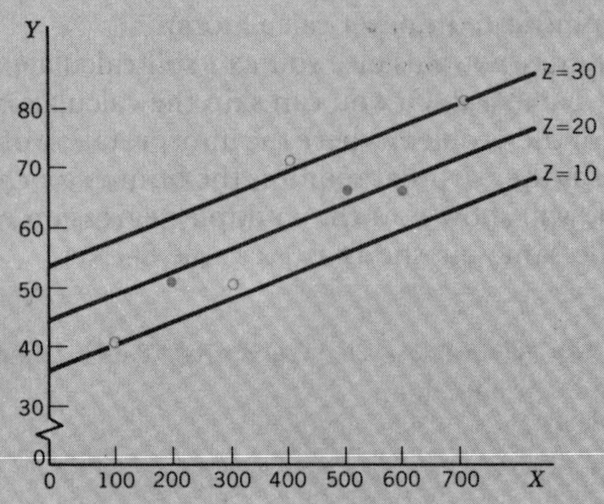

FIGURE 8-3 Multiple regression fits the data with parallel lines.

Note that the three separate lines calculated earlier in Example 8-1 fit the data a little closer. However, they are harder to summarize: In Example 8-1(*b*), we calculated the average slope to be about .04, whereas in (8-11), we can immediately see the slope is .038. Similarly, in Example 8-1(*c*), we calculated the average effect of one inch of rain to be about .75, whereas in (8-11) we can immediately see that it is .83.

Another important difference is that in Example 8-1 we had only 2 or 3 observations to fit each line. However, in estimating the multiple regression (8-11), we used all 7 observations; this puts us in a stronger position to test hypotheses or construct confidence intervals.

PROBLEMS

8-1 Suppose that a random sample of 5 families yielded the following data (S and X are measured in thousands of dollars annually):

Family	*Saving S*	*Income X*	*Children N*
A	2.1	15	2
B	3.0	28	4
C	1.6	20	4
D	2.1	22	3
E	1.2	10	2

(a) Calculate the simple regression equation of S on X.

(b) The multiple regression equation of S on X and N is

$$\hat{S} = .77 + .148X - .52N$$

On the (S, X) plane, graph the following:

(i) The 5 given points (and tag each point with its value of N, as in Figure 8-1).

(ii) The simple regression line.

(iii) The 3 lines you get from the multiple regression, by setting $N = 2, 3$, and 4.

(c) Explain why the simple and multiple regression equations have different coefficients of X.

(d) For the Stein family with 4 children and an income of 25 thousand dollars, predict their saving using the graph. Check your answer using the equation.

(e) Suppose the Tucker family has the same number of children as the Stein family, but 1 thousand dollars more in income. Estimate how much more the Tuckers save annually.

*(f) Using a computer, or a tableau like Table 8-2, calculate the multiple regression coefficients and verify that they agree with the equation given in (b).

8-2 In the midterm U.S. congressional elections (between presidential elections), the party of the President usually loses seats in the House of Representatives. To measure this loss concretely, we take as our base the average congressional vote for the President's party over the previous 8 elections; the amount that the congressional vote drops in a given midterm election, relative to this base, will be our *standardized vote loss* Y.

Y depends on several factors, two of which seem important and easily measurable: X = Gallup poll rating of the President at the time of the election (percent who approved of the way the President is handling his job) and Z = change over the previous year in the real disposable annual income per capita.

Year	Y = *Standardized Vote Loss*	X = *President's Gallup Rating*	Z = *Change in Real Income Over Previous Year*
1946	7.3%	32%	− $40
1950	2.0	43	100
1954	2.3	65	−10
1958	5.9	56	− 10
1962	−.8	67	60
1966	1.7	48	100

From the above data (Tufte, 1974), the following multiple regression equation was computed:

$$\hat{Y} = 10.9 - .13X - .034Z$$

(a) On the (X, Y) plane, graph the 6 points (and tag each point with its Z value, as in Figure 8-1).

(b) Graph the grid of 4 lines you get from the regression equation by setting Z = 100, 50, 0, and −50.

(c) From the graph, find the fitted 1946 vote loss Y, given that X = 32% and Z = \$−40. Confirm it exactly from the regression equation. Compared to the actual vote loss Y = 7.3%, what is the error?

(d) Now consider a real prediction. Put yourself back in time, just before the 1970 election, when President Nixon's rating was X = 56, and the change in real income over the previous year was Z = 70. What would you have predicted for the 1970 vote loss? (Find it on the graph, and confirm it exactly from the regression equation.) It turns out that the actual vote loss Y was 1.0%; what therefore was the prediction error?

*(e) Using a computer or a tableau like Table 8-2, calculate the multiple regression coefficients, and verify that they agree with the given equation.

8-3 In the estimating equation (8-4), suppose it is known a priori that Y does not depend on Z, that is, c = 0. When you solve for b, what do you get?

8-4 CONFIDENCE INTERVALS AND STATISTICAL TESTS

(a) Standard Error

As in simple regression, the true relation of Y to X is measured by the unknown population slope β; we estimate it with the sample slope b. Whereas the true β is fixed, the estimate b varies randomly from sample to sample, fluctuating around its target β with an approximately normal distribution. The standard deviation or standard error of b is customarily computed at the same time as b itself, as shown in Table 8-4. The meaning and use of the standard error are quite analogous to the simple regression case. For example, the standard error forms the basis for confidence intervals and tests.

TABLE 8-4

Computed Standard Errors and t Ratios
(Continuation of Table 8-3)

COLUMN	COEFFICIENT	ST. DEV. OF COEF.	T-RATIO = COEF/S.D.
--	28.095	2.491	11.28
FERT	0.03810	0.00583	6.53
RAIN	0.833	0.154	5.40

(b) Confidence Intervals

The formula for the confidence interval is of the standard form:

$$\boxed{\beta = b \pm t_{.025}\text{SE}} \tag{8-15}$$

like (7-17)

When there are k regressors, the degrees of freedom for the t table are

$$\boxed{\text{d.f.} = n - k - 1} \tag{8-16}$$

like (7-19)

Example 8-3

From the computer output in Table 8-4, calculate a 95% confidence interval for each regression coefficient.

Solution

From (8-16), d.f. = 7 − 2 − 1 = 4, so that Appendix Table V gives $t_{.025} = 2.776$. Also substitute b and its SE from Table 8-4, and then (8-15) gives, for the fertilizer coefficient,

$$\begin{aligned} \beta &= .03810 \pm 2.776\,(.00583) \\ &\simeq .038 \pm .016 \end{aligned} \tag{8-17}$$

And for the rainfall coefficient,

$$\begin{aligned} \gamma &= .833 \pm 2.776\,(.154) \\ &\simeq .83 \pm .43 \end{aligned} \tag{8-18}$$

(c) Prob-Value

The t ratio to test $\beta = 0$ is, as usual,

$$\boxed{t = \frac{b}{\text{SE}}} \tag{8-19}$$

like (7-23)

Example 8-4

(a) From the computer output in Table 8-4, calculate the prob-value for the null hypothesis $\beta = 0$ (fertilizer doesn't increase yield).

(b) Repeat for the null hypothesis $\gamma = 0$ (rainfall doesn't increase yield).

Solution

(a) From (8-19),

$$t = \frac{.03810}{.00583} = 6.53$$

Or, equivalently, this same t ratio can be read from the last column of Table 8-4. In any case, we refer to Appendix Table V, scanning the row where d.f. = 4. We find that the observed t value of 6.53 lies beyond $t_{.0025} = 5.598$. Thus

$$\text{Prob-value} < .0025 \tag{8-20}$$

With such little credibility, the null hypothesis can be rejected; we conclude that yield is indeed increased by fertilizer.

(b) For the null hypothesis $\gamma = 0$, (8-19) gives

$$t = \frac{.833}{.154} = 5.40$$

From Appendix Table V, once again scanning the row where d.f. = 4, we find the observed t value of 5.40 lies beyond $t_{.005} = 4.604$. Thus

$$\text{Prob-value} < .005 \tag{8-21}$$

With such little credibility, this null hypothesis can also be rejected; we conclude that yield is increased by rainfall as well.

It is customary to summarize the calculations of Examples 8-3 and 8-4, by arranging them in equation form as follows:

(8-22)

YIELD = 28 +	.038 FERTILIZER +	.83 RAINFALL
Standard error	.0058	.154
95% CI	± .016	± .43
t ratio	6.5	5.4
Prob-value	< .0025	< .005

PROBLEMS

8-4 The congressional vote loss of the President's party in midterm elections (*Y*) was related to the President's Gallup rating *X* and change in real income over the previous year (*Z*). Using the same 6 years' data as in Problem 8-2, the following regression was calculated:

$\hat{Y} = 10.9$	$-\ .13X$	$-\ .034Z$
Standard error	(.046)	(.010)
95% CI	( )	( )
t ratio	( )	( )
Prob-value	( )	( )

(a) Fill in the blanks.

(b) What assumptions were you making in (a)? How reasonable are they?

8-5 The following regression was calculated for a class of 66 students of nursing (Snedecor and Cochran, 1967):

$\hat{Y} = 3.1$	$+\ .021X_1$	$+\ .075X_2$	$+\ .043X_3$
Standard error	(.019)	(.034)	(.018)
95% CI	( )	( )	( )
t ratio	( )	( )	( )
Prob-value	( )	( )	( )

where Y = student's score on a theory examination
X_1 = student's rank (from the bottom) in high school
X_2 = student's verbal aptitude score
X_3 = a measure of student's character

(a) Fill in the blanks.

(b) What assumptions were you making in (a)? How reasonable are they?

8-5 REGRESSION COEFFICIENTS AS MAGNIFICATION FACTORS

(a) Simple Regression

The coefficients in a linear regression model have a very simple but important interpretation. Recall the simple regression model,

$$Y = a + bX \tag{8-23}$$

like (7-1)

The coefficient b is the slope:

$$\frac{\Delta Y}{\Delta X} = b \qquad \text{(8-24) like (B-2)}$$

where ΔX is any change in X, and ΔY is the corresponding change in Y. (Slopes and lines are reviewed in Appendix B at the back of the book.) We can rewrite (8-24) in another form:

$$\Delta Y = b\,\Delta X \qquad \text{(8-25)}$$

Since this is so important, we write it verbally:

$$\boxed{\text{Change in } Y = b \text{ (change in } X)} \qquad \text{(8-26)}$$

For example, consider the fertilizer-yield example,

$$Y = 36 + .06X \qquad \text{like(7-7)}$$

How much higher would yield Y be, if fertilizer X were 5 pounds higher? From (8-26) we find,

$$\text{Change in yield} = .06(5) = .30 \text{ bushel} \qquad \text{(8-27)}$$

That is, whatever change occurs in X, it is multiplied by b in order to find the corresponding change in Y. We might call b the *magnification factor*. An interesting special case occurs when $\Delta X = 1$. Then (8-26) becomes

$$\text{Change in } Y = b \qquad \text{(8-28)}$$

This provides such an important interpretation of b that we emphasize it:

$$\boxed{b = \text{change in } Y \text{ that accompanies a unit change in } X} \qquad \text{(8-29) like (7-8)}$$

Thus, for example, the coefficient $b = .06$ means that a change of .06 bushel in yield Y accompanies a 1-pound change in fertilizer X. Figure 8-4 illustrates this, as well as (8-27).

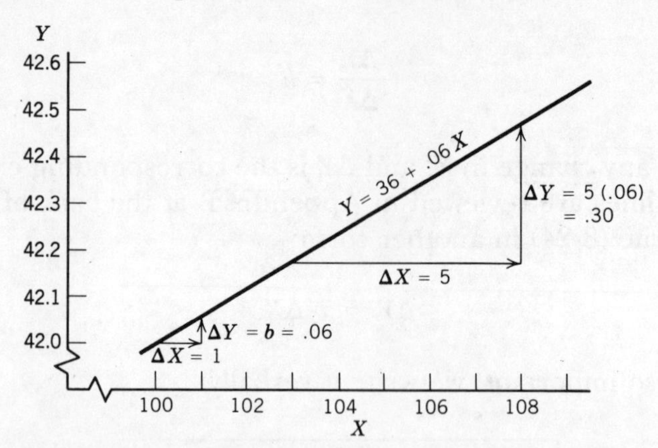

FIGURE 8-4 Interpretation of slope: b = change in Y that accompanies a unit change in X.

We must emphasize that in uncontrolled observational studies, the simple regression slope b does not prove anything about causation. The increase in Y that accompanies a unit change in X reflects not only the effect of X but also the effect of all the extraneous variables that are changing simultaneously. To help sort out how X causes Y, we must therefore turn to multiple regression.

(b) Multiple Regression: "Other Things Being Equal"

Consider next the multiple regression model

$$Y = a + bX + cZ \tag{8-30}$$

like (8-3)

If Z remains constant, it is still true that:

$$\Delta Y = b\,\Delta X \tag{8-31}$$

Equation (8-31) is so important that it is worthwhile giving a simple proof: If we keep Z constant, while we increase X to $(X + \Delta X)$, then from (8-30),

$$\begin{aligned} \text{initial } Y &= a + bX + cZ && \text{(8-32)} \\ \text{new } Y &= a + b(X + \Delta X) + cZ \\ \text{difference: } \Delta Y &= b\,\Delta X && \text{(8-31) proved} \end{aligned}$$

To emphasize its importance, we write (8-31) verbally:

$$\boxed{\begin{array}{c}\text{If the other regressor } Z \text{ remains constant} \\ \text{Change in } Y = b \text{ (change in } X)\end{array}} \tag{8-33}$$

For example, consider the wheat-yield example,

$$Y = 28 + .038X + .83Z \qquad \text{like (8-11)}$$

How much would yield Y increase if fertilizer X were increased 5 pounds, and *rainfall did not change?* From (8-33),

$$\text{Change in yield} = .038(5) = .19 \text{ bushel} \tag{8-34}$$

An interesting special case occurs when $\Delta X = 1$. Then (8-33) becomes

$$\boxed{\begin{array}{ll} b = & \text{change in } Y \text{ that accompanies a unit change in } X, \\ & \text{while the other regressor remains constant}\end{array}} \tag{8-35}$$

It is easy to generalize (8-35) when there are k regressors:

$$\boxed{\begin{array}{ll} & \text{If } Y = a + b_1X_1 + b_2X_2 + \ldots + b_kX_k \\ \text{then } b_1 = & \text{change in } Y \text{ that accompanies a unit} \\ & \text{change in } X_1\text{, while all the other} \\ & \text{regressors remain constant}\end{array}} \tag{8-36}$$

What happens if all the regressors X change simultaneously? Just as we proved (8-31), we could now show that the change in Y is just the sum of the individual changes:

$$\boxed{\begin{array}{l} \text{If } Y = a + b_1X_1 + b_2X_2 + \ldots + b_kX_k \\ \text{then } \Delta Y = b_1\Delta X_1 + b_2\Delta X_2 + \ldots + b_k\Delta X_k\end{array}} \tag{8-37}$$

To contrast the different uses of multiple and simple regression, we end with an example.

Example 8-5

The simple and multiple regressions of yield on fertilizer and rainfall (obtainable from Table 8-2) are as follows:

$$\text{YIELD} = 36 + .059\ \text{FERT} \tag{8-38}$$

$$\text{YIELD} = 30 + 1.50\ \text{RAIN} \tag{8-39}$$

$$\text{YIELD} = 28 + .038\ \text{FERT} + .83\ \text{RAIN} \tag{8-40}$$

(a) If a farmer adds 100 more pounds of fertilizer per acre, how much can he expect his yield to increase?
(b) If he irrigates with 3 inches of water, how much can he expect his yield to increase?
(c) If he simultaneously adds 100 more pounds of fertilizer per acre, and irrigates with 3 inches of water, how much can he expect his yield to increase?
(d) We have already remarked that high fertilizer application tends to be associated with high rainfall, in the data on which these three regression equations were calculated. If this same tendency persisted, how much more yield would you expect on an acre that gets 3 more inches of water than another acre?

Solution

(a) When the farmer adds fertilizer, he will not be changing the rainfall that occurs on his farm. Therefore, it is the multiple regression that matters, and his expected increase in yield is found [according to (8-33)] by multiplying the 100 pounds by the multiple regression coefficient (magnification factor) of .038:

$$.038(100) = 3.8 \text{ bushels}$$

(b) When he adds water through irrigation, he will not be changing the fertilizer. Therefore, it is the multiple regression again that matters, and his expected increase in yield is found by multiplying the 3 inches by the multiple regression coefficient (magnification factor) of .83:

$$.83(3) = 2.5 \text{ bushels}$$

(c) From (8-37),

$$\begin{aligned} \Delta Y &= .038(100) + .83(3) \\ &= 3.8 + 2.5 \\ &= 6.5 \text{ bushels} \end{aligned} \tag{8-41}$$

(d) Now we are not holding fertilizer constant, but letting it vary as rainfall varies, in the same pattern that produced (8-39) (i.e., a pattern where, for example, farmers that get more rainfall are more prosperous and can afford more fertilizer). So we do *not* want to use the coefficient of .83 in (8-40), because this shows how yield rises with rainfall alone (with fertilizer constant). Instead, we go back to (8-39), whose coefficient of 1.50 shows how yield rises with rainfall when fertilizer is changing too:

$$1.50(3) = 4.5 \text{ bushels}$$

To sum up: This is larger than our answer in part (b)—because this simple regression coefficient of 1.50 shows how yield is affected by rainfall *and* the associated fertilizer increase.

Remarks. We assumed that irrigation water had the same effect as rain water. If this assumption is unwarranted, the predictions in (b) and (c) might be far off.

PROBLEMS

8-6 To determine the effect of various factors on land value in Florida, the sale price of residential lots in the Kissimmee River Basin was regressed on various factors. With a data base of $n = 316$ lots, the multiple regression was calculated (Conner and others, 1973; via Anderson and Sclove, 1978):

$$\hat{Y} = 10.3 + 1.5X_1 - 1.1X_2 - 1.34X_3 + \ldots$$

where

$$\begin{aligned} Y &= \text{price per front foot} \\ X_1 &= \text{year of sale } (X_1 = 1, \ldots, 5 \text{ for } 1966, \ldots, 1970) \\ X_2 &= \text{lot size (acres)} \\ X_3 &= \text{distance from nearest paved road (miles)} \end{aligned}$$

(a) Other things being equal (year, distance to nearest paved road, etc.), was the price of a 5 acre lot more or less than a 2 acre lot, per front foot? How much?
(b) Other things being equal, how much higher was the price if the lot was $\frac{1}{2}$ mile closer to the nearest paved road?
(c) Was the trend of prices over time up or down? How much?

8-7 Each year for 20 years, the average values of hay yield Y, temperature T, and rainfall R were recorded in England (Hooker, 1907; via Anderson, 1958), so that the following regressions could be calculated:

$$\hat{Y} = 40.4 - .208T$$
Standard error .112

$$\hat{Y} = 12.2 + 3.22R$$
Standard error .57

$$\hat{Y} = 9.14 + .0364T + 3.38R$$
Standard error .090 .70

Estimate the average increase in yield from one year to the next:
(a) If rainfall increases 3, and temperature remains the same.
(b) If temperature increases 10, and rainfall remains the same.
(c) If rainfall increases 3, and temperature increases 10.
(d) If rainfall increases 3, and we don't know how much temperature changes (although we know it likely will drop, since wet seasons tend to be cold).
(e) If rainfall increases 3, and temperature decreases 13.
(f) If temperature increases 10, and we don't know how much rainfall changes (although we know it will likely fall, since hot seasons tend to be dry).

8-8 In Problem 8-7:
(a) What yield would you predict if $T = 50$ and $R = 5$?
(b) What yield would you predict if $T = 65$ and $R = 7$?
(c) By how much has yield increased in (b) over (a)? Confirm this answer using (8-37).

*8-9 In Problem 8-7(a) and (b), put 95% confidence intervals on your answers, assuming the 20 years formed a random sample. What is the population being sampled?

CHAPTER 8 SUMMARY

8-1 To show how a response Y is related to two (or more) factors, X and Z say, we can fit an equation called the multiple regression, $\hat{Y} = a + bX + cZ$. The coefficient b shows how Y is related to X if Z were constant. Multiple regression is therefore especially valuable in observational studies, where an uncontrolled extraneous factor (Z let us say) would cause bias in the simple regression of Y on X alone.

8-2 The underlying population regression as usual is denoted by Greek letters: $Y = \alpha + \beta X + \gamma Z$. This relation is represented geometrically as a plane. β is the slope of Y in the X direction, and γ is the slope of Y in the Z direction.

8-3 The regression coefficients a, b, and c can be calculated from the data, again using the criterion of least squares. For these calculations there are standard computer packages, which become invaluable when there are many regressors.

8-4 The computer package also calculates the standard errors for b and c. We can therefore construct a confidence interval for β or γ, or calculate the prob-value for $\beta = 0$ or $\gamma = 0$.

8-5 In the model $Y = a + bX + cZ$, the coefficient b is interpreted as the change in Y that accompanies a unit change in X, while Z remains constant.

REVIEW PROBLEMS

8-10 A multiple regression of lung capacity Y (in milliliters), as a function of age (in years) and height (in inches) and amount of smoking (in packs per day), was computed for a sample of 20 men randomly sampled from a large population of workers (Lefcoe and Wonnacott, 1974):

Variable	*Coefficient*	*Standard Error*	*t ratio*	*Prob-value*
AGE	− 39.0	1.88		
HEIT	+ 98.4	7.54		
SMOKIN	−180.0	40.81		

(a) Fill in the blanks.

(b) If the whole population were run through the computer, rather than a mere sample of 20, in what range would you expect to find each of the computed coefficients?

(c) Other things being equal, what would you estimate is the effect of:
  (i) Smoking 1 pack per day?
  (ii) Being 5 years older?
  (iii) Smoking 2 packs per day and being 10 years older?

(d) As far as lung capacity is concerned, the effect of smoking one pack per day is equivalent to aging how many years?

(e) Your answer to (d) is only approximate. What are the sources of error?

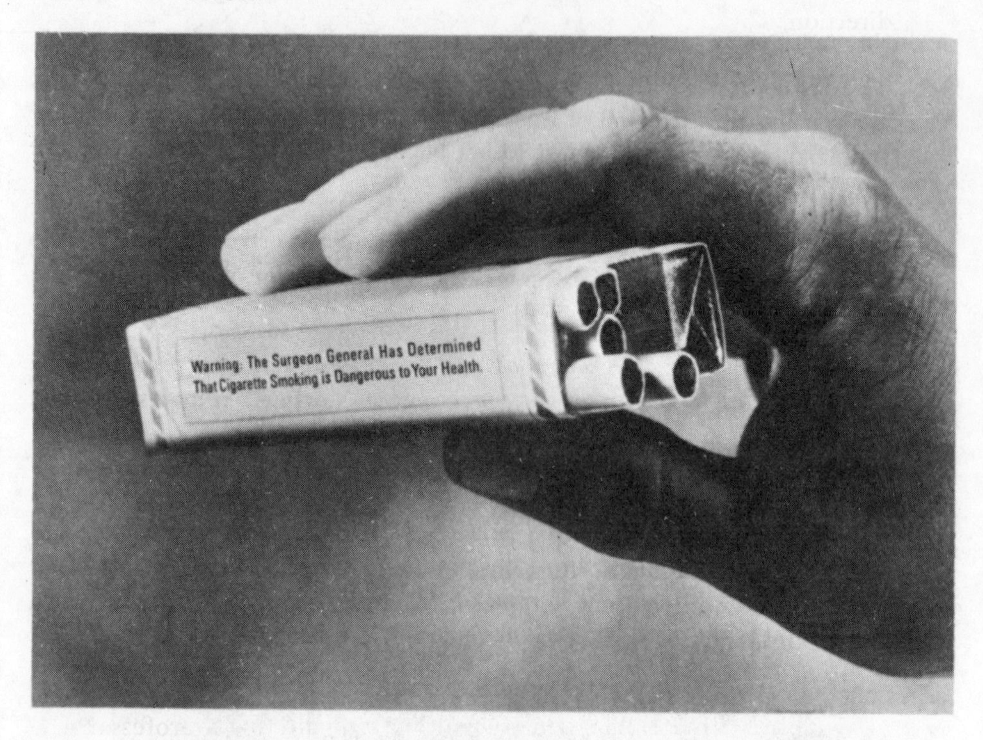

8-11 A simple regression of life expectancy Y (in years) on smoking X (in packs per day) for U.S. males would yield the following approximate regression equation (U.S. Surgeon-General, 1979):

$$\hat{Y} = 70 - 5X$$

Is it fair to say that smoking cuts 5 years off the end of your life, for each pack smoked? If so, why? If not, how *would* you estimate the effect of smoking 1 pack per day?

8-12 A study of several hundred professors' salaries in a large American university in 1969 yielded the following multiple regression. (From Katz, 1973. These same professors were discussed in Problems 2-1 and 5-13. But now we are using more factors than just sex. In order to be brief, however, we do not write down all the factors that Katz included.)

$$\hat{S} = 230B + 18A + 100E + 490D + 190Y + 50T + \ldots$$

Standard error	(86)	(8)	(28)	(60)	(17)	(370)
95% CI	( )	( )	( )	( )	( )	( )
t ratio	( )	( )	( )	( )	( )	( )
Prob-value	( )	( )	( )	( )	( )	( )

where S = the professor's annual salary (dollars)
B = number of books he has written
A = number of ordinary articles
E = number of excellent articles
D = number of Ph.D's supervised
Y = number of years' experience
T = teaching score as measured by student evaluations, severely rounded: the best half of the teachers were rounded up to 100% (i.e., 1); the worst half were rounded down to 0.

(a) Fill in the blanks below the equation.

(b) For someone who knows no statistics, briefly summarize the influences on professors' incomes, by indicating where strong evidence exists and where it does not.

(c) Answer True or False; if False, correct it.

(i) The coefficient of B is estimated to be 230. Other social scientists might collect other samples from the same population and calculate other estimates. The distribution of these estimates would be centered around the true population value of 230.

(ii) Other things being equal, we estimate that a professor who has written one or more books earns \$230 more annually. Or, we might say that \$230 estimates the value (in terms of a professor's salary) of writing one or more books.

(iii) Other things being equal, we estimate that a professor who is 1 year older earns \$190 more annually. In other words, the annual salary increase averages \$190.

(d) Similarly, interpret all the other coefficients for someone who knows no statistics.

CHAPTER 9

Regression Extensions

Happy the man who has been able to learn the causes of things.

Virgil.

In this chapter we will see how versatile a tool multiple regression can be, when it is given a few clever twists. It can be used on categorical data (such as yes-no responses) as well as the numerical data dealt with in Chapter 8. And it can be used to fit curves as well as straight lines.

9-1 DUMMY (0-1) VARIABLES

(a) Parallel Lines for Two Categories

In Chapter 4 we introduced dummy variables for handling data that came in two categories (such as Democrat versus Republican, or treatment versus control). By associating numbers (0 and 1) with the two categories, a dummy variable ingeniously transformed the problem into a numerical one, and so made it amenable to all the standard statistical tools. (For example, standard errors and confidence intervals could be constructed.) Now we shall see how a dummy variable can be equally useful in regression analysis.

Example 9-1

A certain drug (drug A) is suspected of having the unfortunate side effect of raising blood pressure. To test this suspicion, a sample of 10 women were found, 6 of whom had the drug once a day, and 4 of whom had no drug, and hence served as the control group. To transform the treatment-control dichotomy into numerical form, for each patient we let:

$$Z = \text{number of doses of drug } A \text{ she takes daily} \quad (9\text{-}1) \text{ like } (4\text{-}12)$$

that is,

$$\left.\begin{aligned} Z &= 1 \text{ if the patient took drug } A \\ &= 0 \text{ if the patient was a control} \end{aligned}\right\} \quad (9\text{-}2) \text{ like}(4\text{-}13)$$

In the form (9-1), it is clear that Z is a variable that can be run through a regression computer program like any other variable. In the form (9-2), it is clear that Z is a 0-1 variable that clearly distinguishes between the two groups; and this is the form most commonly used.

We want to investigate how treatment affects blood pressure, allowing for extraneous influences such as age. So Table 9-1 lists blood pressure Y, age X, and treatment Z for the 10 women in the sample. In practice, the regression coefficients would be computed with a standard multiple regression package. To emphasize the similarity to the ordinary multiple regression in Table 8-2, however, we carry out the computations in Table 9-1. They yield the fitted equation:

$$\hat{Y} = a + bX + cZ \quad (9\text{-}3)$$

$$\hat{Y} = 69.5 + .44\,X + 4.7\,Z \quad (9\text{-}4)$$

Standard error (.073) (1.9)

(a) Graph the relation of Y to X given by (9-4), when Z has the constant value:

(i) $Z = 0$

(ii) $Z = 1$

(b) What is the meaning of the coefficient of Z? Answer using the graph in (a). Also answer using the fundamental interpretation (8-36).

TABLE 9-1

Multiple Regression of Blood Pressure Y on a Dummy Variable Z and Age X

	Data			Deviations			Products				
Patient	Y	X	Z	$y = Y - \bar{Y}$	$x = X - \bar{X}$	$z = Z - \bar{Z}$	xy	zy	x^2	z^2	xz
Astor	85	30	0	−5	−10	−.6	50	3	100	.36	6
Braun	95	40	1	5	0	.4	0	2	0	.16	0
Cowles	90	40	1	0	0	.4	0	0	0	.16	0
Downes	75	20	0	−15	−20	−.6	300	9	400	.36	12
Ephim	100	60	1	10	20	.4	200	4	400	.16	8
Felllighi	90	40	0	0	0	−.6	0	0	0	.36	0
Grant	90	50	0	0	10	−.6	0	0	100	.36	−6
Howlett	90	30	1	0	−10	.4	0	0	100	.16	−4
Johnson	100	60	1	10	20	.4	200	4	400	.16	8
Klein	85	30	1	−5	−10	.4	50	−2	100	.16	−4
	$\bar{Y} = 90$	$\bar{X} = 40$	$\bar{Z} = .6$	0✓	0✓	0✓	800	20	1600	2.40	20

Estimating equations (8-4) and (8-5) $\begin{cases} 800 = 1600\,b + 20\,c \\ 20 = 20\,b + 2.4\,c \end{cases}$

Solution $\begin{cases} b = .44 \\ c = 4.7 \end{cases}$

From (8-6), $a = 90 - .44\,(40) - 4.7\,(.6)$

$a = 69.5$

(c) Construct a 95% confidence interval for the (true) population coefficient of Z. Is it discernible at the 5% level? What assumptions are required for this confidence interval to be valid?

Solution

(a) Equation (9-3) takes on two values, as $Z = 0$ or 1:

$$\hat{Y} = a + bX \qquad (\text{if } Z = 0) \tag{9-5}$$

$$= a + bX + c \quad (\text{if } Z = 1)$$

$$= (a + c) + bX \tag{9-6}$$

Compared to (9-5), the line (9-6) is seen to be a line with the same slope b, but an intercept that is c units higher. That is, the line where $Z = 1$ is parallel and c units higher.

For the specific equation (9-4), the two lines are plotted in Figure 9-1. Note that the treated group where $Z = 1$ is fitted with a line that is indeed $c = 4.7$ units higher.

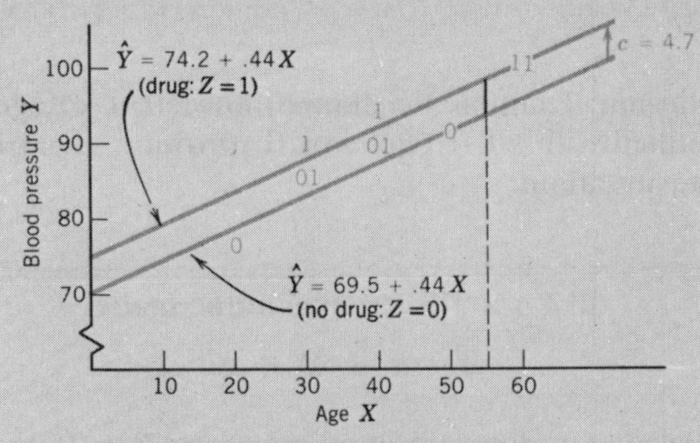

FIGURE 9-1 Graph of equation (9-4) relating blood pressure to age and treatment ($Z = 1$ for drug, or $Z = 0$ for no drug).

(b) On the graph, suppose we keep age constant, at $X = 55$ for example. The difference the drug makes is shown as the vertical distance between the lines, $c = 4.7$.

Alternatively, we could use the fundamental interpretation for any regression coefficient: The coefficient of Z is the change in Y that accompanies a unit change in Z (while X remains constant). This unit change in Z can only be from 0 to 1, that is, from no drug to drug. Thus there is an increase in blood pressure of 4.7 as we go from a woman without the drug to a woman with the drug (of the same age). This agrees with the interpretation in the previous paragraph, of course.

(c) We construct the 95% confidence interval from the estimate and its standard error given in (9-4); for $t_{.025}$, we use d.f. = $n - k - 1 = 10 - 2 - 1 = 7$:

$$\gamma = c \pm t_{.025}\,\text{SE} \qquad \text{like(8-15)}$$

$$= 4.7 \pm 2.365\,(1.9)$$

$$\gamma = 4.7 \pm 4.5 \qquad (9\text{-}7)$$

Since the estimate 4.7 stands out above the sampling allowance 4.5, γ is statistically discernible at the 5% level. Of course, we must remember that the 10 women are assumed to be a *random sample* (VSRS) drawn from a population (with parameter γ).

In conclusion, Example 9-1 showed how a two-category factor can be nicely handled with a 0-1 regressor. It provides an extremely useful graphical interpretation:

> If Z is a 0-1 variable in the model
>
> $$\hat{Y} = a + bX + cZ \qquad (9\text{-}8)$$
>
> then relative to the reference line where $Z = 0$, the line where $Z = 1$ is parallel and c units higher.

In addition, the multiple regression with a 0-1 regressor gives the standard statistical tests and confidence intervals.

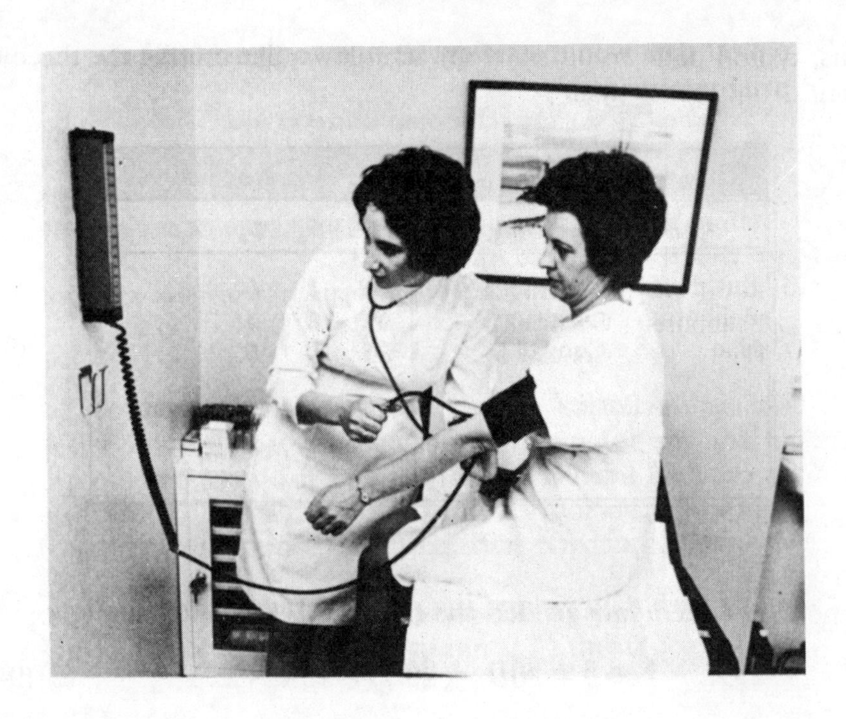

(b) Several Categories

So far we have considered a factor that has only two categories—treatment and control. The dummy variable D measured the effect of the treatment ($D = 1$) relative to the reference control ($D = 0$).

What happens if a factor has three categories? For example, in order to clinically test two drugs A and B against a control C, suppose a pharmaceutical company samples 30 patients. For each patient, as well as measuring the response Y and extraneous variables $X_1, X_2 \ldots$ (such as age, weight, etc.), we could measure drug by means of two dummy variables:

$$D_A = 1 \text{ if drug } A \text{ given; 0 otherwise} \tag{9-9}$$

$$D_B = 1 \text{ if drug } B \text{ given; 0 otherwise} \tag{9-10}$$

Thus, typical data would start out as follows (illustrating the three different drug treatments):

Comments		*Data*			
Person	*Drug*	Y	D_A	D_B	$X_1, X_2 \ldots$
Koval	On drug *A*	60	1	0	...
Bellhouse	On drug *B*	92	0	1	...
Haq	Control *C*	81	0	0	...
.	.	.	.	.	...
.	.	.	.	.	...
.	.	.	.	.	...

(9-11)

Suppose that such data yielded the multiple regression equation:

$$\hat{Y} = 5 + 30D_A + 20D_B - 13X_1 + \ldots \tag{9-12}$$

This means that the effect of drug A (relative to the reference category—control group C) is to raise Y by 30. The effect of drug B is to raise Y by 20. Thus drug A increases Y by 10 units more than does drug B.

Note that we needed only 2 dummy variables to handle 3 categories—since one category (in our example, the control group) was the reference category against which the other two were measured. This is generally true for any number of categories: We need one less dummy than there are categories.

PROBLEMS

9-1 On the (X, Y) plane, graph each of the following equations, clearly marking which lines correspond to which drugs.

(a) In a large-scale study involving men 20 to 50 years of age, blood pressure Y was related to age X and intake of a certain drug A. Using the appropriate dummy variable ($D_A = 1$ if drug A taken, 0 otherwise), the regression equation was

$$\hat{Y} = 70 + .40X + 10D_A$$

(b) The study was enlarged to include some patients taking a second drug. Using the appropriate dummy variable (D_B = 1 if drug B taken, 0 otherwise), the regression equation became

$$\hat{Y} = 66 + .50X + 10D_A + 4D_B$$

(c) In model (b), how much higher does drug A raise blood pressure than drug B? Show this graphically.

9-2 (a) From the values of Σxy and Σx^2 already available in Table 9-1, calculate the simple regression of Y on X.

(b) Compare the simple and multiple regression coefficients. Which do you think is the appropriate measure of "the effect of age on blood pressure?"

(c) By how much was the simple regression thrown off by omitting the regressor Z?

9-3 (a) From the values of Σzy and Σz^2 already available in Table 9-1, calculate the simple regression of Y on Z.

(b) Compare the simple and multiple regression coefficients. Which do you think is the appropriate measure of "the effect of the drug on blood pressure"?

(c) By how much was the simple regression thrown off by omitting the regressor X?

*9-4 Continuing Problem 9-3(a), for the simple regression of Y on Z, construct a 95% confidence interval for the population regression coefficient. State in words what it means.

*9-5 Consider the data of Table 9-1. Ignore X, and use Z to sort Y into two samples, treatment and control. Then calculate the confidence interval (5-24) for the difference between the treatment and control means. State in words what it means.

How does your answer compare to Problem 9-4?

*9-6 To generalize Problem 9-5, answer True or False; if False, correct it.

(a) The two-sample t test is equivalent to simple regression on a 0-1 regressor.

(b) However, it is much better to use multiple regression to avoid the bias that occurs in simple regression (or the t test) when there are uncontrolled extraneous variables.

*9-2 ANALYSIS OF VARIANCE (ANOVA)

(a) One-Factor ANOVA

In Example 9-1, we were mainly concerned with the effect of the dummy regressor (drug), and introduced the numerical regressor (age) primarily to keep this extraneous variable from biasing the estimate of the drug effect. Used in this way, multiple regression is sometimes called *analysis of covariance* (ANOCOVA).

If the numerical regressors are omitted entirely so that dummy regressors alone are used, then the regression equation such as (9-12) reduces to a form like

$$\hat{Y} = 5 + 30D_A + 20D_B \tag{9-13}$$

A regression of this form is equivalent to a traditional technique called Analysis of Variance (ANOVA)—a method introduced by R.A. Fisher about 50 years ago when calculations were done by hand. Although the computational savings are not so important today, traditional ANOVA still provides certain conceptual advantages, especially in large models. In fact, it is a topic so rich that it deserves a whole course—or at least a chapter. Accordingly we devote Chapter 10 to it. And for now, we simply observe that traditional ANOVA is equivalent to a regression with dummies, such as (9-13).

(b) Two-factor ANOVA

It is possible to use dummy variables to introduce a second factor. For example, in the drug study, suppose we want to see the effect of another factor, sex. We can take care of it with one dummy variable:

$$M = 1 \text{ if male; } 0 \text{ if female} \tag{9-14}$$

Then the fitted regression equation, instead of (9-13), would be something like

$$\hat{Y} = 5 + 30D_A + 20D_B + 15M \tag{9-15}$$

Estimating this regression equation is equivalent to traditional *two-factor* ANOVA. One convenient way to graph (9-15) is like (9-4) was graphed in Figure 9-1—with M playing the former role of age X, and

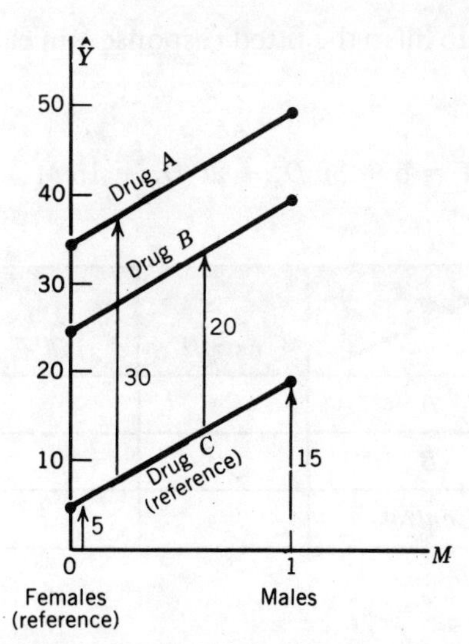

FIGURE 9-2 Graph of two-factor ANOVA.

D_A and D_B playing the former role of the dummy Z. This graph is shown in Figure 9-2.

We can of course introduce into the model as many factors as desired. Soon the model may grow so complex that the simpler calculations of traditional ANOVA may make it a valuable alternative to regression with dummies.

*PROBLEMS

9-7 Suppose three drugs (A, B, and control C) were given to two different age-groups (infant and adult). Let us define two dummies for drugs (D_A and D_B, as usual) and define one dummy for age:

$$I = 1 \text{ if infant; } 0 \text{ otherwise}$$

Use a diagram such as Figure 9-2 to graph the response Y for each of the following equations:

(a) $\hat{Y} = 110 + 20I + 30D_A - 40D_B$

(b) $\hat{Y} = 30 - 10I + 5D_A + 25D_B$

9-8 Use equation (9-15) to fill in the fitted response $\hat{Y}$ in each cell of the following table:

$$Y = 5 + 30\,D_A + 20\,D_B + 15\,M \qquad \text{(9-15) repeated}$$

Drug \ Sex	Female	Male
A		
B		
Control		

Check that this table agrees with Figure 9-2. And note that the improvement of males over females is the same for all drugs.

9-3 SIMPLEST NONLINEAR REGRESSION

Straight lines (and planes) are called *linear* functions. They are characterized by a simple equation ($Y = a + bX$) where the independent variable appears just as X, rather than in some more complicated way such as X^2, $\sqrt{X}$, or $1/X$. In this section we will look at one of these nonlinear functions.

Let us, for example, reconsider how wheat yield depends on fertilizer. Vast amounts of fertilizer eventually begin to burn the crop, causing the yield to fall, as the data in Table 9-2 and Figure 9-3 illustrate. An appropriate model might therefore be a parabola,

$$Y = a + bX + cX^2 \qquad (9\text{-}16)$$

where Y = yield (bushels per acre) and X = fertilizer (thousands of pounds per acre). How do we fit equation (9-16)—that is, find the parabola that best fits the observations shown in Figure 9-3? The answer is,

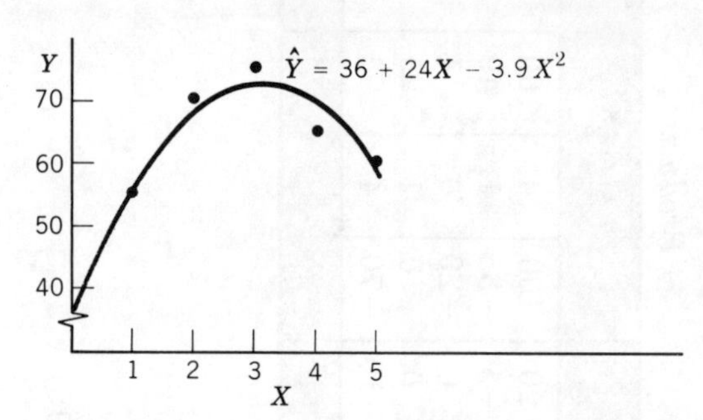

FIGURE 9-3 How yield depends on very large amounts of fertilizer—a parabolic relation.

by ordinary multiple regression with an ingenious twist. We simply define:

$$Z \equiv X^2 \tag{9-17}$$

Then (9-16) takes on the familiar form

$$Y = a + bX + cZ \tag{9-18}$$

like (8-3)

Y is now related to *X* and *Z* in the standard way, and so the regression coefficients (*a*, *b*, and *c*) can be computed with a standard multiple regression package. To emphasize the similarity to the ordinary multiple regression in Table 8-2, however, we carry out the actual computations in Table 9-2. They confirm the computer output:

$$\hat{Y} = 36 + 24X - 3.9X^2 \tag{9-21}$$

Standard error	(6.5)	(1.1)
t ratio	(3.7)	(−3.5)

In Figure 9-3, the graph of this function displays a good fit to the given data. Thus we have shown how ordinary multiple regression can be used to estimate a nonlinear function.

TABLE 9-2

Multiple Regression Fit of a Parabola

Data		*New Regressor*	*Deviations*			*Products*				
Yield (bu/acre)	*Fertilizer X (1000 lb/acre)*	$Z = X^2$	$y = Y - \overline{Y}$	$x = X - \overline{X}$	$z = Z - \overline{Z}$	xy	zy	x^2	z^2	xz
55	1	1	−10	−2	−10	20	100	4	100	20
70	2	4	5	−1	−7	−5	−35	1	49	7
75	3	9	10	0	−2	0	−20	0	4	0
65	4	16	0	1	5	0	0	1	25	5
60	5	25	−5	2	14	−10	−70	4	196	28
$\overline{Y} = 65$	$\overline{X} = 3$	$\overline{Z} = 11$	0√	0√	0√	5	−25	10	374	60

$$\text{Estimating equations} \begin{cases} 5 = 10b + 60c \\ -25 = 60b + 374c \end{cases}$$

$$\text{Solution} \begin{cases} b = 24 \\ c = -3.9 \end{cases} \tag{9-19}$$

$$a = \overline{Y} - b\overline{X} - c\overline{Z} = 36$$

$$\text{Thus } \hat{Y} = 36 + 24\,X - 3.9\,X^2 \tag{9-20}$$

PROBLEMS

9-9

Output X (Thousands)	*Marginal Cost Y*
1	32
2	20
3	20
4	28
5	50

From the above data, the following quadratic polynomial was fitted:

$$\hat{Y} = 55 - 28.2X + 5.4X^2$$

(a) Graph the 5 points.
(b) Calculate and graph $\hat{Y}$ for $X = 1, 2, 3, 4, 5$. Then sketch the fitted parabola that passes through these 5 points.
(c) From (b), what is the estimated increase in Y when X increased 1 unit, from 4 to 5? Is this the same as the coefficient of X? Is this consistent with the fundamental interpretation (8-29)?
*(d) Using a computer, or a tableau like Table 9-2, calculate the regression coefficients and verify that they agree with the given equation.

9-10 In Figure 9-3 and equation (9-21), is the parabolic model really necessary, or would a straight-line fit be adequate? Discuss, in relation to:
(a) Prior belief about what is a reasonable model.
*(b) The statistical evidence (discernibility).

CHAPTER 9 SUMMARY

9-1 A two-category factor (such as male-female or treatment-control) can be handled with a dummy (0-1) variable. Similarly, any number of categories can be handled with one less dummy than there are categories.

*9-2 When all factors are categorical and there are no numerical factors, the regression is called ANOVA.

9-3 Nonlinear functions such as parabolas can be fitted by using a simple twist on standard multiple regression.

REVIEW PROBLEMS

9-11 A regression equation related personal income Y (annual, in \$1000) to education E (in years) and geographical location, measured with dummies as follows:

$$D_S = 1 \text{ if in the south; 0 otherwise}$$

$$D_W = 1 \text{ if in the west; 0 otherwise}$$

The remaining region (northeast) is left as the reference region. Suppose the fitted regression was

$$\hat{Y} = 4.5 + 0.5E - 1.0D_S + 1.5D_W$$

Graph the estimated income $\hat{Y}$ as a function of E, for all 3 regions, with E running from 8 to 16.

9-12 Now we can consider more precisely the regression equation for several hundred professors' salaries in Problem 8-12 by including some additional regressors (Katz, 1973):

$$\hat{S} = 230B + \cdots + 50T - 2400X + 1900P + \cdots$$

Standard error	(370)	(530)	(610)
95% CI	( )	( )	( )
t ratio	( )	( )	( )
Prob-value	( )	( )	( )

where S = the professor's annual salary (dollars)
T = 1 if the professor received a student evaluation score above the median; 0 otherwise
X = 1 if the professor is female; 0 otherwise
P = 1 if the professor has a Ph.D.; 0 otherwise

(a) Fill in the blanks in the equation.

(b) Answer True or False; if False, correct it.

(i) A professor with a Ph.D. earns annually $1900 more than one without a Ph.D.

(ii) Or, we might say that $1900 estimates the value (in terms of a professor's salary) of one more unit (in this case, a Ph.D.).

(iii) The average woman earns $2400 more than the average man.

(c) Give an interpretation of the coefficient of T.

9-13 For the raw data of Problem 9-12, the mean salaries for male and female professors were $16,100 and $11,200 respectively. By referring to the coefficient of X in Problem 9-12, answer True or False; if False, correct it:

After holding constant all other variables, women made $2400 less than men. Therefore, $2400 is a measure of the extent of sex discrimination, and $2500 (16,100 − 11,200 − 2400) is a measure of the salary differential due to other factors, for example, productivity and experience.

*9-14 Comparing Problems 9-12 and 8-12, note that the same variable T appears in two different forms. In the form of Problem 8-12, it is apparent that it involves a severe degree of rounding. What are the advantages and disadvantages of such rounding?

9-15

	Years of Formal Education E	*Father's Income F*
Urban sample	15	$16,000
	18	22,000
	12	18,000
	16	24,000
Rural sample	13	$10,000
	10	6,000
	11	12,000
	14	20,000

Suppose the above data (hypothetical) is a random sample of eight 25-year-old women. Let us define a dummy variable $D = 1$ if rural, 0 if urban. Then the multiple regression equation relating each woman's education to her father's income and urban-rural background is

$$\hat{E} = 9.42 + .00029\,F - .917D$$

(a) On the *EF* plane, graph this regression equation, and the 8 data points (8 women).

*(b) Using a computer, or a tableau like Table 9-1, calculate the regression coefficients and verify that they agree with the equation above.

CHAPTER 10

Analysis of Variance (ANOVA)

Variability is the law of life.

Sir William Osler

ANOVA has already been discussed in Section 9-2, where we showed how it could be done with dummy variable regression. In this chapter we will give the traditional view of ANOVA, following the lines of its development some 50 years ago.

We study this traditional approach because: (1) It provides the basic language for understanding a large body of applied statistical analysis; (2) it frequently has computational advantages; (3) it is so intuitively appealing that it provides a very helpful viewpoint for the applied statistician who analyzes or designs experiments; and (4) it provides a vehicle for a very simple introduction to Bayesian statistics in Chapter 12.

10-1 ONE-FACTOR ANALYSIS OF VARIANCE

(a) Variation Between Machines

In Chapter 5, we made inferences about one population mean, and then compared two means. Now we will compare several means. For these ANOVA techniques to be exactly valid, we require the populations to have a common variance and normal shape, just as we did in comparing two means in Chapter 5. However, ANOVA techniques remain approximately valid under broader conditions, and are therefore called *robust.*

As an illustration, suppose that three machines are to be compared. Because these machines are operated by people, and because of other, inexplicable reasons, output per hour is subject to chance fluctuation. In the hope of "averaging out" and thus reducing the effect of chance fluctuation, a random sample of 5 different hours is obtained from each machine and set out in Table 10-1, where each sample mean is calculated. (The bottom of this table can be ignored for now.)

The first question is, "Are the machines really different?" That is, are the sample means $\overline{X}$ in Table 10-1 different because of differences in the underlying population means μ (where μ represents the lifetime performance of a machine)? Or may these differences in $\overline{X}$ be reasonably attributed to chance fluctuations alone? To illustrate, suppose that we collect three samples from just *one* machine, as shown in Table 10-2. As

TABLE 10-1
Sample Outputs of Three Machines

	Machine 1	*Machine 2*	*Machine 3*	
	47	55	54	
	53	54	50	
	49	58	51	
	50	61	51	
	46	52	49	
$\overline{X}$	$\overline{X}_1 = 49$	$\overline{X}_2 = 56$	$\overline{X}_3 = 51$	$\overline{\overline{X}} = 52$
$(\overline{X} - \overline{\overline{X}})$	−3	4	−1	$\Sigma(\overline{X} - \overline{\overline{X}}) = 0\checkmark$
$(\overline{X} - \overline{\overline{X}})^2$	9	16	1	$\Sigma(\overline{X} - \overline{\overline{X}})^2 = 26$

expected, sampling fluctuations cause small differences in the $\overline{X}$'s even though the μ's in this case are identical. So the question may be rephrased, "Are the differences in the $\overline{X}$'s of Table 10-1 of the same order as those of Table 10-2 (and thus attributable to chance fluctuation), or are they large enough to indicate a difference in the underlying μ's?" The latter explanation seems more plausible; but how do we develop a formal test?

TABLE 10-2
Three Sample Outputs of the Same Machine

Sample 1	*Sample 2*	*Sample 3*
49	52	55
55	51	51
51	55	52
52	58	52
48	49	50
$\bar{X}_1 = 51$	$\bar{X}_2 = 53$	$\bar{X}_3 = 52$

As usual, the hypothesis of "no difference" in the population means is called the null hypothesis:

$$H_0: \mu_1 = \mu_2 = \mu_3 \tag{10-1}$$

A test of this hypothesis first requires a numerical measure of how much the sample means differ. We therefore take the three sample means $\bar{X}$ and calculate their variance at the bottom of Table 10-1, obtaining

$$s_{\bar{X}}^2 = \frac{26}{2} = 13 \tag{10-2}$$

The variance formula we used here, of course, is (2-10)—where $\bar{X}$ is substituted for X, and c (the number of sample means or columns) is substituted for n, so that

$$s_{\bar{X}}^2 = \frac{1}{c-1}\Sigma(\bar{X} - \bar{\bar{X}})^2 \tag{10-3}$$

where $\bar{\bar{X}}$ is the average of the $\bar{X}$, and is called the *grand mean.*

(b) Variation Within Machines

The variance between machines that we have just calculated does not tell the whole story. For example, consider the data of Table 10-3, which has the same $s_{\bar{X}}^2$ as Table 10-1, yet more erratic machines that produce large chance fluctuations within each column. The implications

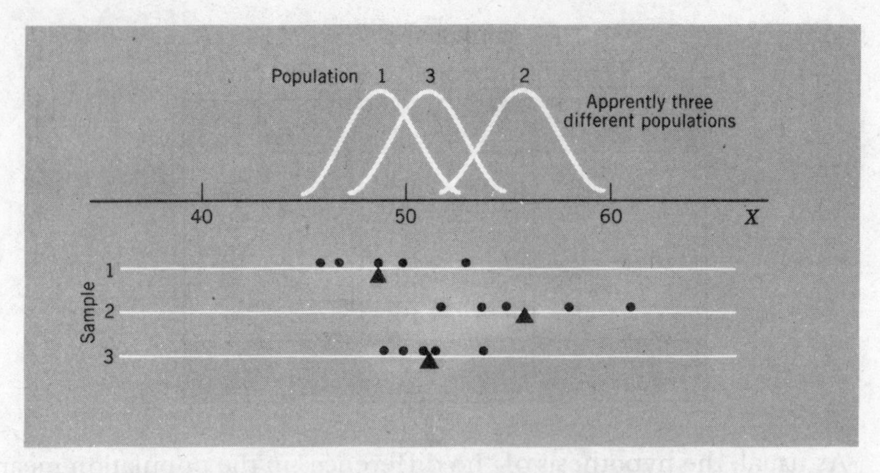

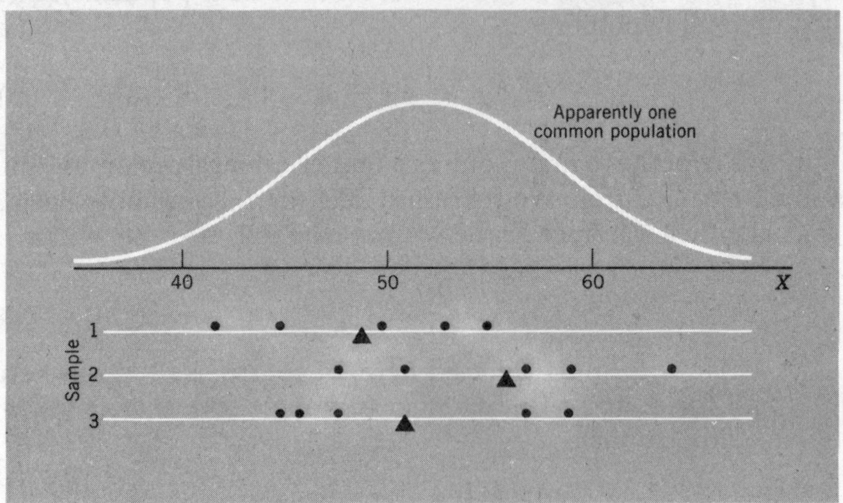

FIGURE 10-1 (*a*) Outputs of 3 relatively predictable machines (data from Table 10-1) (*b*) Outputs of 3 erratic machines (data from Table 10-3). They have the same sample means $\overline{X}$, hence the same $s_{\overline{X}}^2$, as the machines in (*a*).

of this are shown in Figure 10-1. In panel (*b*), the machines are so erratic that all samples could be drawn from the same population. That is, the differences in sample means may be explained by chance. On the other hand, the same differences in sample means can hardly be explained by chance in panel (*a*), because the machines in this case are *not* so erratic.

TABLE 10-3
Sample Outputs of Three Erratic Machines

Machine 1	*Machine 2*	*Machine 3*
50	48	57
42	57	59
53	65	48
45	59	46
55	51	45
$\overline{X}_1 = 49$	$\overline{X}_2 = 56$	$\overline{X}_3 = 51$

We now have our standard of comparison. In panel (*a*) we conclude that the μ's are different—reject H_0—because the variance in sample means ($s_{\overline{X}}^2$) is large *relative* to the chance fluctuation.

How can we measure this chance fluctuation? Intuitively, it seems to be the spread (or variance) of observed values *within* each sample. Thus we compute the squared deviations within the first sample in Table 10-1:

$$\sum (X_1 - \overline{X}_1)^2 = (47 - 49)^2 + (53 - 49)^2 + \ldots = 30$$

Similarly, we compute the squared deviations within the second and third samples, and add them all up. Then we divide by the total d.f. in all three samples ($n - 1 = 4$ in each). We thus obtain the pooled variance s_p^2 (just as we did for the two-sample case in Chapter 5):

$$s_p^2 = \frac{30 + 50 + 14}{4 + 4 + 4} = \frac{94}{12} = 7.83 \tag{10-4}$$

The generalization is easy to see. When there are c columns of data, each having n observations, then

$$s_p^2 = \frac{\sum(X_1 - \overline{X}_1)^2 + \sum(X_2 - \overline{X}_2)^2 + \ldots}{c(n - 1)} \tag{10-5}$$

like (5-25)

(c) The F Ratio

The key question now can be stated. Is $s_{\bar{X}}^2$ large relative to s_p^2? That is, what is the ratio $s_{\bar{X}}^2/s_p^2$? It is customary to examine a slightly modified ratio:

$$\boxed{F = \frac{ns_{\bar{X}}^2}{s_p^2}} \tag{10-6}$$

where n has been introduced into the numerator to make it equal on average to the denominator, when H_0 is true. Then the F ratio will fluctuate around 1.

If H_0 is not true (and the μ's are not the same), then $ns_{\bar{X}}^2$ will be relatively large compared to s_p^2, and the F ratio in (10-6) will tend to be much greater than 1. The larger is F, therefore, the less credible is the null hypothesis.

To numerically measure the credibility of H_0, as usual we find its prob-value—in this case, the probability in the tail of the F distribution beyond the observed value. We can estimate the prob-value from Appendix Table VI, which lists the critical values of the F distribution when H_0 is true (just like Table V lists the critical values of t). To use Table VI, we have to recognize that the F distribution depends on the degrees of freedom in the numerator variance $(c - 1)$, and in the denominator variance $[c(n - 1)]$. This is written briefly as

$$F \text{ has } (c - 1) \text{ and } c(n - 1) \text{ d.f.} \tag{10-7}$$

An example is the easiest way to demonstrate the actual calculation of the prob-value.

Example 10-1

For the data in Table 10-1, we have already calculated how much variance there is *between* the 3 sample means:

$$s_{\bar{X}}^2 = 13 \qquad \text{(10-2)repeated}$$

and how much residual variance there is *within* the 3 samples (of 5 observations each):

$$s_p^2 = 7.83 \qquad \text{(10-4)repeated}$$

(a) Calculate the F ratio.
(b) Calculate the degrees of freedom for F.
(c) Find the prob-value for H_0 (no difference in population means).

Solution

(a)
$$F = \frac{ns_{\bar{X}}^2}{s_p^2} \qquad \text{(10-6)repeated}$$
$$= \frac{5(13)}{7.83} = 8.3$$

(b)
$$\text{d.f.} = (c - 1) \text{ and } c(n - 1) \qquad \text{(10-7)repeated}$$
$$= (3 - 1) \text{ and } 3(5 - 1)$$
$$= 2 \text{ and } 12 \qquad \text{(10-8)}$$

(c) We look up Table VI where d.f. = 2 and 12, and find 5 critical values listed in a column that we scan down—till we find the observed F value of 8.3 lies beyond $F_{.01} = 6.93$. As the diagram below shows, we conclude that

$$\text{Prob-value} < .01 \qquad \text{(10-9)}$$

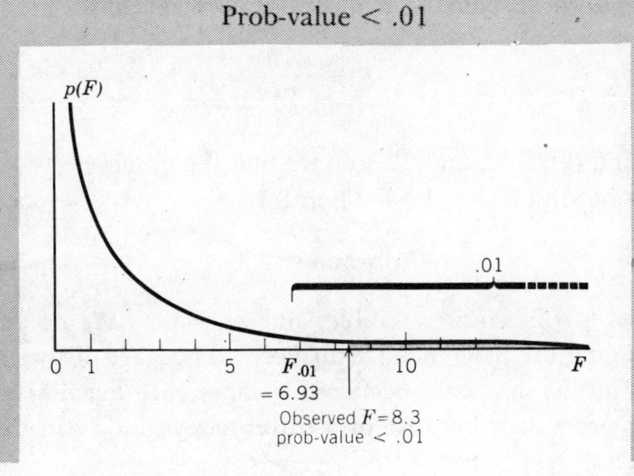

This means that if H_0 were true, there is less than a 1% chance of getting sample means that differ so much. It seems that the 3 machines in Table 10-1 really are different.

Let us see what the F test shows for the other cases in Tables 10-2 and 10-3.

Example 10-2

(a) Calculate the prob-value for H_0 using the data in Table 10-2, which showed 3 samples from the same machine. Thus $s_{\bar{X}}^2$ was only 1.0, while s_p^2 was 7.83.

(b) Calculate the prob-value for H_0, using the data in Table 10-3, which showed 3 erratic machines. Thus s_p^2 took on the large value 39.0, while $s_{\bar{X}}^2$ was 13.0.

Solution

(a)

$$F = \frac{5(1)}{7.83} = .64 \tag{10-10}$$

Using d.f. = 2 and 12 again, we find the observed F value of .64 falls far short of the first critical value $F_{.25} = 1.56$. Therefore,

$$\text{Prob-value} \gg .25$$

Since the prob-value is much greater than .25, H_0 is very credible. This is the correct conclusion, of course, since we generated these 3 samples in Table 10-2 from the same machine.

(b)

$$F = \frac{5(13)}{39} = 1.67 \tag{10-11}$$

Using d.f. = 2 and 12 still, we find the observed F value of 1.67 lies beyond $F_{.25} = 1.56$. Therefore,

$$\text{Prob-value} < .25$$

That is, H_0 is quite credible, and we really have no grounds for claiming the machines are different. The large difference in sample means may have occurred because each machine is erratic—not necessarily because of a difference in long-run machine averages.

(d) The ANOVA Table

The calculations we have done so far can be followed more easily if we summarize them in a table of standard form, called the ANOVA table. As Table 10-4a shows, it is mostly a bookkeeping arrangement, with the first row showing calculations of the numerator of the F ratio, and the second row the denominator. In Table 10-4(b) we evaluate the specific example of the three machines in Table 10-1.

In addition, the ANOVA table provides a handy check on our calculations. For example, in Table 10-4(b) consider the sum of squares (SS, also called *variations*) in the second column. At the bottom we list the total SS, which is obtained by ignoring the column-by-column structure of the data. That is, we just take all 15 numbers within Table 10-1 and calculate how much each deviates from the overall mean $\overline{\overline{X}}$. All the resulting deviations $(X - \overline{\overline{X}})$ are then squared and summed, to produce the total variation SS_t. To indicate that we must sum over the whole table, both rows and columns, we write the Σ sign twice:

$$SS_t = \Sigma\Sigma(X - \overline{\overline{X}})^2 \tag{10-12}$$

$$= (47 - 52)^2 + (53 - 52)^2 + \ldots + (49 - 52)^2$$

$$= 25 + 1 + \ldots + 9 = 224 \tag{10-13}$$

Then we note that the SS in the second column add up to this total, that is,

SS_t	=	SS_b	+	SS_w
Total variation	=	variation *between* columns	+	variation *within* columns

(10-14)

In addition, the degrees of freedom in Table 10-4 should add up in the same way. In Table 10-4(b) we indicate with a check (√) that the variations and degrees of freedom do indeed total properly.

When each variation is divided by its appropriate degrees of freedom, we obtain the *variance* in Table 10-4. The variance between columns is "explained" by the fact that the columns may come from different parent populations (machines that perform differently). The variance within columns is "unexplained" because it is the random or chance variance that cannot be systematically explained (by differences in machines). Thus their ratio F is sometimes referred to as the variance ratio:

$$F = \frac{\text{explained variance}}{\text{unexplained variance}} \tag{10-15}$$

PROBLEMS

For Problems 10-1 to 10-4, calculate the ANOVA table, including the approximate prob-value for the null hypothesis.

10-1 Twelve plots of land are divided randomly into 3 groups. The first is held as a control group, while fertilizers A and B are applied to the other two groups. Yield is observed to be:

Control C	*A*	*B*
60	75	74
64	70	78
65	66	72
55	69	68

10-2 In a large American university in 1969, the male and female professors were sampled independently, yielding the following annual salaries (in thousands of dollars, rounded. From Katz, 1974):

Men	*Women*
12	9
11	12
19	8
16	10
22	16

TABLE 10-4

(a) ANOVA Table, General

Source of Variation	*Variation; Sum of Squares (SS)*	*d.f.*	*Variance; Mean Sum of Squares (MSS)*	*F Ratio*
Between columns, due to differences in $\overline{X}$	$SS_b = n\left[\begin{matrix}(\overline{X}_1 - \overline{\overline{X}})^2 + \\ (\overline{X}_2 - \overline{\overline{X}})^2 \ldots\end{matrix}\right]$	$(c - 1)$	$MSS_b = SS_b/(c-1) = ns^2_{\overline{X}}$	$F = \dfrac{MSS_b}{MSS_w} = \dfrac{ns^2_{\overline{X}}}{s^2_p}$
Within columns, the residual variation, due to chance fluctuation	$SS_w = \Sigma(X_1 - \overline{X}_1)^2 + \Sigma(X_2 - \overline{X}_2)^2 + \ldots$	$c(n - 1)$	$MSS_w = SS_w/c(n-1) = s^2_p$	
Total	$SS_t = \Sigma\Sigma(X - \overline{\overline{X}})^2$	$(nc - 1)$		

(b) ANOVA Table, for Observations Given in Table 10-1

Source of Variation	*Variation*	*d.f.*	*Variance*	*F Ratio*	*Prob-value*
Between machines	130	2	65	$\dfrac{65}{7.83} = 8.3$	$p < .01$
Within machines	94	12	7.83		
Total	224√	14√			

10-3 A sample of American Homeowners in 1970 reported the following home values (in thousands of dollars) by region:

Region	*Population Size*	*Sample of Home Values*	$\bar{X}$	$\Sigma(X - \bar{X})^2$
Northeast	16,000,000	23, 18, 12, 15, 27	19	146
Northcentral	19,000,000	17, 32, 12, 13, 11	17	302
South	21,000,000	13, 10, 16, 12, 19	14	50
West	12,000,000	13, 38, 17, 25, 17	22	396

10-4 In 1977, 50 full-time working women were sampled from each of three educational categories, and their incomes (in thousands of dollars, annually) were reported as follows:

Years of School Completed	*Mean Income* $\bar{X}$	$\Sigma(X - \bar{X})^2$
Elementary school (8 years)	7.8	1835
High school (12 years)	9.7	2442
College (16 years)	14.0	4707

*10-2 EXTENSIONS

(a) Unequal Sample Sizes

In Table 10-1 we took the same number of observations (5) on each machine. Often this is the most efficient way to collect observations: Make all samples the same size n. However, when there are different sample sizes, $n_1, n_2, n_3, \ldots$, it is easy to modify the ANOVA calculations appropriately. The total number of observations is now $n_1 + n_2 + \ldots$ instead of nc. Consequently, the grand mean of all the numbers in the table is

$$\bar{\bar{X}} = \frac{\Sigma\Sigma X}{n_1 + n_2 + \ldots}$$

Alternatively, $\overline{\overline{X}}$ can be written in a form that clearly shows that it is still an average (but now a weighted average) of the machine means:

$$\overline{\overline{X}} = \frac{n_1\overline{X}_1 + n_2\overline{X}_2 + \ldots}{n_1 + n_2 + \ldots} \tag{10-16}$$

The variation between columns must be changed accordingly, to

$$SS_b = n_1(\overline{X}_1 - \overline{\overline{X}})^2 + n_2(\overline{X}_2 - \overline{\overline{X}})^2 + \ldots \tag{10-17}$$

Corresponding changes must also be made in the d.f. Instead of $nc - 1$, the total d.f. is now

$$\text{Total d.f.} = n_1 + n_2 + \ldots - 1 \tag{10-18}$$

And instead of $c(n - 1)$, the d.f. within columns is now

$$\text{Within columns d.f.} = (n_1 - 1) + (n_2 - 1) + \ldots \tag{10-19}$$

With these changes, the ANOVA table is exactly the same as Table 10-4.

(b) Two-factor ANOVA

So far we have studied how a response such as production depends on one factor such as machine type. This is commonly called *one-factor* ANOVA. However, in practice a response may depend on two, three, or more factors. For example, production may depend not only on machine type but also on operator experience, and even quality of the input material. The appropriate analysis in this case is to extend the ANOVA Table 10-4 to include all the other factors as sources of explained variation too. The table is then called two-factor ANOVA (or three-factor ANOVA, etc.)

*PROBLEMS

10-5 A sample of 11 American families in 1971 reported the following annual incomes (in thousands of dollars) by region.

Northeast	*North Central*	*South*	*West*
8	13	7	7
14	9	14	7
		8	16
		7	

Construct the ANOVA table, including the prob-value for H_0.

10-6 Verify that each formula in Section 10-2 reduces to its counterpart in Section 10-1, when all the sample sizes $n_1, n_2, \ldots$ are equal to n. Specifically:

(a) Show that (10-16) reduces to $(\overline{X}_1 + \overline{X}_2 + \ldots)/c$.
(b) Show that (10-17) reduces to $n[(\overline{X}_1 - \overline{\overline{X}})^2 + (\overline{X}_2 - \overline{\overline{X}})^2 + \ldots]$.
(c) Show that (10-18) reduces to $nc - 1$.
(d) Show that (10-19) reduces to $c(n - 1)$.

CHAPTER 10 SUMMARY

10-1 When more than two populations are to be compared, we need an extension of the two-sample t test. This is provided by the F ratio, which compares the explained variance between the sample means with the unexplained variance within the samples. The ANOVA table provides an orderly way to calculate F, step by step.

*10-2 ANOVA can be extended in many ways. For example, it can handle unequal sample sizes, or more than one factor.

REVIEW PROBLEMS

(10-7)

Yarn Types			
A	*B*	*C*	*D*
35	42	41	40
39	40	41	34
34	35	44	31

Four types of yarn were each randomly sampled 3 times, and their breaking strengths (pounds) were recorded. Calculate the ANOVA table.

*10-8 Consider once again the data in Problem 10-2 giving annual salaries (in thousands of dollars, rounded):

Men	12, 11, 19, 16, 22
Women	9, 12, 8, 10, 16

Denote income by Y, and sex by a dummy variable X having $X = 0$ for men, $X = 1$ for women. Then:

(a) Graph Y against X.

(b) Estimate by eye the regression line of Y on X. (*Hint:* Where will the line pass through the men's salaries? Through the women's salaries?)

(c) Calculate the regression line of Y on X. How well does your eyeball estimate in (b) compare?

(d) Construct a 95% confidence interval for the coefficient of X. Explain what it means in simple language.

(e) At the 5% error level, test whether income is unrelated to sex (i.e., whether the null hypothesis lies within the confidence interval).

(f) How well does (d) measure the university's discrimination against women?

*10-9 In Problems 5-13, 10-2, and 10-8 we analyzed the same data, using, respectively, t, ANOVA, and regression. Now answer True or False; if False, correct it.

(a) In all 3 cases we got the same test result: At the 5% level, H_0 cannot quite be rejected.

(b) Comparing t with regression, we also got the same estimate for the difference between men's and women's salaries and, in fact, the very same confidence interval.

CHAPTER 11

Correlation

Truth is stranger than fiction, but not so popular.

O. W. Holmes

This chapter will use some of the insights gained in ANOVA to shed new light on regression. The main tool for doing this will be the *correlation coefficient*, a close relative of the regression coefficient.

11-1 SIMPLE CORRELATION

The simple regression coefficient b showed us *how* one variable Y was related to (or could be predicted from) another variable X. The correlation coefficient r gives this a slight twist, and shows us *how closely* two variables are related.

(a) Sample Correlation r

Recall how the regression coefficient of Y on X was calculated: We first expressed X and Y in deviation form (x and y), and then calculated

$$b = \frac{\Sigma xy}{\Sigma x^2} \tag{11-1}$$

(7-5) repeated

TABLE 11-1
Math Score (X) and Corresponding Verbal Score (Y) of a Sample of Eight Students Entering College

Data		*Deviation Form*		*Products*		
X	Y	$x = X - \overline{X}$	$y = Y - \overline{Y}$	xy	x^2	y^2
80	65	20	15	300	400	225
50	60	−10	10	−100	100	100
36	35	−24	−15	360	576	225
58	39	−2	−11	22	4	121
72	48	12	−2	−24	144	4
60	44	0	−6	0	0	36
56	48	−4	−2	8	16	4
68	61	8	11	88	64	121
$\overline{X} = 60$	$\overline{Y} = 50$	0✓	0✓	$\sum xy = 654$	$\sum x^2 = 1304$	$\sum y^2 = 836$

The correlation coefficient r uses the same quantities Σxy and Σx^2, and uses Σy^2 as well:

Correlation of X and Y

$$r \equiv \frac{\Sigma xy}{\sqrt{\Sigma x^2}\,\sqrt{\Sigma y^2}} \qquad (11\text{-}2)$$

However x appears in this formula, now y appears symmetrically in the same way. Thus the correlation r does not make a distinction between the response y and the regressor x, the way the regression coefficient b did. [In the denominator of the regression formula (11-1), the regressor X appeared, but the response Y did not.]

To illustrate, how are math and verbal scores related? A sample of 8 college students is given in the first two columns of Table 11-1. In the succeeding columns we calculate the deviations, and then the sums Σxy, Σx^2 and Σy^2. If we wanted to measure how Y can be predicted from X, we would calculate the regression coefficient:

$$b = \frac{\Sigma xy}{\Sigma x^2} = \frac{654}{1304} = .50 \tag{11-3}$$

On the other hand, if we wanted to measure *how much* X and Y are related, we would calculate the correlation coefficient:

$$r = \frac{\Sigma xy}{\sqrt{\Sigma x^2}\sqrt{\Sigma y^2}} = \frac{654}{\sqrt{1304}\sqrt{836}} = .63 \tag{11-4}$$

(b) *r* Measures the Degree of Relation

We have claimed that the correlation r measures how much X and Y are related. Now let's support that claim, by analyzing what the formula (11-2) really means. First, recall how we interpreted a deviation from the mean:

$$x \equiv X - \overline{X} \qquad \text{(7-4) repeated}$$

In Figure 2-6, we saw that the deviation x tells how far we are from the mean $\overline{X}$. Similarly, the deviation y tells us how far we are from the

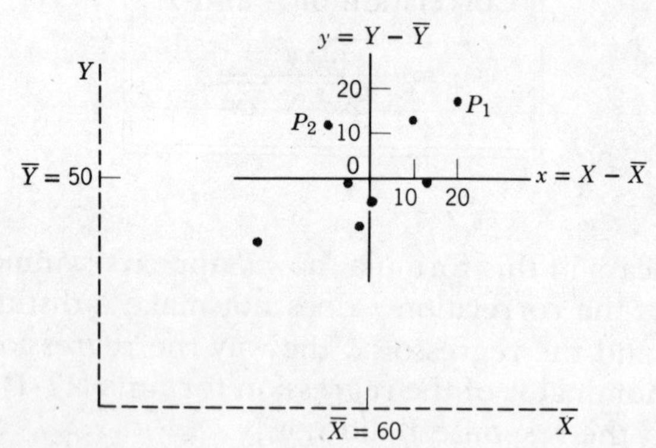

FIGURE 11-1 Scatter of math and verbal scores, from Table 11-1.

mean $\overline{Y}$. When we plot the pair (x, y) in two dimensions, therefore, we see how far we are from the center of the data $(\overline{X}, \overline{Y})$. Figure 11-1 illustrates this for the math and verbal scores.

Suppose we multiply the x and y values for each student, and sum them to get Σxy. This gives us a good measure of how math and verbal scores tend to move together. We can see this by referring to Figure 11-1: For any observation such as P_1 in the first or third quadrant, x and y agree in sign, so their product xy is positive.[1] Conversely, for any observation such as P_2 in the second or fourth quadrant, x and y disagree in sign, so their product xy is negative. If X and Y move together, most observations will fall in the first and third quadrants; consequently most products xy will be positive, as will their sum—a reflection of the positive relationship between X and Y. But if X and Y are negatively related (i.e., when one rises the other falls), most observations will fall in the second and fourth quadrants, yielding a negative value for our Σxy index. We conclude that as an index of correlation, Σxy at least carries the right sign. Moreover, when there is no relationship between X and Y, with the observations distributed evenly over the four quadrants, positive and negative terms will cancel, and Σxy will be 0.

To measure how X and Y vary together, Σxy suffers from just one defect: It depends on the units that X and Y are measured in. For example, suppose in Table 11-1 that X were marked on a different scale—specifically, a scale that is 10 times larger (in other words, suppose X is marked out of 1000 instead of 100). Then every deviation x would be 10 times larger and so, therefore, would the whole sum Σxy.

We would like a measure of relation that is not so fickle, one that remains "invariant to scale" even if we decide to use an X scale that is 10 times larger. How can we adjust Σxy to obtain such a measure? First, note that the quantity $\sqrt{\Sigma x^2}$ would also change by the same[2] factor 10. So if

---

[1]The point P_1 is given in the first line of Table 11-1. It represents a student with excellent scores, $X = 80$ and $Y = 65$—or in deviation form, $x = 20$ and $y = 15$. Note that the product xy is indeed positive (+ 300).

For the point P_2, $x = -10$ and $y = +10$. So now the product xy is negative (− 100).

[2]This is because each x^2 and consequently the whole sum Σx^2 would be 100 times larger. But when we take the square root, $\sqrt{\Sigma x^2}$ is reduced to being 10 times larger.

we divide Σxy by $\sqrt{\Sigma x^2}$, the factor of 10 cancels out, and we are left with an invariant measure. Of course, to protect against changes in the Y scale, we should divide by $\sqrt{\Sigma y^2}$ too. The result would be

$$\frac{\Sigma xy}{\sqrt{\Sigma x^2}\sqrt{\Sigma y^2}}$$

Thus our search for an appropriate measure of how closely two variables are related has indeed led us to the correlation coefficient (11-2).

To get a further idea of the meaning of r, in Figure 11-2 we have used the computer to plot various scatters and their correlation coefficients. In panel (*a*), for example, we show a larger sample than the 8 points in Figure 11-1. Nevertheless, the outline of the scatter displays about the same degree of relation, so it is no surprise that r is about the same value .60.

In panel (*b*) of Figure 11-2 there is a perfectly positive association, so that the product xy is always positive. Consequently r takes on its largest possible value, which turns out to be $+1$. Similarly, in panel (*d*), where there is a perfectly negative association, r takes on its most negative possible value, which turns out to be -1. We therefore conclude

$$\boxed{-1.00 \leq r \leq 1.00} \qquad (11\text{-}5)$$

Finally consider the symmetric scatters shown in panels (*e*) and (*f*) of Figure 11-2. The calculation of r in either case yields 0, because each product xy is offset by a corresponding product xy of the opposite sign in the opposite quadrant (to the left or right). Yet these two scatters show quite different patterns: In (*e*) there is no relation between X and Y; in (*f*), however, there is a strong relation (knowledge of X will tell us a great deal about Y). A zero value for r therefore does not necessarily mean "no relation." Instead, it means "no *linear* relation" (no straight-line relation). Thus

$$\boxed{r \text{ is a measure of linear relation only}} \qquad (11\text{-}6)$$

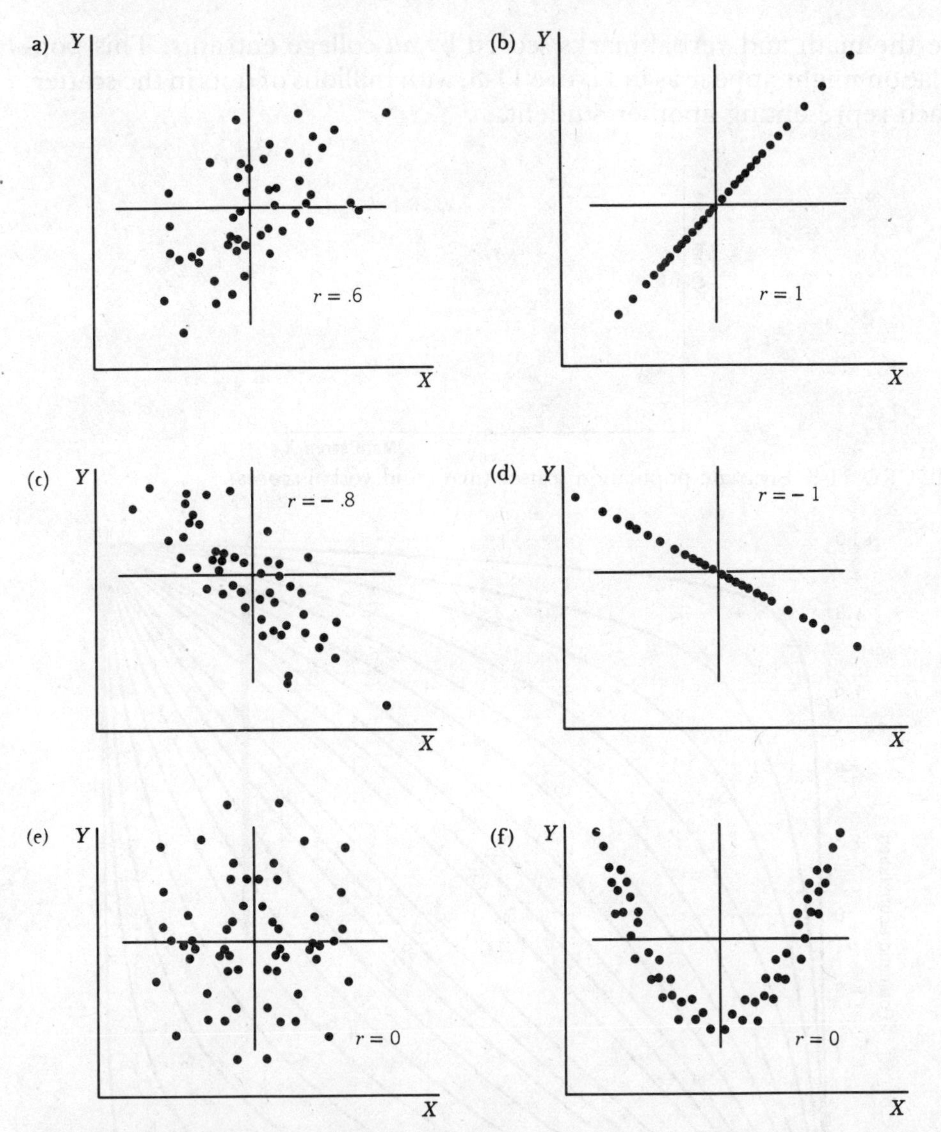

FIGURE 11-2 Various correlations illustrated: The vaguer is the scatter, the closer r is to 0.

(c) Population Correlation ρ

Once we have calculated the sample r, how can it be used to make inferences about the underlying population? In our example, this would be the math and verbal marks scored by *all* college entrants. This population might appear as in Figure 11-3, with millions of dots in the scatter, each representing another student.

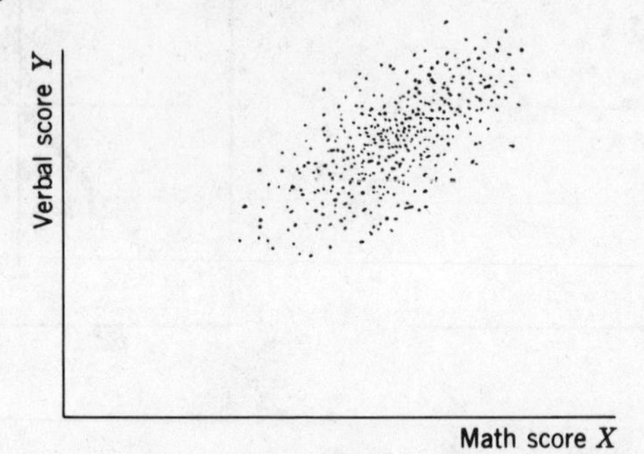

FIGURE 11-3 Bivariate population scatter (math and verbal scores).

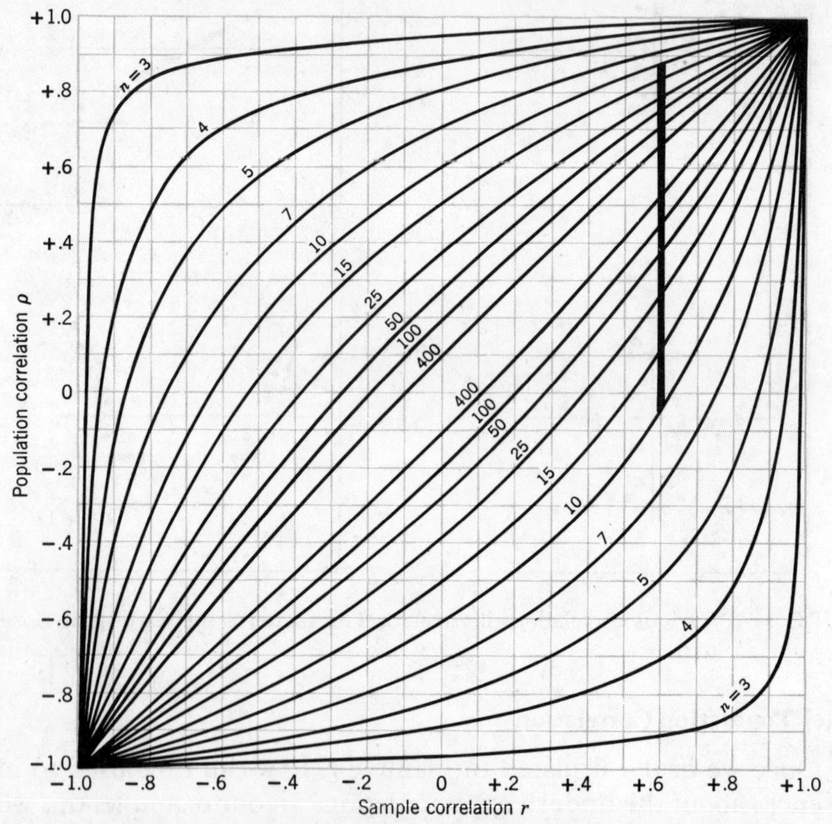

FIGURE 11-4 95% confidence bands for correlation ρ in a bivariate normal population, for various sample sizes n.

If we calculate (11-2) using all the points in the population, we call it the *population correlation* ρ. (*rho* is the Greek equivalent of r. We use a Greek letter, as usual, to denote a population parameter). Of course, as always in statistics, the population is unknown, and the problem is to infer ρ from an observed sample r. To do so, recall how we used P to make an inference about π in Figure 5-5. Similarly, in Figure 11-4 we can use r to make an inference[3] about ρ. For example, if a random sample of 10 students has $r = .60$, the 95% confidence interval for ρ is read vertically as

$$-.05 < \rho < .87 \tag{11-7}$$

as shown by the heavy blue line.

PROBLEMS

11-1 A random sample of 6 states gave the following figures for X = annual per capita cigarette consumption and Y = annual death rate per 100,000, from lung cancer (Fraumini, 1968):

State	X	Y
Delaware	3400	24
Indiana	2600	20
Iowa	2200	17
Montana	2400	19
New Jersey	2900	26
Washington	2100	20
Averages	2600	21

(a) Calculate the sample correlation r. A correlation calculated from aggregate data such as this is called an *ecological* correlation.

(b) Find a 95% confidence interval for the population correlation ρ.

---

[3]For the inference to be strictly valid, we assume the population is *bivariate normal.* Among other things, this means that if the X scores and also the Y scores were tabulated and graphed, each distribution would be normally shaped.

(c) At the 5% error level, test whether cigarette consumption and lung cancer are unrelated (i.e., whether the null hypothesis is in the confidence interval).

11-2

Son's Height (inches)	*Father's Height (inches)*
68	64
66	66
72	71
73	70
66	69
69 = average	68 = average

From the random sample of 5 son and father pairs given above:

(a) Calculate r.

(b) Find the 95% confidence interval for ρ.

(c) Calculate the regression of son's height on father's height, and find a 95% confidence interval for β.

(d) Graph the 5 pairs and the estimated regression line.

(e) At the 5% error level, can you reject:

(i) The null hypothesis $\beta = 0$?

(ii) The null hypothesis $\rho = 0$?

11-2 CORRELATION AND REGRESSION

(a) Relation of Regression Slope b and Correlation r

As we mentioned when we first set out their formulas, b and r are very similar. In fact, it may be easily shown (in Problem 11-5) that we can write b explicitly in terms of r as

$$\boxed{b = r\frac{s_Y}{s_X}} \tag{11-8}$$

Thus, for example, if either b or r is 0, the other will also be 0. Similarly, if either of the population parameters β or ρ is 0, the other will also be 0. Thus it is no surprise that in Problem 11-2, the tests for $\beta = 0$ and for $\rho = 0$ were equivalent ways of examining "no linear relation between X and Y."

To see how (11-8) works in other cases, let us give an example.

Example 11-1

Verify (11-8) for the data in Table 11-1. You may use $b = .50$ and $r = .63$ already calculated [in (11-3) and (11-4)].

Solution

To calculate s_Y and s_X, we merely substitute Σx^2 and Σy^2 from Table 11-1 into the formula (2-10):

$$s_X^2 = \frac{1}{n-1}\sum(X-\overline{X})^2 = \frac{1}{n-1}\sum x^2 \tag{11-9}$$

$$= \frac{1}{(8-1)}1304 = 186.3 \tag{11-10}$$

$$s_Y^2 = \frac{1}{(8-1)}836 = 119.4 \tag{11-11}$$

Substitute all these into (11-8):

Left side : $b = .50$

$$\text{Right side} : r\frac{s_Y}{s_X} = .63\frac{\sqrt{119.4}}{\sqrt{186.3}} = .50 = \text{left side} \tag{11-12}$$

(b) Explained and Unexplained Variation

The sample of math (X) and verbal (Y) scores from Figure 11-1 is reproduced in Figure 11-5, along with the fitted regression line. Suppose we want to predict the verbal Y score of a given student—to be concrete, the right-most student in Figure 11-5.

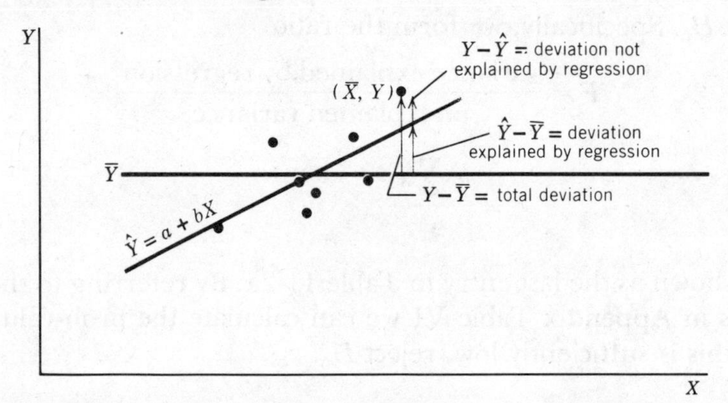

FIGURE 11-5 The value of regression in reducing variation in the fitted response.

If the math score X were not known, the only available prediction is to use the sample average $\overline{Y}$. Then the prediction error would be $Y - \overline{Y}$. In Figure 11-5, this appears as the large error shown by the longest blue arrow.

However, if X is known we can do better: We predict Y to be $\hat{Y}$ on the regression line. Note how this reduces our error, since a large part of our deviation $(\hat{Y} - \overline{Y})$ is now explained. This leaves only a relatively small unexplained deviation $(Y - \hat{Y})$. The total deviation of Y is the sum:

$$\begin{array}{ccccc} (Y - \overline{Y}) & = & (\hat{Y} - \overline{Y}) & + & (Y - \hat{Y}) \\ \text{total deviation} & = & \text{explained deviation} & + & \text{unexplained deviation} \end{array} \tag{11-13}$$

A similar breakdown can be proved to hold true when these deviations are squared and summed. Thus we obtain a result very similar to ANOVA in Chapter 10:

$$\begin{array}{ccccc} \sum (Y - \overline{Y})^2 & = & b^2 \sum x^2 & + & \sum (Y - \hat{Y})^2 \\ \text{total variation} & = & \text{variation explained by } X & + & \text{unexplained variation} \end{array} \tag{11-14}$$

like (10-14)

This procedure of analyzing or decomposing total variation into its components is called *analysis of variance* (ANOVA) applied to regression. The components may be displayed in an ANOVA table such as Table 11-2(a). From this, we may formulate a test of the null hypothesis $\beta = 0$. Just as in the standard ANOVA test in Chapter 10, the question is whether the ratio of explained variance to unexplained variance is large enough to reject H_0. Specifically, we form the ratio

$$F = \frac{\text{variance explained by regression}}{\text{unexplained variance}} \tag{11-15}$$

$$= \frac{b^2 \sum x^2}{s^2} \tag{11-16}$$

This is shown as the last entry in Table 11-2a. By referring to the critical F values in Appendix Table VI we can calculate the prob-value for H_0 and, if this is sufficiently low, reject H_0.

TABLE 11-2
ANOVA Table for Linear Regression
(*a*) General

Source of Variation	*Variation*	*d.f.*	*Variance = Variation/d.f.*	*F ratio*
Explained (by regression)	$\sum(\hat{Y} - \overline{Y})^2$ or $b^2 \sum x^2$	1	$\frac{b^2\sum x^2}{1}$	$\frac{b^2 \sum x^2}{s^2}$
Unexplained (residual)	$\sum(Y - \hat{Y})^2$	$n - 2$	$s^2 = \frac{\sum(Y - \hat{Y})^2}{n - 2}$	
Total	$\sum(Y - \overline{Y})^2$	$n - 1$		

(*b*) For Sample of Math and Verbal Scores (in Table 11-1)

Source of Variation	*Variation*	*d.f.*	*Variance*	*F ratio*	*Prob-value*
Explained (by regression)	328	1	328	3.87	$p < .10$
Unexplained (residual)	508	6	84.7		
Total	836√	7√			

The F test is just an alternative way of testing the null hypothesis that $\beta = 0$. The first method, using the t ratio in (7-23), is preferable if a confidence interval also is desired. The two tests are equivalent because the t statistic is related to F (with 1 d.f. in the numerator) by

$$\boxed{t^2 = F} \tag{11-17}$$

To sum up, there are three equivalent ways of testing the null hypothesis that X has no relation to Y: the F test and the t test of $\beta =$

0, and the test of $\rho = 0$ by using a confidence interval constructed from Figure 11-4. All three will now be illustrated in an example.

Example 11-2

(a) Analyze the math and verbal scores of Table 11-1 in an ANOVA table, including the prob-value and a test of the null hypothesis $\beta = 0$ at the 5% level.
(b) Test the same null hypothesis by alternatively using the t confidence interval.
(c) Test the equivalent null hypothesis $\rho = 0$ using the confidence interval for ρ (based on our observed $r = .63$).

Solution

(a) The ANOVA table is set out in Table 11-2b, yielding an F ratio of 3.87. We refer this to Table VI where d.f. = 1 and $n - 2 = 6$: We scan down the column of critical values till we find that the observed F value of 3.87 lies beyond $F_{.10} = 3.78$. Thus

$$\text{prob-value} < .10 \tag{11-18}$$

Since the prob-value for H_0 is more than 5%, we cannot reject H_0.

(b) We use the confidence interval

$$\beta = b \pm t_{.025} \frac{s}{\sqrt{\Sigma x^2}} \qquad \text{(7-18) repeated}$$

$$\begin{aligned} \text{where } b &= .50 \text{ from (11-3)} \\ t_{.025} &= 2.447 \text{ from Appendix Table V} \\ s^2 &= 84.7 \text{ from Table 11-2b} \\ \Sigma x^2 &= 1304 \text{ from Table 11-1} \end{aligned}$$

$$\text{Thus } \beta = .50 \pm 2.447 \frac{\sqrt{84.7}}{\sqrt{1304}}$$

$$= .50 \pm 2.447(.254) \tag{11-19}$$

$$= .50 \pm .62 \tag{11-20}$$

Since $\beta = 0$ is included in the confidence interval, we cannot reject the null hypothesis at the 5% level.

(c) In Figure 11-4, we must interpolate to find $n = 8$ and $r = .63$. This yields the approximate 95% confidence interval:

$$-.15 < \rho < +.90 \tag{11-21}$$

Since $\rho = 0$ is included in the confidence interval, we cannot reject the null hypothesis at the 5% level. This agrees with the conclusions in (a) and (b).

(c) Coefficient of Determination, r^2

The variations in Y in the ANOVA table can be related to r:

$$r^2 = \frac{\text{explained variation of } Y}{\text{total variation of } Y} \tag{11-22}$$

This equation provides a clear intuitive interpretation of r^2. Note that this is the *square* of the correlation coefficient r, and is often called the *coefficient of determination. It is the proportion of the total variation in Y explained by fitting the regression.* Since the numerator cannot exceed the denominator, the maximum value of the right-hand side of (11-22) is 1; hence the limits on r are $\pm$ 1. These two limits were illustrated in Figure 11-2: in panel *(b)*, $r = +1$ and all observations lie on a positively sloped straight line; in panel *(d)*, $r = -1$ and all observations line on a negatively sloped straight line. In either case, a regression fit will explain 100% of the variation in Y.

At the other extreme, when $r = 0$, then the proportion of the variation of Y that is explained is $r^2 = 0$, and a regression line explains nothing. That is, when $r = 0$, then $b = 0$; again note that these are just two equivalent ways of formally stating "no observed linear relation between X and Y."

(d) Multiple R^2

There are several kinds of correlation coefficient for multiple regression, analogous to the correlation coefficient r for simple regression that we have studied so far. Space permits us to mention only the most important: Like r^2 in (11-22), the *multiple coefficient of determination* R^2 gives the proportion of the total variation in Y that is explained by the *whole set of regressors* in a multiple regression.

(e) Correlation or Regression?

Both the regression and correlation models require that Y be a random variable. But the two models differ in the assumptions made about X. The regression model makes few assumptions about X, but the more restrictive correlation model of this chapter requires that X be a random variable, as well as Y. We therefore conclude that the regression model has wider application. It may be used for example to describe the fertilizer-yield problem in Chapter 7 where X was fixed at prespecified levels, or the population of X and Y in this chapter; however, the correlation model describes only the latter.

In addition, regression answers more interesting questions. Like correlation, it indicates if two variables move together; but it also estimates how, and provides the equation for prediction. Moreover, the key question in correlation analysis (whether or not any relationship exists between the two variables) is answered by testing the null hypothesis $\rho = 0$; but it can alternatively be answered directly from regression analysis by testing the equivalent null hypothesis $\beta = 0$. If this is the only correlation question, then there is no need to introduce correlation analysis at all.

In conclusion, regression answers a broader and more interesting set of questions (including some correlation questions as well), so it is usually the preferred technique. Correlation is used primarily as an aid to understanding regression.

(f) Spurious Correlation

Even though correlation (or simple regression) may have established that two variables move together, no claim can be made that this necessarily indicates cause and effect. For example, the correlation of teachers' salaries and the consumption of liquor over a period of years turned out to be .9. This does not prove that teachers drink; nor does it prove that liquor sales increase teachers' salaries. Instead, both variables moved together, because both are influenced by a third variable—long-run growth in national income and population. (To establish cause-and-effect, extraneous factors like this would have to be kept constant, as in a randomized controlled study—or their effects allowed for, as in multiple regression.)

Such correlations are often called *spurious* or *nonsense* correlations. It would be more accurate to say that the correlation is real enough, but any naive inference of cause and effect is nonsense.

PROBLEMS

11-3 A random sample of 7 U.S. women gave the following data on X = age (years) and Y = concentration of cholesterol in blood (grams/liter). The correlation is $r = .693$.

	X	Y
	30	1.6
	60	2.5
	40	2.2
	20	1.4
	50	2.7
	50	1.6
	30	2.0
Average	40	2.0
Variation	1200	1.46
Variance	200	.243
Standard deviation	14.1	.493

(a) Calculate the regression line of Y on X. [*Hint:* It is easiest to use (11-8).] Graph the regression line, along with the 7 given points.

(b) Write out the ANOVA table for the regression of Y on X; the following order is the easiest way to obtain the variations column:

  (i) Calculate the explained variation, using b found in part (a), and $\Sigma x^2 = 1200$ given in the table above.

  (ii) Copy down the total variation $\Sigma(Y - \overline{Y})^2 = 1.46$ given in the table above.

  (iii) Find the residual variation by subtraction.

Carry through the ANOVA table as far as the prob-value for H_0, using the F test. Can you reject H_0, at the 5% error level?

(c) Using the slope calculated in (a) and the residual variance calculated in (b), calculate the 95% confidence interval for β. Can you reject H_0, at the 5% error level?

(d) Find the 95% confidence interval for ρ. Can you reject H_0, at the 5% error level?

(e) Do you get consistent answers in (b), (c), and (d) for the question "Are X and Y linearly related?"

(f) From the ANOVA table in (b), find the proportion of the variation that is explained by the regression. Does it agree with r^2?
Also find the proportion left unexplained. Does it agree with $(1 - r^2)$?

11-4 Repeat Problem 11-3 for the following sample, where $r = .690$:

	X	Y
	60	2.9
	20	2.0
	50	1.7
	20	1.5
	50	2.4
Average	40	2.1
Variation	1400	1.26
Variance	350	.315
Standard deviation	18.7	.561

11-5 Here is a proof of (11-8):

(a) Show that dividing (11-1) by (11-2) gives

$$\frac{b}{r} = \frac{\sqrt{\Sigma y^2}}{\sqrt{\Sigma x^2}}$$

(b) On the right side of this equation, divide numerator and denominator by $\sqrt{n-1}$. Show that this gives

$$\frac{b}{r} = \frac{s_Y}{s_X}$$

(c) Solve this equation for b. Do you get (11-8)?

11-6 The explained variation can be calculated two ways, as Table 11-2*a* indicates: either as $\Sigma(\hat{Y} - \overline{Y})^2$, or $b^2\ \Sigma x^2$. For the data in Table 11-1, show that these two numbers are the same (both equal 326, within rounding error).

CHAPTER 11 SUMMARY

11-1 The correlation coefficient r measures how closely two variables move together. Its value lies between -1 and $+1$.

11-2 Correspondingly, the coefficient of determination r^2 lies between 0 and 1 (0% and 100%). And r^2 gives the percent of the variation in one variable that is explained by regressing it on the other variable. There are other close connections between correlation and regression; for example, the regression slope b is easily expressed in terms of r. Finally, the t test on b is equivalent to the F test in the ANOVA table.

REVIEW PROBLEMS

11-7 Suppose men always married women who were 3 years younger than themselves. What would be the correlation between husbands' and wives' ages?

11-8 (a) Referring to the math and verbal scores of Table 11-1, suppose that only the students with math score exceeding 65 were admitted to college. For this subsample of 3 students, calculate the correlation of X and Y.

(b) For the other subsample of the 5 remaining students, calculate the correlation of X and Y.

(c) Are these 2 correlations in the subsamples greater or less than the correlation in the whole sample? Do you think this will be generally true?

11-9 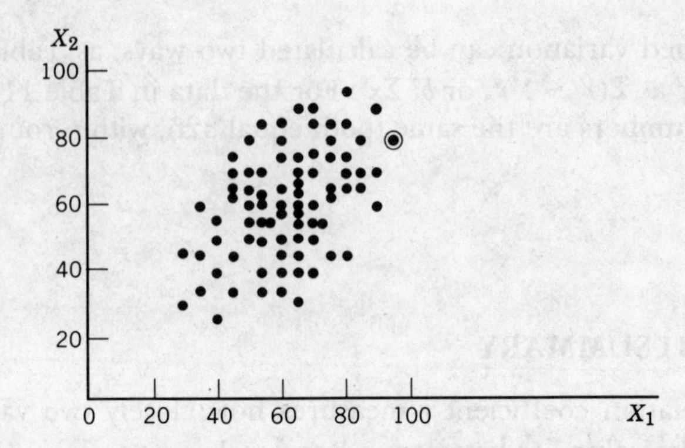

In an experimental program, 80 pilot trainees were drawn at random, and each given 2 trial landings. Their scores X_1 and X_2 were recorded in the above graph, and the following summary statistics were calculated:

$$\bar{X}_1 = 62, \quad \bar{X}_2 = 61,$$

$$\Sigma(X_1 - \bar{X}_1)^2 = 18{,}000, \quad \Sigma(X_2 - \bar{X}_2)^2 = 21{,}000,$$

$$\Sigma(X_1 - \bar{X}_1)(X_2 - \bar{X}_2) = 11{,}000$$

(a) By comparison with Figure 11-2, guess the approximate correlation of X_1 and X_2. Then calculate it.

(b) Draw in the line of equivalent performance, the line $X_2 = X_1$. For comparison, calculate and then graph the regression line of X_2 on X_1.

(c) What would you predict would be a pilot's second score X_2, if his first score was $X_1 = 90$? and $X_1 = 40$?

(d) On the figure below we graph the distribution of X_1 and of X_2. The arrow indicates that the pilot who scored $X_1 = 95$ later scored $X_2 = 80$. (This is the pilot in the original graph who is

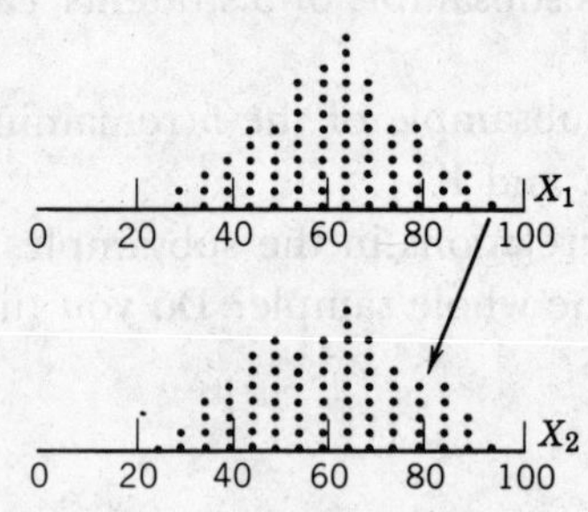

circled on the extreme right.) Draw in similar arrows for all three pilots who scored $X_1 = 90$. What is the mean of their three X_2 scores? How does it compare to the answer in (c)?

(e) Repeat (d), for all the pilots who scored $X_1 = 40$.

(f) Answer True or False; if False correct it.

The pilots who scored very well or very badly on the first test X_1 were closer to the average on the second test X_2.

One possible reason for this "regression toward the mean" is that a very high score probably represents some good luck as well as good skill. On another test, the good luck may not persist. So the second score will generally be not so good as the first, although still better than average.

11-10 Repeat Problem 11-9, interchanging X_1 and X_2 everywhere.

11-11 (a) When the flight instructors in Problem 11-9 graded the pilots, they praised them if they did well, or criticized them if they did poorly. They noticed that the pilots who has been praised did worse the next time, while the pilots who had been criticized did better; so they concluded that criticism works better than praise. Comment.

(b) The instructors therefore decided, for a fresh sample of 80 pilots, to criticize *every* pilot no matter how well or poorly the pilot did on the first test. They were disappointed to find that this uniform criticism made the second test scores no better, on average. Comment.

PART IV

FURTHER TOPICS

CHAPTER 12

An Introduction to Bayes Estimation

All virtue is a compromise.

William Godwin

In the first 11 chapters, we have shown how to use a sample to estimate a population parameter. Sometimes there are other sources of information as well as the sample, sources ranging all the way from good intuition to a wide variety of experience with similar samples. This kind of information, that is available even before the particular sample at hand is collected, is customarily called *prior information.* How can it be combined with the sample to provide an overall final estimate?

This final estimate is also called the *posterior* estimate, or *Bayes* estimate, after the English mathematician Rev. Thomas Bayes, 1702–1761. We will derive it for just the simplest case, where the prior information is that the unknown parameter is in the neighborhood of the null hypothesis. That is, in this chapter, we will regard the null hypothesis as a vague prior estimate to be modified in light of the data.

12-1 RESOLVING THE CLASSICAL DILEMMA

(a) Difficulty with Classical ANOVA

To be concrete, let us look at an example where classical estimation involves a dilemma that is very clear. In Table 12-1, suppose we need an estimate of μ_2, the long-run mean output of the second machine. What do we use?

TABLE 12-1
Production of 3 Machines
(repeat of Table 10-3)

Machine 1	*Machine 2*	*Machine 3*
50	48	57
42	57	59
53	65	48
45	59	46
55	51	45
$\bar{X}_1 = 49$	$\bar{X}_2 = 56$	$\bar{X}_3 = 51$
Grand mean $\bar{\bar{X}} = 52$		
$F = \frac{65}{39.0} = 1.67$		

The obvious estimate of the long-run production of the second machine is 56 (the second sample mean $\bar{X}_2$). But what if H_0 is true (i.e., there is no difference in machines, so that all 15 observations come from the same population)? Then a better estimate would be 52 (the grand mean $\bar{\bar{X}}$, based on all the relevant data).

The problem in a nutshell is this. *We don't know for sure which of these two estimates to choose because we don't know for certain whether H_0 or H_1 is true:* If H_0 is true, the better estimate is $\bar{\bar{X}} = 52$; but if H_1 is true (and machines do differ) then the better estimate is $\bar{X}_2 = 56$. The classical way out of this dilemma is to use the F test at some customary level, say 5%. In our example in Table 12-1, F was calculated to be 1.67. Since this is less than the critical $F_{.05} = 3.89$, we cannot reject H_0; so we use the H_0 estimate of 52.

But imagine this was only one experiment in a whole sequence. In addition to the value $F = 1.67$, suppose we had also, from the other experiments, calculated $F = 8.26$, and a whole sequence of F values in between. In each of the cases where F is less than the critical value $F_{.05} = 3.89$, no matter how slightly, we would use $\bar{\bar{X}}$. But whenever F

is above 3.89, no matter how slightly, we would use $\overline{X}_2$. That is, our estimate could only take on two values ($\overline{\overline{X}}$ and $\overline{X}_2$), and would jump from one to the other because of a trivial change in the data (F changing, for example, from 3.88 to 3.90). This is clearly unsatisfactory. Instead, we should develop an estimate that changes *continuously* between $\overline{\overline{X}}$ and $\overline{X}_2$, depending on the strength of the sample message we get from F. The more strongly F confirms H_1 (i.e., the larger the value of F), the closer our estimate should come to the H_1 value $\overline{X}_2 = 56$.

(b) Bayes Solution in General

The dilemma of whether to choose H_0 or H_1 is thus resolved: Compromise. Whichever hypothesis is more credible should have the greater weight. So what we need is a measure of credibility, and the F statistic provides it. To the extent that the F statistic is large, H_1 is more credible; to the extent that F is small, H_0 is more credible. The Bayes solution (derived, for example, in Box and Tiao, 1973) makes this idea precise:

Bayes estimate (for $F \geq 1$)

$$\text{Give } H_0 \text{ a weight} = \frac{1}{F} \qquad \qquad (12\text{-}1)$$

$$\text{Give } H_1 \text{ the remaining weight} = 1 - \frac{1}{F}$$

For $F < 1$, we would run into trouble if we tried to use (12-1). H_0 would get a weight of more than 1, and H_1 would get a negative weight. The Bayes estimate is therefore defined, for $F < 1$, by

$$\left.\begin{array}{l}\text{Give } H_0 \text{ a weight} = 1\\ \text{Give } H_1 \text{ a weight} = 0\end{array}\right\} \qquad (12\text{-}2)$$

In the rest of this chapter, we will simply show how formula (12-1) works in specific cases.

(c) Bayes Estimate for ANOVA

For ANOVA, we saw that the null hypothesis H_0 used the grand mean $\overline{\overline{X}}$ and the alternative hypothesis H_1 used the specific machine mean $\overline{X}_2$. If we denote the Bayes estimate of μ_2 by BE (μ_2), then (12-1) becomes

Bayes estimate of the second population mean in ANOVA

$$\text{BE}(\mu_2) = \frac{1}{F}\overline{\overline{X}} + \left(1 - \frac{1}{F}\right)\overline{X}_2 \tag{12-3}$$

A similar formula holds for the first mean (or any other mean), of course. An example will show how the formula works, and how it can be interpreted graphically.

Example 12-1

Recall that in Table 12-1, from a sample of $n = 5$ observations on each machine, the three machine means were 49, 56, and 51. The grand mean was $\overline{\overline{X}} = 52$, and the F statistic was 1.67.

(a) Calculate the Bayes estimate of the second machine mean, and then the first and third.

(b) On a graph, contrast the Bayes and classical estimates.

Solution

(a) The Bayes estimate is a compromise between $\overline{\overline{X}} = 52$ and $\overline{X}_2 = 56$, with the weights determined by $F = 1.67$. When these numbers are substituted into (12-3), we obtain

$$\text{BE}(\mu_2) = \frac{1}{1.67}\overline{\overline{X}} + \left(1 - \frac{1}{1.67}\right)\overline{X}_2$$

$$= .60(52) + .40(56) = 53.6 \tag{12-4}$$

The Bayes estimate 53.6 is thus a 60-40 compromise. For the other machines, the compromise is still 60-40, because the F statistic is still 1.67. Therefore,

$$\text{BE}(\mu_1) = .60(52) + .40(49) = 50.8 \tag{12-5}$$

$$\text{BE}(\mu_3) = .60(52) + .40(51) = 51.6 \tag{12-6}$$

(b) As Figure 12-1 shows, the Bayes estimates are shrunk toward the grand mean $\overline{\overline{X}}$. So they are sometimes called *shrinkage* estimates.

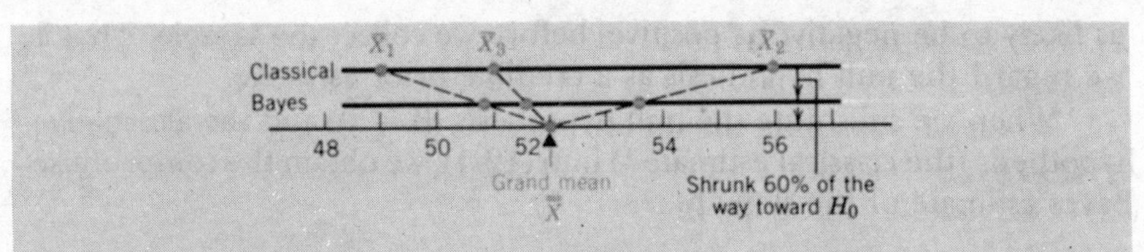

FIGURE 12-1 The Bayes estimates are classical estimates shrunk toward H_0. (In our example, with $F = 1.67$, there is a 60% shrinkage.)

PROBLEMS

12-1 Recall that in Table 10-1, a sample of $n = 5$ observations from each machine gave 3 machine means: 49, 56, and 51. The grand mean was 52 and the F statistic was 8.3.

(a) Calculate the Bayes estimates for all 3 machine population means.

(b) On a graph, contrast the Bayes and classical estimates.

(c) Comparing the graph in (b) to the graph in Figure 12-1:

(i) Which graph shows more shrinkage toward the null hypothesis value $\bar{\bar{X}}$?

(ii) In which case was there more evidence that H_0 was true, as given by the classical test?

12-2 In Problem 10-7, a sample of $n = 3$ was taken from each of four yarn types. The four sample means were: 36, 39, 42, and 35. The grand mean was 38, and the F statistic was 2.73.

(a) Calculate the Bayes estimates for the four population means.

(b) On a graph, contrast the Bayes and classical estimates.

12-3 Repeat Problem 12-2, for the data slightly changed: Suppose the four sample means are now 36, 39, 37, and 36. The grand mean is then 37, and the F statistic .75.

12-2 BAYES ESTIMATES IN OTHER CASES

Let us apply our basic Bayes solution to two more cases—regression and the comparison of two means.

(a) Regression Slope β

As usual, we suppose that our prior knowledge leads us to expect the parameter is somewhere in the neighbourhood of 0; the slope β is as likely to be negative as positive, before we collect the sample. That is, we regard the null hypothesis as a credible prior estimate.

When we substitute the null hypothesis ($\beta = 0$) and the alternative hypothesis (the classical estimate b) into (12-1), we obtain the compromise Bayes estimate of the slope β:

$$\begin{aligned} \text{BE}(\beta) &= \left(\frac{1}{F}\right)0 + \left(1 - \frac{1}{F}\right)b \\ &= \left(1 - \frac{1}{F}\right)b \end{aligned} \tag{12-7}$$

where F is the customary F ratio for testing the null hypothesis:

$$F = t^2 = \left(\frac{b}{\text{SE}}\right)^2 \tag{12-8}$$

(11-17) repeated

Example 12-2

The Statistics 200 course at a large university has many sections, and has been offered for many years. Every year, each section evaluates its instructor on his "effectiveness" (or popularity?) X. In turn, the students are all given a common final exam, and each section average Y is calculated. How are X and Y related? Of the hundreds of sections evaluated over the years, a random sample of 10 was selected, and plotted in Figure 12-2. Rather surprisingly, the fitted least squares line showed a negative slope ($b = -2.4$)—due perhaps to the large degree of scatter, and subsequent high standard error (SE = 1.94).

(a) What is the prob-value for H_0 ($\beta = 0$, i.e., no relation of Y to X)?

(b) Suppose we felt, before looking at the data, that a positive relation between Y and X was about as likely to occur as a negative relation. This means that (12-7), the Bayes compromise with 0, is legitimate. So calculate it.

Solution

(a) From (12-8),

$$F = \left(\frac{b}{\text{SE}}\right)^2 = \left(\frac{-2.4}{1.94}\right)^2 = 1.53$$

We refer to Appendix Table VI, with F having d.f. $= 1$ and $n - 2 = 8$. The observed value $F = 1.53$ is close to $F_{.25} = 1.54$, so that we conclude:

$$\text{Prob-value} \simeq .25$$

(b) Because H_0 had substantial credibility in part (a), we expect the Bayes compromise slope to be shrunk substantially toward 0. Formula (12-7) gives it explicitly:

$$\begin{aligned} \text{BE}(\beta) &= \frac{1}{1.53}(0) + \left(1 - \frac{1}{1.53}\right)(-2.4) \\ &= .65(0) + .35(-2.4) \\ &= -.8 \end{aligned} \tag{12-9}$$

Equation (12-9) states explicitly that 0 gets 65% of the weight (because F is so small). As Figure 12-2 shows, this means that the classical slope b is shrunk (flattened) 65% of the way toward 0 (and thus is shrunk from -2.4 to $-.8$).

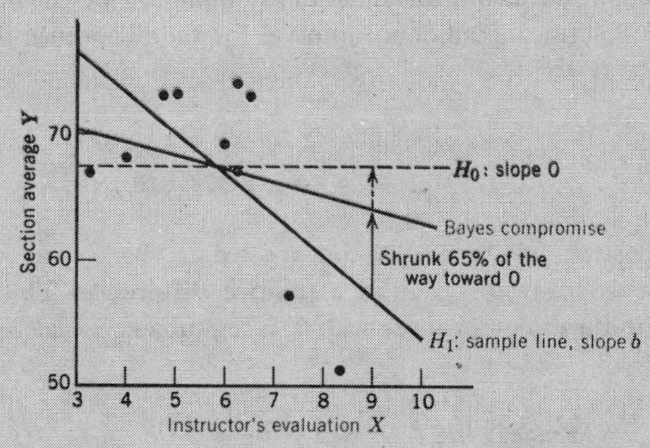

FIGURE 12-2 The Bayes line is the classical line shrunk toward H_0. (In our example, with $F = 1.53$ there is a 65% shrinkage.)

(b) The Difference in Two Means $(\mu_1 - \mu_2)$

When we substitute the null hypothesis ($\mu_1 - \mu_2 = 0$) and the alternative hypothesis (the classical estimate $\bar{X}_1 - \bar{X}_2$) into (12-1), we obtain the compromise Bayes estimate:

$$\begin{aligned} \text{BE}(\mu_1 - \mu_2) &= \left(\frac{1}{F}\right) 0 + \left(1 - \frac{1}{F}\right)(\bar{X}_1 - \bar{X}_2) \\ &= \left(1 - \frac{1}{F}\right)(\bar{X}_1 - \bar{X}_2) \end{aligned} \tag{12-10}$$

where F is the customary F ratio for testing H_0,

$$F = t^2 = \left(\frac{\bar{X}_1 - \bar{X}_2}{\text{SE}}\right)^2 \tag{12-11}$$

like (6-19)

We use whatever SE (standard error) is appropriate, depending on whether the two samples are independent or paired. Since proportions are special cases of means, they also are covered by (12-10) and (12-11). An example will illustrate these formulas.

Example 12-3

From a large class, a random sample of 4 grades was drawn: 64, 66, 89, and 77. From a second large class, a random sample of 3 grades was drawn: 56, 71, 53. From this data, in Example 5-3 we calculated the following 95% classical confidence interval for the difference in population means:

$$\begin{aligned} \mu_1 - \mu_2 &= (\bar{X}_1 - \bar{X}_2) \pm t_{.025}\,\text{SE} \\ &= 14 \pm 2.571(8.26) \end{aligned}$$

Suppose we felt, before looking at the data, that a negative difference was about as likely to occur as a positive difference. This means that (12-10), the Bayes compromise with 0, is legitimate. So calculate it.

Solution

First, we calculate the F ratio. From (12-11),

$$F = t^2 = \left(\frac{14}{8.26}\right)^2 = 2.87$$

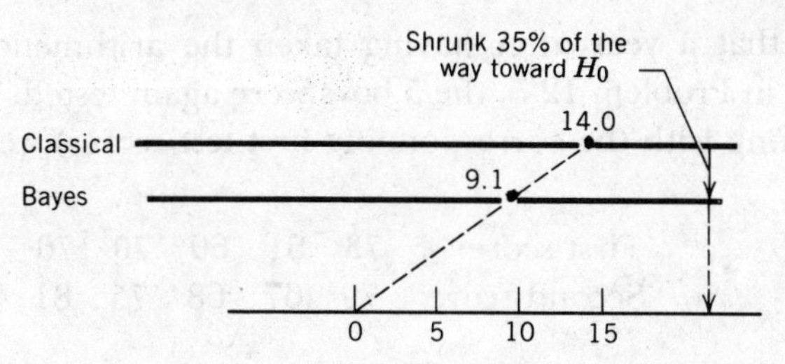

FIGURE 12-3 The Bayes estimate is the classical estimate shrunk toward H_0. (In our example, with $F = 2.87$, there is a 35% shrinkage.)

Substitute this into (12-10):

$$\begin{aligned} BE(\mu_1 - \mu_2) &= \frac{1}{2.87}(0) + \left(1 - \frac{1}{2.87}\right)(14) \\ &= .35(0) + .65(14) \\ &= 9.1 \end{aligned} \tag{12-12}$$

As Figure 12-3 shows, this means the Bayes estimate is shrunk 35% of the way toward 0.

The past few examples have illustrated Bayes estimation in several contexts. We could give further examples, and they would all show a shrinkage of the classical estimate toward the prior null hypothesis.

PROBLEMS

12-4 A random sample of 7 women showed a positive slope of blood cholesterol level on age. If the least squares slope was .024 with a standard error of .010, what is the Bayes slope?

12-5 An arithmetic reasoning test was given to a random sample of 5 boys and an independent random sample of 5 girls in a large third-grade class. The scores obtained were:

Boys:	73	61	60	70	76
Girls:	71	85	72	93	84

For the difference between the boys' and girls' scores (average over the whole class):

(a) Calculate the classical estimate.

(b) Calculate the Bayes estimate.

12-6 Suppose that a year after having taken the arithmetic reasoning test discussed in Problem 12-5, the 5 boys were again tested. The second test scores (along with the corresponding first test scores) were as follows:

First score:	73	61	60	70	76
Second score:	79	67	68	75	81

For the improvement in score (average over the whole class):
(a) Calculate the classical estimate.
(b) Calculate the Bayes estimate.

*12-7 In 1962, a sample of 2500 women in a region of Taiwan showed 14.2% pregnant. A year later, another sample of 2500 women showed 11.4% pregnant (Berelson and Freedman, 1964). For the drop in pregnancy rate in the whole population:
(a) Calculate the classical estimate.
(b) Calculate the Bayes estimate.

12-3 CONCLUSIONS

(a) Classical and Bayes Estimates Compared

Broadly speaking, a classical hypothesis test is a simple approach that sees only black and white. It leads to a simple conclusion: Accept or reject H_0. And the classical estimates are correspondingly simple and uncomprising.

The Bayes estimate is a more subtle one that sees all shades of gray. It is a compromise between the prior hypothesis H_0 and the alternative H_1—with the relative weights depending on how strongly the data supports H_1 (i.e., how large is the calculated value of F).

If the sample is large, the standard error in the denominator of F is small. This in turn makes F large, and hence the shrinkage factor $1/F$ small. With so little shrinkage, the Bayes estimate is very close to the classical estimate. Thus we conclude

> For large samples, the Bayes estimate is about the same as the classical estimate. (12-13)

In other words, if the sample is large, there is no need to fuss with the Bayes estimate—it would just be like the classical estimate anyhow. The Bayes estimate is valuable, however, for small samples—where the classical estimate is unreliable, and therefore needs to be substantially shrunk toward the prior belief H_0.

(b) Some Other Approaches

There are ways other than Bayes to modify classical estimates. They range from the very theoretical to the very practical, and are known by such names as *James-Stein estimation, ridge regression,* and *cross validation.* They all share a common goal of minimizing estimation error, and all produce roughly the same answer—the Bayes estimate given in (12-1). They therefore confirm the reasonableness of the Bayes approach.

(c) Limitations of Bayes Estimates

Classical estimates have no tendency to overestimate, or to underestimate. That is, they are unbiased. For example, the classical least squares slope estimate b fluctuates equally above and below its target β. So the Bayes estimate, which is moved toward the prior hypothesis, must be a biased estimate. This bias is the cost of Bayes estimation.

What are the benefits? Variance is reduced, and reduced more than enough to compensate for the bias introduced. If we have to put our judgement on the line, and want our estimate to be as close as possible to the target, this one particular time, the Bayes estimate works well.

On the other hand, what if our estimate is just one of many in a long sequence of estimates that are to be averaged in a general scientific context? Over many estimates, random variability cancels out. But bias, which is in the *same* direction every time, will *not* cancel out in the long run. Instead it will persist, and bias the overall estimate by the same amount. This is the case against any kind of bias, and a very strong case it is. In particular, it is important to avoid the bias of naively averaging the Bayes estimates. It is better to average the unbiased classical estimates instead. (The best solution of all would be a sophisticated Bayes estimate based on *all the data* pooled from the whole sequence of studies. This would make one huge sample, with perhaps a very large F statistic, and hence very little shrinkage. Thus the Bayes estimate might be practically the same as the classical.)

In conclusion, Bayes estimates are useful when all three of the following conditions are met:

1. We want an estimate that is close to the target one particular time

(as opposed to an estimate that forms part of a long sequence that will be naively averaged).

2. The null hypothesis H_0 (toward which the classical estimate is to be shrunk) is credible, in the sense that values above and below H_0 are judged to be about equally likely.[1]
3. The sample is small enough that the Bayes estimate really differs from the classical estimate. (Recall that when sample size is very large, the two estimates are practically the same.)

*12-4 BAYES CONFIDENCE INTERVALS

In Section 12-1 we have seen how the classical ANOVA estimate ($\overline{X}_2 = 56$) is adjusted by Bayes analysis to use *all* the data—that is, all 15 of the observations used to calculate $\overline{\overline{X}}$. There is a further benefit of using all the information that has been collected: The variability of the estimate is reduced. In fact, *when F is large* ($F > 3$, say) the variance of the estimate is reduced by the same shrinkage factor $1 - (1/F)$ that appears in the Bayes estimate. Consequently, the standard error of the estimate is reduced by the square root $\sqrt{1 - (1/F)}$. We can therefore obtain a Bayes confidence interval from a classical confidence interval in two steps:

> *Bayes confidence interval (for $F > 3$)*
>
> Shrink the estimate by the factor $1 - (1/F)$
>
> Shrink the standard error by the factor $\sqrt{1 - (1/F)}$ (12-14)

*PROBLEMS

12-8 Calculate the Bayes 95% confidence intervals for the data in Problems 12-4 to 12-7.

---

[1]In fact, if we *define* the null hypothesis H_0 as the most credible hypothesis a priori, and make this apropriate substitution in (12-1), then the Bayes estimate is quite generally valid.

CHAPTER 12 SUMMARY

12-1 When we have prior knowledge that a parameter is in the neighborhood of the null hypothesis H_0, how can we modify the sample estimate in this direction? The Bayes estimate gives H_0 a weight of $1/F$, and the sample estimate the remaining weight $1 - (1/F)$. For ANOVA, this shrinks each sample mean toward the grand mean.

12-2 The Bayes estimate involves this same sort of shrinking in other cases too. For example, the regression slope, or the difference in two means.

12-3 Although Bayes estimates are often attractive, there are certain assumptions that have to be satisfied before they should be used.

*12-4 In addition to shrinking estimates by the factor $1 - (1/F)$, Bayes analysis shrinks the confidence allowance by the factor $\sqrt{1 - (1/F)}$.

REVIEW PROBLEMS

12-9 An experiment to compare three treatments was conducted. Eighteen rats were assigned at random, 6 to each of the 3 groups. Their lengths of survival (in weeks) after being injected with a carcinogen were as follows:

	T_1	T_2	T_3
	13	3	10
	9	6	7
	16	6	8
	7	5	12
	11	7	14
	10	9	9
Mean $\overline{X}$	11	6	10
$\Sigma(X - \overline{X})^2$	55	20	34

On a graph, contrast the classical and Bayes estimates of the mean survival time for the 3 treatments.

CHAPTER 13

Nonparametric Statistics

Let all things be done decently and in order.

St. Paul

The classical statistics we have so far considered (like t) require the assumption of population normality. If this assumption is not approximately correct, we should look for techniques that are free of this distribution requirement. Such techniques are called *distribution free* or *nonparametric,* and they are preferred for two reasons:

1. The classical confidence interval may be invalid (i.e., its confidence level may not actually be as high as 95%).
2. But even where the classical confidence interval is reasonably valid, a nonparametric statistic may provide a narrower, more precise confidence interval. In fact, this is the more important reason for using nonparametric statistics.

For nearly every classical test (such as the t test or ANOVA) there are usually several corresponding nonparametric tests; a description of all of these would fill several volumes. Of this great variety, we select two of the simplest and most useful to discuss in this chapter: the sign test, which corresponds to the one-sample t (5-13); and the W test, which corresponds to the two-sample t (5-24).

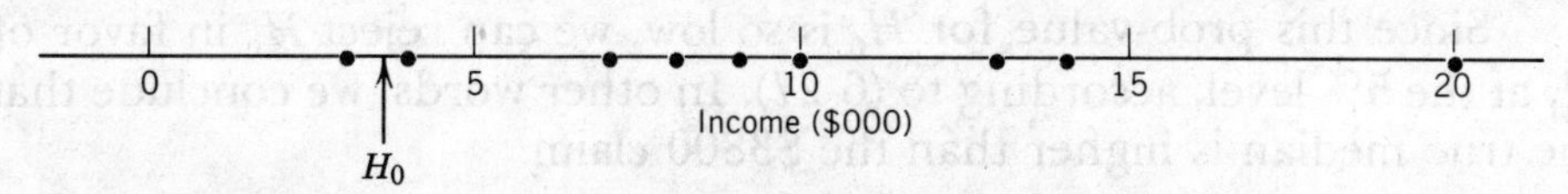

FIGURE 13-1 A sample of 9 ordered observations of income.

13-1 THE SIGN TEST

(a) Single Sample

Suppose the median 1971 family income in the U.S. South was claimed to be $3800. But in a random sample of 9 families, shown in Figure 13-1, 8 have an income above $3800, while only 1 has an income below. Does this evidence allow us to reject the claim?

The null hypothesis may be stated formally as:

$$H_0\text{:population median} = \$3800 \tag{13-1}$$

That is, half the population incomes lie above $3800; or, if an observation is randomly drawn, the probability that it lies above $3800 is

$$H_0\text{:}\pi = 1/2 \tag{13-2}$$

We recognize (13-2) as being just like the hypothesis that a coin is fair. To state it more explicitly, we have two events that are mathematically equivalent:

"Random observation will fall above the median" is equivalent to "A coin will turn up heads"	(13-3)

If H_0 is true, the sample of $n = 9$ observations is just like tossing a coin 9 times. The total number of successes S (families above $3800) will have the binomial distribution, and H_0 may be rejected if S is too far away from its expected value to be reasonably explained by chance.

The alternative hypothesis, of course, is that the true median is higher than the $3800 claim. To judge between these two competing hypotheses, we calculate as usual the prob-value for H_0, that is, the probability (assuming H_0 is true) that we would observe an S of 8 or more. Thus

$$\text{Prob-value} = \Pr(S \geq 8) \tag{13-4}$$

from Appendix Table IIIc,

$$= .0195 \tag{13-5}$$

Since this prob-value for H_0 is so low, we can reject H_0 in favor of H_1 at the 5% level, according to (6-27). In other words, we conclude that the true median is higher than the $3800 claim.

(b) Two Paired Samples

With a little imagination, we can use the sign test for two paired samples [just as we used the t test in (5-28)].

Example 13-1

Suppose a small sample of 8 men had their lung capacity measured before and after a certain treatment; the results are shown in Table 13-1. Use the binomial distribution to calculate the prob-value for the null hypothesis that the treatment is ineffective.

Solution

TABLE 13-1

Lung Capacity of 8 Patients, Before and After Treatment

X (Before)	*Y (After)*	*Difference* $D = Y - X$
750	850	+100
860	880	+ 20
950	930	− 20
830	860	+ 30
750	800	+ 50
680	740	+ 60
720	760	+ 40
810	800	− 10

The original matched pairs can be forgotten, once the differences (improvements) have been found in the third column of Table 13-1. These differences D form a single sample, to which we can apply the sign test. The null hypothesis (that the treatment provides no improvement) is

$$H_0\text{:median, } \nu = 0$$
$$\text{that is, } \pi = \Pr(\text{observing positive } D) = 1/2$$

The question is: are the 6 positive D's (i.e., 6 "heads") observed in a sample of 8 observations ("tosses") consistent with H_0? The probability of this is found in Appendix Table IIIc:

$$\begin{aligned} \text{Prob-value} &= \Pr(S \geq 6) \\ &= .1445 \end{aligned}$$

Since this prob-value is so high, we cannot reject H_0 (that the treatment is ineffective) at the 5% level.

The origin of the term sign test should now be clear. It is a test based on the binomial distribution, where the test statistic is the number of positive differences (heads).

Incidentally, if an observation occurs right on the hypothetical median $D = 0$, it is rather like a coin falling on its edge. It is best to discard it, and not count it as an observation at all.

PROBLEMS

13-1 A random sample of annual incomes (thousands of dollars) of 10 brother-sister pairs was ordered according to the man's income as follows:

Brother's Income (M)	*Sister's Income (W)*
9	14
14	10
16	8
16	14
18	13
19	16
22	12
23	40
25	13
78	24

Calculate the prob-value for the following null hypotheses:

(a) That the male median income is as low as 15.

(b) That the female median income is as high as 15.

(c) That on the whole, men earn no more than women.

13-2

Random Sample of Heights (Inches) of 8 Brother-Sister Pairs

Brother's Height (M)	*Sister's Height (W)*
65	63
67	62
69	64
70	65
71	68
73	66
76	71
77	69

Calculate the prob-value for the following null hypotheses:

(a) That the male median height is as low as 66 inches.

(b) That the female median height is as low as 63 inches.

(c) That on the whole, men are no taller than women.

*13-3 In a sample of 25 bolts from a production line, only 6 were above 10.0 cm. What is the prob-value for the hypothesis that the median length is as high as 10.0 centimeters?

*13-2 CONFIDENCE INTERVALS FOR THE MEDIAN

The sign test of the previous section can be given a nice twist, to provide a confidence interval for the population median. To emphasize how similar this will be to constructing a confidence interval for the population mean μ, we denote the population median by the Greek letter ν, (nu, pronounced "new").

(a) Example

Suppose the claimed median of $3800 in Section 13-1 had been suspected of being simply inaccurate, possibly on the high side as well as the low side. It would then have been appropriate to calculate the two-sided prob-value for $H_0(\nu = \$3800)$. From (13-5), therefore,

$$\text{Prob-value} = 2(.0195) = .039 \tag{13-6}$$

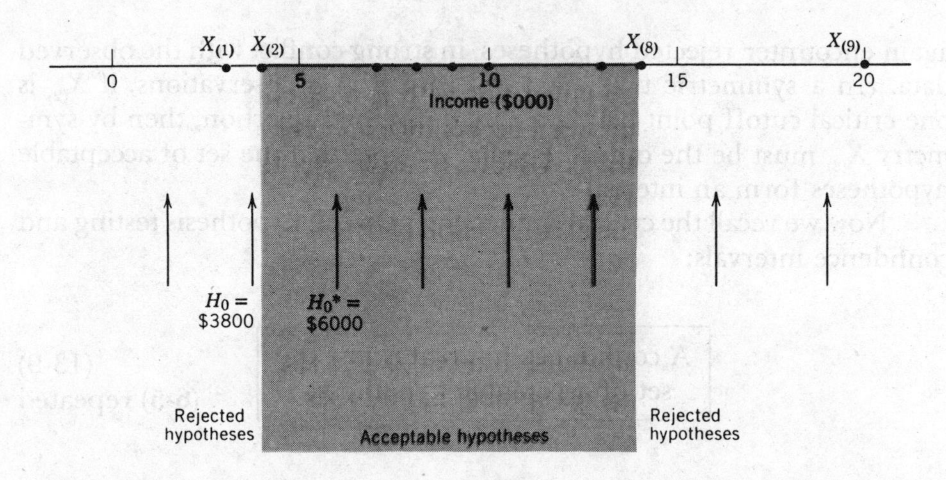

FIGURE 13-2 Nonparametric confidence interval for the population median ν.

Now suppose we wished to make a classical test of $H_0(\nu = \$3800)$ at the level of, say, $\alpha = .040$. According to (6-27), with the prob-value of .039 given in (13-6), we could just barely reject H_0. This is shown in Figure 13-2, which introduces an important convention: *Ordered observations* are denoted by bracketed subscripts. Thus the second smallest observation, for example, is $X_{(2)} = \$4000$. It is evident that every hypothesis below $X_{(2)}$ would be rejected for the same reasons as we rejected $\nu = \$3800$: Its prob-value would be below α. Such rejected hypotheses are shown as black arrows in Figure 13-2.

On the other hand, consider an hypothesis above $X_{(2)} = \$4000$ such as

$$H_0^*: \nu = \$6000 \tag{13-7}$$

Since there are now 7 rather than 8 heads (incomes above this median),

$$\begin{aligned} \text{Prob-value} &= 2 \Pr(S \geq 7) \\ &= 2\,(.0898) = .1796 \end{aligned} \tag{13-8}$$

Since this now exceeds $\alpha = .04$, H_0^* is an acceptable hypothesis. By the same logic, even higher hypotheses such as $\nu = \$8000$ or $\nu = \$11{,}000$ are also acceptable. These acceptable hypotheses are shown as colored arrows in Figure 13-2. Of course, at the very high end of the range, we again encounter rejected hypotheses, in strong conflict with the observed data. (In a symmetric two-tailed test with $n = 9$ observations, if $X_{(2)}$ is one critical cutoff point between acceptance and rejection, then by symmetry $X_{(8)}$ must be the other). Finally, we note that the set of acceptable hypotheses form an interval.

Now we recall the crucial connection between hypothesis testing and confidence intervals:

A confidence interval is just the set of acceptable hypotheses.

(13-9)
(6-5) repeated

Thus in Figure 13-2 we have constructed a confidence interval for the population median, the level of confidence being $1 - \alpha = 96\%$. In algebraic language, the 96% confidence interval for the population median, when $n = 9$, is

$$X_{(2)} \leq \nu \leq X_{(8)} \tag{13-10}$$

that is,

$$4000 \leq \nu \leq 14000 \tag{13-11}$$

(b) Generalization

To generalize, consider a sample of n ordered observations $X_{(1)}, X_{(2)}, \ldots, X_{(n)}$ from a population with unknown median ν. The confidence interval is defined by counting off an equal number of observations from each end,

$$\boxed{X_{(q)} \leq \nu \leq X_{(r)}} \tag{13-12}$$

where q and r are symmetrically chosen. For example, when $n = 9$, q and r could be 1 and 9, or 2 and 8. And in general,

$$q + r = n + 1 \tag{13-13}$$

The problem then is to solve for the level α, which may be obtained from the binomial distribution, using $\pi = 1/2$ in Table IIIc. For S successes in n trials,

$$\boxed{\alpha = 2\,\Pr(S \geq r)} \tag{13-14}$$

Finally, this α allows us to specify the level of confidence $(1 - \alpha)$ for the confidence interval (13-12).

*PROBLEMS

13-4 In Problem 13-1, construct a 98% nonparametric confidence interval for:
(a) Median men's income.
(b) Median women's income.
(c) Median difference in income.

13-5 In Problem 13-2, construct a 93% nonparametric confidence interval for:
(a) Median men's height.
(b) Median women's height.
(c) Median difference in heights.

13-6 (a) Referring to Problem 13-1, do you think that the population of men's incomes is normally distributed?
(b) Do you think the nonparametric confidence interval in Problem 13-4(a) would be more or less precise than the classical confidence interval based on t?
(c) To check your conjecture in (b), go ahead and calculate the 98% confidence interval of the form (5-13).

13-7 (a) Referring to Problem 13-2, do you think the population of men's heights is normally distributed?
(b) Do you think the nonparametric confidence interval in Problem 13-5(a) would be more or less precise than the classical confidence interval based on t?

(c) To check your conjecture in (b), go ahead and calculate the 93% confidence interval of the form (5-13). (*Hint:* More extensive t tables than Table V are available, and give $t_{.035} = 2.138$ for 7 d.f.)

13-8 Write a summary of what you learned from Problems 13-6 and 13-7.

13-3 THE W TEST FOR TWO INDEPENDENT SAMPLES

(a) Small Sample

The Wilcoxon-Mann-Whitney (W-M-W, or W) test is used for two independent samples (in contrast to the sign test, which used matched pairs). The objective again is to detect whether the two underlying populations are centered differently. For example, suppose independent random samples of family income were taken from two different regions of the United States in 1971, and then ordered as in Table 13-2.

TABLE 13-2

Two Samples of Income

South X	*Northeast Y*
4,000	5,000
8,000	10,000
9,000	11,000
14,000	12,000
	15,000
	22,000
Size $m = 4$	$n = 6$

Let us test the null hypothesis that the two underlying populations are identical. Suppose the alternate hypothesis is that the South is poorer than the Northeast, so that the usual one-sided test is appropriate.

We first rank the combined X and Y observations, as shown in Table 13-3.

TABLE 13-3
Combined Ranking Yields the W Statistic

Combined Ordered Observations		*Combined Ranks*	
X	Y	X	Y
4,000		1	
	5,000		2
8,000		3	
9,000		4	
	10,000		5
	11,000		6
	12,000		7
14,000		8	
	15,000		9
	22,000		10
$m = 4$	$n = 6$	$W = 16$	

The actual income levels are now discarded in favor of this ranking. (This provides a test that is not affected by skewness, or any other peculiarity of the population—in other words, a distribution-free test.) Then W is defined as the sum of all the X ranks; in this case,

$$W = 1 + 3 + 4 + 8 = 16$$

For $W = 16$, in Appendix Table VIII we can find the tail probability:

$$\text{Prob-value} = .129 \simeq 13\% \tag{13-15}$$

This provides very weak evidence of a lower family income in the South; we can't reject H_0 at a 5% level.

*(b) Generalization

In general, suppose there are m observations in the smaller sample (call them $X_1, X_2, \ldots, X_m$) and $n \geq m$ observations in the larger sample (call them $Y_1, Y_2, \ldots, Y_n$). Start counting from the end where the X values predominate. This keeps the X ranks low. Adding up all the X ranks yields the rank sum W.

Appendix Table VIII gives the corresponding prob-value, for $n \leq 6$ and prob-value $\leq .25$. For samples larger than those covered by this table, W is approximately normal, with

$$\text{Expected value of } W = \frac{1}{2} m (m + n + 1)$$
$$\text{Standard deviation of } W = \sqrt{\frac{1}{12} m\, n\, (m + n + 1)} \tag{13-16}$$

Thus we can use the normal tables to find the prob-value. Since it is the small values of W that mean the X values are all bunched up at the low end (and hence H_0 is incredible), it is in the left-hand tail that we calculate the tail probability. An example will illustrate.

Example 13-2

In Table 13-4, suppose we have acquired larger samples of income than in Table 13-3. Calculate the prob-value for H_0 in the light of this increased information.

TABLE 13-4

W Statistic, Large-Sample

Combined Ordered Observations		*Combined Ranks*	
South X	*Northeast* Y	*South* X	*Northeast* Y
4,000		1	
	5,000		2
6,000		3	
	7,000		4
8,000		5	
9,000		6	
	10,000		7
	11,000		8
	12,000		9
14,000		10	
	15,000		11
	21,000		12
	22,000		13
$m = 5$	$n = 8$	$W = 25$	

Solution

Since $m = 5$ and $n = 8$ exceeds the capacity of Table VIII, we resort to the normal approximation. Substituting $n = 8$ and $m = 5$ into (13-16),

$$\text{Expected value of } W = 35$$
$$\text{Standard deviation of } W = \sqrt{46.7} = 6.83$$

To calculate the prob-value, we first must standardize the observed value $W = 25$:

$$Z = \frac{W - \mu}{\sigma} \qquad \text{like (3-24)}$$
$$= \frac{25 - 35}{6.83} = -1.46$$

Thus

$$\text{Prob-value} \equiv \Pr(W \leq 25)$$
$$= \Pr(Z \leq -1.46)$$
$$= .0735 \simeq 7\%$$

This provides stronger evidence than in (13-15) of a lower family income in the South. In fact, at the level $\alpha = 10\%$, this result is statistically discernible.

Of course, the approximately normal distribution of W must not be confused with the population of the X and Y values, which may be very nonnormal. In fact the W test, like all nonparametric tests, is specifically designed for very nonnormal populations.

PROBLEMS

13-9 A random sample of 6 men's heights and an independent random sample of 6 women's heights were observed, and ordered as follows:

Men's Heights (M)	*Women's Heights (W)*
65	62
67	63
69	64
70	66
73	68
75	71

Calculate the prob-value for the null hypothesis H_0 that on the whole, men are no taller than women.

13-10 Two makes of cars were randomly sampled, to determine the mileage (in thousands) until the brakes required relining. Calculate the prob-value for the null hypothesis that make *A* is no better than *B*.

Make A	*Make B*
30	22
41	26
48	32
49	39
61	

*13-11 Two samples of children were randomly selected to test two art education programs, A and B. At the end, each child's best painting was collected, to be judged by an independent artist. In terms of creativity, she ranked them as follows:

Rank of Child	1	2	3	4	5	6	7	8	9	10	11	12	13	14	15	16
Art Program	B	A	B	B	B	A	B	B	A	B	A	B	A	A	A	A

Test whether program B (the intensive program) is better than A.

*13-12 Recalculate the prob-value in Problem 13-10:

(a) Using the large sample formula (13-16), just to see how well it works.

(b) Using the large sample formula (13-16), with continuity correction (wcc). Do you think the cc is worthwhile?

CHAPTER 13 SUMMARY

13-1 Nonparametric tests are often quick and convenient, and require no assumption that the population is normal. The simplest is the sign test for two matched samples (or one single sample).

*13-2 Pursuing the logic of the sign test, a nonparametric confidence interval can be constructed for the population median. We simply count off an appropriate number of observations from each end of the sample.

13-3 Another very useful nonparametric test is the W test for two independent samples, which is based on combined ranks.

REVIEW PROBLEMS

13-13 A random sample of 8 pairs of identical twins was used in a certain study. In each pair, one twin was chosen at random for an enriched education program E, and the other twin was given the standard program S. At the end of the year, the performance scores were as follows:

Twin's Surname	*Jones*	*Chan*	*Ryan*	*Mario*	*Strom*	*Lee*	*Roy*	*Stein*
S group	57	91	68	75	82	47	63	73
E group	64	93	72	71	90	52	79	81

Calculate the prob-value for the null hypothesis that the enriched program is no better:

(a) Assuming the scores are normally distributed (t test).
(b) Assuming the scores are quite nonnormal (nonparametric test).

13-14 Suppose the 16 numbers in Problem 13-13 had been obtained differently, from 16 unmatched students. Now answer the same questions.

CHAPTER 14

Chi-Square Tests

It is better to know nothing than to know what ain't so.

H. W. Shaw

14-1 χ^2 TEST FOR GOODNESS OF FIT

Because it is so easy to understand, chi-square (χ^2) is a very popular form of hypothesis testing. It provides a simple test based on the difference between observed and expected frequencies. Let us illustrate with an extended example.

Suppose that we wish to test the null hypothesis that births in Sweden occur equally often throughout the year. And suppose the only available data is a random sample of 88 births distributed over the year, grouped into seasons of differing length. These observed frequencies O are given in column (2) of Table 14-1.

How well does the data fit the null hypothesis? The notion of goodness of fit may be developed in several steps:

1. First consider the implications of H_0, the null hypothesis that every birth is likely to occur in any given season with a probability proportional to the length of that season. Spring is defined to have 3 months, or 91 days; thus the probability (given H_0) of a

TABLE 14-1
Distribution of n = 88 Births Among 4 Cells
(Seasons of the Year)

Given		*χ^2 calculations*			
(1) *Season*	*(2)* *Observed Frequency* O	*(3)* *Probability (if H_0 true)* π	*(4)* *Expected Frequency* $E = n\pi$	*(5)* *Deviation* $(O - E)$	*(6)* *Deviation Squared and Weighted* $(O - E)^2/E$
Spring Apr–June	27	91/365 = .25	.25(88) = 22.0	27 − 22 = +5.0	25/22 = 1.14
Summer July–Aug	20	.17	15.0	+5.0	1.67
Fall Sept–Oct	8	.167	14.7	−6.7	3.05
Winter Nov–Mar	33	.413	36.3	−3.3	.30
	n = 88	1.00✓	88✓	0✓	χ^2 = 6.16 Prob-value < .25

birth occurring in the spring is 91/365 = .25. Similarly, all the probabilities π are calculated in column (3) of Table 14-1.

2. Now calculate what the expected frequencies in each cell (season) would be if the null hypothesis were true. For example, consider the first cell; if H_0 were true, $\pi = .25 = 25\%$, so that the expected frequency would be 25% of 88 = 22 births. Similarly, the other expected frequencies E are calculated in column (4):

$$\boxed{E = n\pi} \tag{14-1}$$

3. The question now is: By how much does the observed frequency deviate from the expected frequency? For example, in the first cell, this deviation is 27 − 22 = 5. Similarly, all the deviations $(O - E)$ are set out in column (5).
4. The deviations always would sum algebraically to 0, which is not a useful criterion. This problem is avoided by taking the square[1]

---

[1]This squaring of the deviations is a familiar strategy: It also occurred in the least squares criterion, and as far back as the MSD.

of the deviation $(O - E)^2$. Then, to show its relative importance, each squared deviation somehow must be compared with the expected frequency in its cell, E. Therefore, we calculate the ratio $(O - E)^2/E$, as shown in column (6). Finally, we sum the contributions from all cells to obtain an overall measure of deviation, called chi-square:

$$\boxed{\chi^2 = \sum \frac{(O - E)^2}{E}} \tag{14-2}$$

$$= 6.16 \tag{14-3}$$

This whole argument obviously has strong similarities to the one that we used in developing ANOVA; specifically, the χ^2 statistic is very similar to the F statistic (10-6) which also measured discrepancy from a null hypothesis. Hence, the χ^2 statistic may be analyzed in a similar way.

First, χ^2 is a random variable that fluctuates from sample to sample; this fluctuation in χ^2 occurs because of the fluctuation in the observed cell frequencies O that comprise it. In passing, note that these frequencies are not independent: since $O_1 + O_2 + O_3 + O_4 = n$, any one of them may be expressed in terms of the others. For example, $O_4 = n - (O_1 + O_2 + O_3)$; thus, the last cell is determined by the previous three, and does not provide fresh information. Therefore, in this case, χ^2 has only 3 d.f.; and in general, with k cells,

$$\boxed{\text{d.f.} = k - 1} \tag{14-4}$$

If H_0 is true, the χ^2 distribution has the critical points ($\chi^2_{.25}$, $\chi^2_{.10}$, etc.) given in Appendix Table VII (just like the critical points of the t distribution in Table V). Since d.f. = 3, we scan along the third row of Table VII, and find that the observed χ^2 value of 6.16 lies beyond $\chi^2_{.25} = 4.11$. Thus

$$\text{Prob-value} < .25 \tag{14-5}$$

Since this prob-value is so large, the data has indicated that the null hypothesis H_0 (births evenly distributed) is fairly credible.

When all the cells are equally likely, the calculation of χ^2 is simpler, as the next example illustrates.

Example 14-1

A die was cast 1000 times, with the following results:

Face	Relative Frequency
1	.183
2	.161
3	.142
4	.174
5	.181
6	.159

(a) Find the prob-value for H_0 (the die is fair).
(b) Can you reject H_0 at the 10% level?

Solution

(a) To use χ^2, we first must express the data as observed frequencies O by multiplying the relative frequencies by $n = 1000$. Then we calculate the expected frequencies E by stating the null hypothesis:

$$H_0 : \pi_1 = \pi_2 = \cdots = \pi_6 = \frac{1}{6}$$

Thus: $$E = n\pi = 1000\left(\frac{1}{6}\right) \simeq 167$$

Face	O	E	$(O - E)$	$(O - E)^2$	$(O - E)^2/E$
1	183	167	+16	256	1.53
2	161	167	− 6	36	.22
3	142	167	−25	625	3.74
4	174	167	+ 7	49	.29
5	181	167	+14	196	1.17
6	159	167	− 8	64	.38

$\chi^2 = 7.33$

We refer to Table VII with d.f. $= k - 1 = 6 - 1 = 5$. We scan along the fifth row and find that the observed χ^2 value of 7.33 lies beyond $\chi^2_{.25} = 6.63$. Thus

$$\text{Prob-value} < .25 \qquad (14\text{-}6)$$

(b) According to (6-27), since prob-value $> .10$, we cannot reject H_0. That is, the die may well be fair, and the differences in the given relative frequencies may just be due to chance.

These examples have shown how widely the χ^2 test can be applied to data counted off in several cells (such as seasons). This χ^2 test is sometimes called the *multinomial* test to indicate that it is an extension of the *binomial,* where data was counted off in two cells (such as male and female, or Democrat and Republican.)

PROBLEMS

14-1

Period of Day	*Number of Accidents*
8–10 A.M.	17
10–12 A.M.	15
1–3 P.M.	20
3–5 P.M.	28

This table classifies the industrial accidents in an auto plant into 4 time periods.

(a) Find the prob-value for H_0 (that accidents are equally likely to occur at any time of day).

(b) Can you reject H_0 at the 5% error level?

14-2 According to the Mendelian genetic model, a certain garden pea plant should produce offspring that have white, pink, and red flowers, in the long-run proportion 25%, 50%, 25%. A sample of 1000 such offspring were colored as follows:

white, 21%; pink, 52%; red, 27%

(a) Find the prob-value for the Mendelian hypothesis.

(b) Can you reject the Mendelian hypothesis at the 5% level?

14-3 Throw a fair die 30 times (or simulate it with random digits in Table I).
(a) Use χ^2 to calculate the prob-value for H_0 (that it is a fair die).
(b) Do you reject H_0 at the 25% level?
(c) If each student in a large class carries out this test, approximately what proportion will reject H_0?

14-4 Repeat Problem 14-3 for an unfair die. (Since you do not have an unfair die available, use the table of random digits to simulate a die that is biased toward aces; for example, let the digit 0 as well as 1 represent the ace, so that the ace has twice the probability of any other face.)

*14-5 Is there a better test than χ^2 for the die in Problem 14-4 that is suspected of being biased toward aces? If so, use it to recalculate Problem 14-4.

14-2 CONTINGENCY TABLES

(a) Example

Contingency means dependence, so a contingency table is simply a table that displays how two characteristics depend on each other; for example, Table 14-2 shows the dependence of income on region in a sample of 400 U.S. families in 1971. To test the null hypothesis of no dependence in the underlying population, χ^2 again may be used in a goodness-of-fit test. In these special circumstances, it is customarily renamed the χ^2 *test of independence.*

TABLE 14-2
Observed Frequencies O for 400 U.S. Families Classified by Region and Income, 1971

Income ($000) \ *Region*	*South*	*North*	*Total Frequency*
0–5	28	44	72
5–10	42	78	120
10–15	30	78	108
15–over	24	76	100
Total frequency	124	276	400
Relative frequency	.31	.69	1.00

Statistical independence means that the pattern of income should be the same in each region. For example, Table 14-3 displays this independence for a small hypothetical country of population 6000, one-third of which is in the South, two-thirds in the North. You can easily verify that in both regions the same relative frequency distribution occurs: 15%, 35%, 25%, and 25% in the four income cells, totaling 100%.

TABLE 14-3

Population Frequencies for a Small Hypothetical Country Where Income and Region Are Independent (Null Hypothesis H_0)

Region / *Income ($000)*	*South*	*North*	*Total Frequency*
0–5	300	600	900
5–10	700	1400	2100
10–15	500	1000	1500
15–over	500	1000	1500
Total frequency	2000	4000	6000
Relative frequency	1/3	2/3	

Equivalently, independence means that the frequencies in each column (region) are just a proportion of the last column (the whole country). For example, since the South has 1/3 of the population, its frequencies are just 1/3 of the country's frequencies.

If a sample perfectly reflected this independence, we would observe the same pattern: Each column would be a proportion of the last column (the total). Let us see what this would mean for the given sample in Table 14-2. To obtain the South column, we would multiply the Total column by .31 (since 31% of the observations are in the South.) Similarly, to obtain the North column, we would multiply the Total column by .69. We thus obtain the expected frequencies E shown in Table 14-4. Then we can compare them to the observed frequencies O using the χ^2 statistic.

TABLE 14-4

Expected Frequencies E for the Sample in Table 14-2, if It Were from a Population Where Independence Occurs (H_0)

Income ($000) \ *Region*	*South*	*North*
0–5	22.3	49.7
5–10	37.2	82.8
10–15	33.5	74.5
15–over	31.0	69.0

Let us state in general terms what we do to test the null hypothesis of independence:

1. In the given table of observed frequencies, calculate the total frequencies in both margins—at the bottom and at the right. In the bottom margin, go on to calculate the *relative* frequencies.
2. Multiply the right-hand total column by each relative frequency in turn, to generate the columns of expected frequencies E.
3. Finally, compare the expected frequencies E with the observed frequencies O using the χ^2 statistic:

$$\chi^2 \equiv \sum\sum \frac{(O - E)^2}{E} \qquad (14\text{-}7)$$

We have written the Σ sign twice, to indicate that we have summed over the whole table. If we let c denote the number of columns in the table, and r the number of rows, the degrees of freedom are

$$\text{d.f.} = (c - 1)(r - 1) \qquad (14\text{-}8)$$

Then χ^2 is referred to Table VII to find the prob-value for the null hypothesis (independence).

Let us finish these χ^2 calculations for our example of regional income. The observed and expected frequencies of Tables 14-2 and 14-4 are

TABLE 14-5
χ^2 Calculations

(1) *Observed Frequencies O*		(2) *Expected Frequencies E*	
28	44	22.3	49.7
42	78	37.2	82.8
30	78	33.5	74.5
24	76	31.0	69.0

(3) *Deviations (0 − E)*		(4) *(0 − E)²/E*	
5.7	−5.7	1.46	.65
4.8	−4.8	.62	.28
−3.5	3.5	.37	.16
−7.0	7.0	1.58	.71

Sum, $\chi^2 = 5.83$

briefly laid out in Table 14-5, where the χ^2 calculations are completed, yielding

$$\chi^2 = 5.83 \tag{14-9}$$

From (14-8),

$$\text{d.f.} = (2 - 1)(4 - 1) = 3$$

We scan along the third row of Table VII, and find that the observed χ^2 value of 5.83 lies beyond $\chi^2_{.25} = 4.11$ Thus

$$\text{Prob-value} < .25 \tag{14-10}$$

This prob-value is so high that we cannot reject H_0 at the customary 5% level (or even at the 10% level). That is, this test fails to establish any dependence of income on region in the U.S. population as a whole.

(b) Alternative to χ^2: A Confidence Interval Once Again

Since the χ^2 test does not exploit the numerical nature of income, it misses the essential question: *How much* do incomes differ between regions? Even the secondary question of testing may be ineffectively answered by χ^2.

To remedy these shortcomings, the data in Table 14-2 can be reworked. Since income is numerical, we can calculate the mean income in the North ($\overline{X}_1$), and the South ($\overline{X}_2$). Then, as detailed in Problem 14-11, we can find the 95% confidence interval for the difference in the population mean incomes:

$$(\mu_1 - \mu_2) = (\overline{X}_1 - \overline{X}_2) \pm t_{.025}\, s_p \sqrt{\frac{1}{n_1} + \frac{1}{n_2}} \qquad \text{like (5-24)}$$

$$= (10.87 - 9.52) \pm 1.96\sqrt{27.3}\sqrt{\frac{1}{276} + \frac{1}{124}}$$

$$= 1.35 \pm 1.11 \text{ thousand dollars} \qquad (14\text{-}11)$$

That is, the North has a mean income that is \$1350 ± \$1110 higher than the South.

The secondary question of testing H_0 (no difference between regions) now can be answered immediately: At the 5% level, H_0 now can be rejected, since 0 does not lie in the confidence interval (14-11). That is, there is a discernible difference between the two regions. This is a much stronger conclusion than we obtained from the χ^2 test following (14-10), where we failed to find any discernible difference between the two regions.

(c) General Alternatives to χ^2 Tests of Independence

The lesson to be drawn from the preceding example is clear:

> Whenever *numerical* variables appear, they should be analyzed with a tool (such as the 2-sample *t*, ANOVA, or regression) that exploits their numerical nature. A χ^2 test is more appropriate for *categorical* variables.

In fact, even if a variable is not naturally numerical but merely ordered (e.g., a variable such as social class, or degree of success), it is often a good strategy to code the various levels of the variable by 0, 1, 2, 3, . . . , and then proceed with this new numerical variable. Although this may seem arbitrary, it usually will yield a more powerful test of H_0 than χ^2, and also will give at least a rough-and-ready answer to the question, "*How much* do things differ?"

PROBLEMS

In each of the following problems, a random sample of several hundred Americans was classified according to two characteristics.

14-6 Newspaper Read by Social Class

Social Class \ Newspaper	A	B	C
Poor	15	20	13
Lower middle class	27	27	18
Middle class	44	26	14
Rich	22	11	3

(a) State H_0 in words.

(b) Calculate χ^2 and the prob-value for H_0. Is the difference between newspapers discernible at the .01 level?

14-7 Suppose the table in Problem 14-6 had its rows and columns interchanged:

Newspaper \ Social Class	Poor	Lower Middle	Middle Class	Rich
A	15	27	44	22
B	20	27	26	11
C	13	18	14	3

This contains exactly the same information, of course (e.g., both tables show that 13 readers were poor and read newspaper *C*). So it should provide exactly the same value of χ^2. Carry out the χ^2 calculations to show this is indeed so.

14-8 Income, by Sex, 1964

Income ($) \ Sex	Male	Female
Less than 5000	53	89
More than 5000	47	11

(a) State H_0 in words.

(b) Calculate χ^2 and the prob-value for H_0.

14-9 Educational Attainment, by Color, 1964

Color \ *Education*	*Elementary School Attendance*	*High School Attendance*	*University Attendance*
White	127	232	91
Black	23	23	4

(a) State H_0 in words.
(b) Calculate χ^2 and the prob-value for H_0.

14-10 Employment in Various Occupations, by Color, 1965

Color \ *Occupation*	*White Collar*	*Blue Collar*	*Household and Other Services*	*Farm Workers*
White	281	182	37	41
Black	15	25	12	7

(a) State H_0 in words.
(b) Calculate χ^2 and the prob-value for H_0.
(c) Analyze more graphically: For each occupation, calculate the proportion of blacks, including a confidence interval. Graph.

*14-11 Use the following steps to verify (14-11):
(a) In Table 14-2, approximate the incomes from 0 to 5 by the cell midpoint 2.5. Continue for the other cells. (In the last cell, which is open-ended and has no midpoint, use 17.5 as a rough approximation to the typical income within the cell.)
(b) For each region, South and North, calculate $\overline{X}$ and the squared deviations $\Sigma(X - \overline{X})^2 f$.
(c) Substitute into equation (5-25) to find s_p^2. Finally substitute into (5-24) to find $\mu_1 - \mu_2$.

*14-12 (a) Analyze Problem 14-6 in a way that exploits the ordered nature of social class: Since the four social classes are ordered from poor to rich, a reasonable strategy is to number them 1, 2, 3, 4 (call it a *social class score* if you like). With the 240 people all having their social class transformed into a numerical score, it is now possible to calculate the mean score for newspaper A, and compare it to the mean scores for newspapers B and C using ANOVA.

(b) Did you find the difference between newspapers discernible at the .01 level? Is this ANOVA test better than χ^2?

CHAPTER 14 SUMMARY

14-1 χ^2 is an hypothesis test based on the difference between observed values and expected values (expected under the null hypothesis). In the simplest case, χ^2 can be applied to data sorted into several cells, according to a single factor such as season.

14-2 χ^2 can also be applied to data sorted according to two factors such as income and region, to test their independence of each other. Although χ^2 tests are easy to calculate, we can often find more graphic alternatives such as confidence intervals.

REVIEW PROBLEMS

14-13 A random sample of 10,000 American babies born in 1960 gave the following breadown by sex:

Males	5120
Females	4880

(a) Using the χ^2 test, calculate the prob-value for the null hypothesis that boys and girls are equally likely.

(b) Calculate the 95% confidence interval for the probability of a boy.

(c) Which do you think is the more meaningful analysis, (a) or (b)?

14-14 A random sample of 10,000 American babies born in 1960, and another in 1970, gave the following breakdown by sex:

Sex \ Year	1960	1970
Male	5060	5180
Female	4940	4820
Totals	10,000	10,000

(a) Using the χ^2 test, calculate the prob-value for the null hypothesis that the probability of a boy stays constant from 1960 to 1970.
(b) Calculate the 95% confidence interval for the change in the probability of a boy.
(c) Which do you think is the more meaningful analysis, (a) or (b)?

14-15 Repeat Problem 14-14 for the following data:

Total U.S. Births, by Year, and Sex

Sex \ Year	*1960*	*1970*
Male	2,180,000	1,907,000
Female	2,078,000	1,811,000
Totals	4,258,000	3,718,000

14-16 A random sample of 1250 university degrees earned in 1976 gave the following breakdown:

Sex \ Degree	*Bachelors*	*Masters*	*Doctorate*
Male	501	162	27
Female	409	143	8

(a) State the null hypothesis in words.
(b) Calculate the χ^2 prob-value for H_0.
(c) For a more graphic alternative to χ^2, calculate the proportion of women in each of the 3 different degree categories. Include confidence intervals, and a graph.

APPENDIX A

Monte Carlo Using Normal Random Numbers

If the population is approximately normal, (for example, men's heights), then sampling can be speeded up if we bypass the serial numbering of the individuals in the population. Instead, we reverse the standardized normal transformation by rewriting (3-24) as:

$$X = \mu + \sigma Z \qquad \text{(A-1)}$$

Recall that Z has a normally shaped distribution, with mean 0 and standard deviation 1. Then X also has a normally shaped distribution, but with mean μ and standard deviation σ. To use a specific example, suppose

$$X = 69 + 3.2Z \qquad \text{(A-2)}$$

Because the fluctuating Z is multiplied by 3.2, X fluctuates more. In fact, the standard deviation of X is exactly 3.2 times larger than the standard deviation of Z (which was 1). In addition, because we add on 69, the mean of X is 69 more than the mean of Z (which was 0). Thus X has mean 69 and standard deviation 3.2. That is, it may be regarded as drawn from the population of student heights in Example 4-1.

Table II gives a list of random observations from the standard normal distribution of Z. By plugging these into (A-2) we can generate random observations from the normal distribution of X. For example, starting from the first entry in Table II, since $Z = .5$,

$$X = 69 + 3.2(.5) = 70.6 \tag{A-3}$$

Continuing in this way, we can simulate a complete sample of 4 observations of students' heights:

Z	X
.5	70.6
.1	69.3
2.5	77.0
−.3	68.0
	$\bar{X} = 71.2$

Example A-1

Let everyone in the class perform the above sampling experiment (i.e., simulating a random sample of $n = 4$ observations from a normal population with $\mu = 69$ and $\sigma = 3.2$). Let the instructor graph the frequency distribution of all the values of $\bar{X}$. Is it approximately the same sampling distribution as in Example 4-1?

Solution

The sampling distribution of $\bar{X}$ will indeed be similar to the distribution generated in Example 4-1. We have simply done the same problem in two different ways.

In practice, simulation is usually done on a computer, which can do repetitive calculations like (A-3) in a split second. The computer can draw a sample and calculate its mean, then repeat this procedure over and over, very quickly. It takes only a few seconds, therefore, to build up the sampling distribution of $\bar{X}$.

Problems

A-1 (a) In Problem 4-3, graph the given distribution of demands $p(X)$.

(b) Graph the normal distribution with mean $\mu = 31$ and standard deviation $\sigma = 6$; note that this is an excellent approximation to the distribution in (a).

(c) Using the normal distribution in (b), and the normal random numbers of Table II, redo Problem 4-3.

A-2 In a population of working couples, the man's weekly income X and the woman's weekly income Y are statistically independent and normally distributed. X has $\mu = 250$ and $\sigma = 50$, while Y has $\mu = 200$ and $\sigma = 60$. Five couples are sampled randomly in a sociological survey. Simulate this experiment, and calculate:

(a) The average male income.

(b) The average female income.

(c) By how much the husband's income exceeds his wife's, on average.

APPENDIX B

Lines

The definitive characteristic of a straight line is that it continues forever in the same constant direction. In Figure B-1 we make this idea precise. In moving from one point P_1 to another point P_2, we denote the horizontal distance by ΔX (where Δ is the Greek letter D, for difference) and the vertical distance by ΔY. Then the slope is defined as

$$\text{Slope} \equiv \frac{\Delta Y}{\Delta X} \tag{B-1}$$

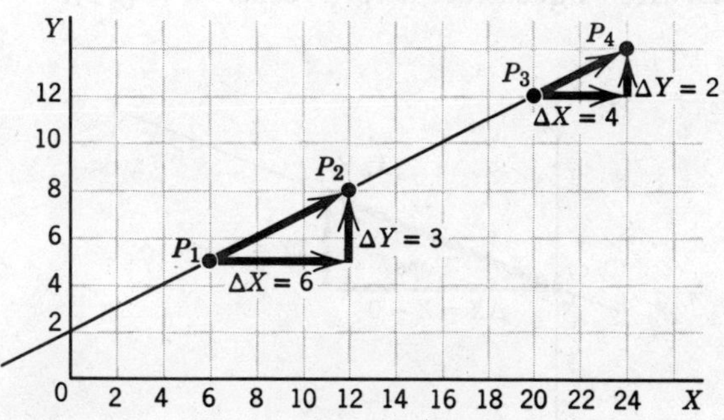

FIGURE B-1 A straight line is characterized by constant slope, $\Delta Y / \Delta X = b$.

Slope is a concept useful in engineering as well as mathematics. For example, if a highway rises 12 feet over a distance of 200 feet, its slope is 12/200 = 6%. The characteristic of a *straight* line is that the slope remains the same everywhere:

$$\boxed{\frac{\Delta Y}{\Delta X} = b \text{ (a constant)}} \tag{B-2}$$

For example, the slope between P_3 and P_4 is the same as between P_1 and P_2, as calculation will verify:

$$P_1 \text{ to } P_2\text{:} \ \frac{\Delta Y}{\Delta X} = \frac{3}{6} = .50 \tag{B-3}$$

$$P_3 \text{ to } P_4\text{:} \ \frac{\Delta Y}{\Delta X} = \frac{2}{4} = .50$$

A very instructive case occurs when X increases just one unit; then (B-2) yields

$$\boxed{\text{When } \Delta X = 1, \ \Delta Y = b} \tag{B-4}$$

In words, "b is the increase in Y that accompanies a unit increase in X," which agrees with the regression interpretation (8-29).

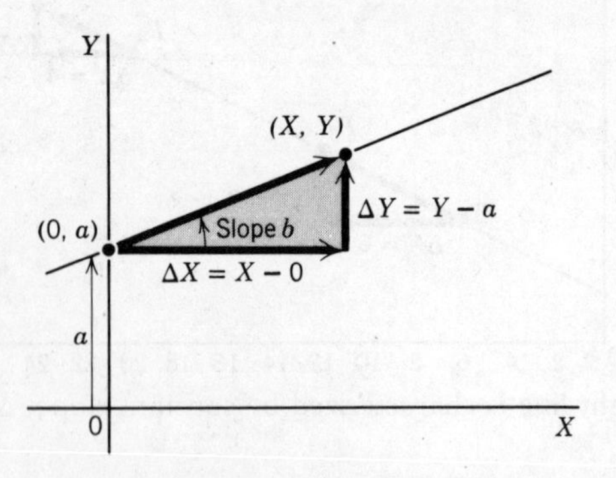

FIGURE B-2 Derivation of the equation of a straight line.

It is now very easy to derive the equation of a line, if we know its slope b and any one point on the line. Suppose the one point we know is the Y intercept—the point where the line crosses the Y axis, at height a, let us say. As Figure B-2 shows, in moving to any other point (X, Y) on the line, we may write

$$\text{Slope, } \frac{\Delta Y}{\Delta X} = \frac{Y - a}{X - 0} \tag{B-5}$$

For the line to be straight, we insisted in (B-2) that this slope must equal the constant b:

$$\frac{Y - a}{X - 0} = b$$

$$Y - a = bX$$

$$\boxed{Y = a + bX} \tag{B-6}$$

(7-1) proved

APPENDIX C

Solution of a Set of Simultaneous Linear Equations

To illustrate, let us take the pair of estimating equations in Table 8-2:

$$16{,}500 = 280{,}000b + 7000c \tag{C-1}$$

$$600 = 7000b + 400c \tag{C-2}$$

To solve a set of simultaneous equations, we eliminate one unknown at a time. To eliminate the first unknown b, we match its coefficient in (C-2) with (C-1), by multiplying (C-2) by 40:

$$24{,}000 = 280{,}000b + 16{,}000c \tag{C-3}$$

When (C-1) is subtracted from (C-3), b is indeed eliminated:

$$7500 = 9000c$$

$$c = \frac{7500}{9000} = .833 \tag{C-4}$$

With c solved, we can use it to solve for the other unknown b. We substitute (C-4) back into (C-1):

$$16{,}500 = 280{,}000b + 7000(.833)$$

$$b = \frac{10{,}667}{280{,}000} = .0381 \tag{C-5}$$

Thus (C-4) and (C-5) are the required solution.

APPENDIX D

Tables

I Random Digits ... 341
II Random Normal Numbers ... 342
III (a) Binomial Coefficients ... 343
(b) Binomial Probabilities, Individual ... 344
(c) Binomial Probabilities, Cumulative ... 346
IV Standard Normal Probabilities, Cumulative ... 348
V t Critical Points ... 349
VI F Critical Points ... 350
VII χ^2 Critical Points ... 352
VIII Wilcoxon-Mann-Whitney Two-Sample Test ... 353

TABLE I

Random Digits
(Blocked merely for convenience)

39 65 76 45 45	19 90 69 64 61	20 26 36 31 62	58 24 97 14 97	95 06 70 99 00
73 71 23 70 90	65 97 60 12 11	31 56 34 19 19	47 83 75 51 33	30 62 38 20 46
72 20 47 33 84	51 67 47 97 19	98 40 07 17 66	23 05 09 51 80	59 78 11 52 49
75 17 25 69 17	17 95 21 78 58	24 33 45 77 48	69 81 84 09 29	93 22 70 45 80
37 48 79 88 74	63 52 06 34 30	01 31 60 10 27	35 07 79 71 53	28 99 52 01 41
02 89 08 16 94	85 53 83 29 95	56 27 09 24 43	21 78 55 09 82	72 61 88 73 61
87 18 15 70 07	37 79 49 12 38	48 13 93 55 96	41 92 45 71 51	09 18 25 58 94
98 83 71 70 15	89 09 39 59 24	00 06 41 41 20	14 36 59 25 47	54 45 17 24 89
10 08 58 07 04	76 62 16 48 68	58 76 17 14 86	59 53 11 52 21	66 04 18 72 87
47 90 56 37 31	71 82 13 50 41	27 55 10 24 92	28 04 67 53 44	95 23 00 84 47
93 05 31 03 07	34 18 04 52 35	74 13 39 35 22	68 95 23 92 35	36 63 70 35 33
21 89 11 47 99	11 20 99 45 18	76 51 94 84 86	13 79 93 37 55	98 16 04 41 67
95 18 94 06 97	27 37 83 28 71	79 57 95 13 91	09 61 87 25 21	56 20 11 32 44
97 08 31 55 73	10 65 81 92 59	77 31 61 95 46	20 44 90 32 64	26 99 76 75 63
69 26 88 86 13	59 71 74 17 32	48 38 75 93 29	73 37 32 04 05	60 82 29 20 25
41 47 10 25 03	87 63 93 95 17	81 83 83 04 49	77 45 85 50 51	79 88 01 97 30
91 94 14 63 62	08 61 74 51 69	92 79 43 89 79	29 18 94 51 23	14 85 11 47 23
80 06 54 18 47	08 52 85 08 40	48 40 35 94 22	72 65 71 08 86	50 03 42 99 36
67 72 77 63 99	89 85 84 46 06	64 71 06 21 66	89 37 20 70 01	61 65 70 22 12
59 40 24 13 75	42 29 72 23 19	06 94 76 10 08	81 30 15 39 14	81 83 17 16 33
63 62 06 34 41	79 53 36 02 95	94 61 09 43 62	20 21 14 68 86	94 95 48 46 45
78 47 23 53 90	79 93 96 38 63	34 85 52 05 09	85 43 01 72 73	14 93 87 81 40
87 68 62 15 43	97 48 72 66 48	53 16 71 13 81	59 97 50 99 52	24 62 20 42 31
47 60 92 10 77	26 97 05 73 51	88 46 38 03 58	72 68 49 29 31	75 70 16 08 24
56 88 87 59 41	06 87 37 78 48	65 88 69 58 39	88 02 84 27 83	85 81 56 39 38
22 17 68 65 84	87 02 22 57 51	68 69 80 95 44	11 29 01 95 80	49 34 35 86 47
19 36 27 59 46	39 77 32 77 09	79 57 92 36 59	89 74 39 82 15	08 58 94 34 74
16 77 23 02 77	28 06 24 25 93	22 45 44 84 11	87 80 61 65 31	09 71 91 74 25
78 43 76 71 61	97 67 63 99 61	80 45 67 93 82	59 73 19 85 23	53 33 65 97 21
03 28 28 26 08	69 30 16 09 05	53 58 47 70 93	66 56 45 65 79	45 56 20 19 47
04 31 17 21 56	33 73 99 19 87	26 72 39 27 67	53 77 57 68 93	60 61 97 22 61
61 06 98 03 91	87 14 77 43 96	43 00 65 98 50	45 60 33 01 07	98 99 46 50 47
23 68 35 26 00	99 53 93 61 28	52 70 05 48 34	56 65 05 61 86	90 92 10 70 80
15 39 25 70 99	93 86 52 77 65	15 33 59 05 28	22 87 26 07 47	86 96 98 29 06
58 71 96 30 24	18 46 23 34 27	85 13 99 24 44	49 18 09 79 49	74 16 32 23 02
93 22 53 64 39	07 10 63 76 35	87 03 04 79 88	08 13 13 85 51	55 34 57 72 69
78 76 58 54 74	92 38 70 96 92	52 06 79 79 45	82 63 18 27 44	69 66 92 19 09
61 81 31 96 82	00 57 25 60 59	46 72 60 18 77	55 66 12 62 11	08 99 55 64 57
42 88 07 10 05	24 98 65 63 21	47 21 61 88 32	27 80 30 21 60	10 92 35 36 12
77 94 30 05 39	28 10 99 00 27	12 73 73 99 12	49 99 57 94 82	96 88 57 17 91

TABLE II

Random Normal Numbers, $\mu = 0$, $\sigma = 1$
(Rounded to 1 Decimal Place)

.5	.1	2.5	−.3	−.1	.3	−.3	1.3	.2	−1.0
.1	−2.5	−.5	−.2	.5	−1.6	.2	−1.2	.0	.5
1.5	−.4	−.6	.7	.9	1.4	.8	−1.0	−.9	−1.9
1.0	−.5	1.3	3.5	.6	−1.9	.2	1.2	−.5	−.3
1.4	−.6	.0	.3	2.9	2.0	−.3	.4	.4	.0
.9	−.5	−.5	.6	.9	−.9	1.6	.2	−1.9	.4
1.2	−1.1	.0	.8	1.0	.7	1.1	−.6	.3	−.7
−1.5	−.5	−.2	−.1	1.0	.2	.4	.7	−.4	−.4
−.7	.8	−1.6	−.3	−.5	−2.1	−.5	−.2	.9	−.5
1.4	.2	.4	.8	.2	−.7	1.0	−1.5	−.3	.1
−.5	1.7	−.1	−1.2	−.5	.9	−.5	−2.0	−2.8	.2
−1.4	−.2	1.4	−.6	−.3	−.2	.2	.8	1.0	−.9
−1.0	.6	−.9	1.6	.1	.4	−.2	.3	−1.0	−1.0
.0	−.9	.0	−.7	1.1	−.1	1.1	.5	−1.7	.4
1.4	−1.2	−.9	1.2	−.2	−.2	1.2	−2.6	−.6	.1
−1.8	−.3	1.2	1.0	−.5	−1.6	−.1	.4	−.6	.6
−.1	−.4	−1.4	.4	−1.0	−.1	−1.7	−2.8	−1.1	−2.4
−1.3	1.8	−1.0	.4	1.0	−1.1	−1.0	.4	−1.7	2.0
1.0	.5	.7	1.4	1.0	−1.3	1.6	−1.0	.5	−.3
.3	−2.1	.7	−.9	−1.1	−1.4	1.0	.1	−.6	.9
−1.8	−2.0	−1.6	.5	.2	−.2	.0	.0	.5	−1.0
−1.2	1.2	1.1	.9	1.3	−.2	.2	−.4	−.3	.5
.7	−1.1	1.2	−1.2	−.9	.4	.3	−.9	.6	1.7
−.4	.4	−1.9	.9	−.2	.6	.9	−.4	−.2	−.1
−1.4	−.2	.4	−.6	−.6	.2	−.3	.5	.7	−.3
.2	.2	−1.1	−.2	−.3	1.2	1.1	.0	−2.0	−.6
.2	.3	.3	.1	−2.8	−.4	−.8	−1.3	−.6	−1.0
2.3	.6	.6	−.7	.2	1.3	.1	−1.8	−.7	−1.3
.0	−.3	.1	.8	−.6	.5	.5	−1.0	.5	1.0
−1.1	−2.1	.9	.1	.4	−1.7	1.0	−1.4	−.6	−1.0
.8	.1	−1.5	.0	−2.1	.7	.1	−.9	−.6	.6
.4	−1.7	−.9	.2	−.7	.3	−.1	−.2	−.1	.4
−.5	−.3	.2	−.7	1.0	.0	.4	−.8	.2	.1
.3	−.5	1.3	−1.2	−.9	.1	−.5	−.8	.0	.5
1.0	3.0	−.6	−.5	−1.1	1.3	−1.4	−1.3	−3.0	.5
−1.3	1.3	−.6	−.1	−.5	−.6	2.9	.5	.4	.3
−.3	−.1	−.3	.6	−.5	−1.2	−1.2	−.3	−.1	1.1
.2	−.9	−.9	−.5	1.4	−.5	.2	−.4	1.5	1.1
−1.3	.2	−1.2	.4	−1.0	.8	.9	1.0	.0	.8
−1.2	−.2	−.3	1.8	1.4	.6	1.2	.7	.4	.2
.6	−.5	.8	.1	.5	.4	1.7	1.2	.9	−.3
.4	−1.9	.2	−.5	.7	−.1	−.1	−.5	.5	1.1
−1.4	.5	−1.7	−1.2	.8	−.7	.1	1.0	−.8	.2
−.2	−.2	−.4	−.8	.3	1.0	1.8	2.9	−.8	−.1
−.3	.5	.4	−1.5	1.5	2.0	−.1	.2	.0	−1.2
.4	−.4	.6	1.0	−.1	.1	.5	−1.3	1.1	1.1
.6	.7	−1.1	−1.4	−1.6	−1.6	1.5	1.3	.7	−.9
.9	−.9	−.1	−.5	.5	1.4	.0	−.3	−.3	1.2
.2	−.6	.0	−.5	−.9	−.4	−.5	1.7	−.2	−1.2
−.9	.4	.8	.8	.4	−.3	−1.1	.6	1.4	1.3

TABLE IIIa

Binomial Coefficients $\binom{n}{S}$

n	S = Number of Successes 0	1	2	3	4	5	6	7	8	9	10
0	1										
1	1	1									
2	1	2	1								
3	1	3	3	1							
4	1	4	6	4	1						
5	1	5	10	10	5	1					
6	1	6	15	20	15	6	1				
7	1	7	21	35	35	21	7	1			
8	1	8	28	56	70	56	28	8	1		
9	1	9	36	84	126	126	84	36	9	1	
10	1	10	45	120	210	252	210	120	45	10	1
11	1	11	55	165	330	462	462	330	165	55	11
12	1	12	66	220	495	792	924	792	495	220	66
13	1	13	78	286	715	1287	1716	1716	1287	715	286
14	1	14	91	364	1001	2002	3003	3432	3003	2002	1001
15	1	15	105	455	1365	3003	5005	6435	6435	5005	3003
16	1	16	120	560	1820	4368	8008	11440	12870	11440	8008
17	1	17	136	680	2380	6188	12376	19448	24310	24310	19448
18	1	18	153	816	3060	8568	18564	31824	43758	48620	43758
19	1	19	171	969	3876	11628	27132	50388	75582	92378	92378
20	1	20	190	1140	4845	15504	38760	77520	125970	167960	184756

Note. $\binom{n}{S} = \dfrac{n(n-1)(n-2)\cdots \text{to } S \text{ factors}}{S(S-1)(S-2)\cdots 3\cdot 2\cdot 1}$

For coefficients missing from the above table, use the relation:

$$\binom{n}{S} = \binom{n}{n-S}, \text{ e.g., } \binom{16}{12} = \binom{16}{4} = 1820$$

TABLE IIIb

Individual Binomial Probabilities $p(S)$

n	S	π .10	.20	.30	.40	.50	.60	.70	.80	.90
1	0	.9000	.8000	.7000	.6000	.5000	.4000	.3000	.2000	.1000
	1	.1000	.2000	.3000	.4000	.5000	.6000	.7000	.8000	.9000
2	0	.8100	.6400	.4900	.3600	.2500	.1600	.0900	.0400	.0100
	1	.1800	.3200	.4200	.4800	.5000	.4800	.4200	.3200	.1800
	2	.0100	.0400	.0900	.1600	.2500	.3600	.4900	.6400	.8100
3	0	.7290	.5120	.3430	.2160	.1250	.0640	.0270	.0080	.0010
	1	.2430	.3840	.4410	.4320	.3750	.2880	.1890	.0960	.0270
	2	.0270	.0960	.1890	.2880	.3750	.4320	.4410	.3840	.2430
	3	.0010	.0080	.0270	.0640	.1250	.2160	.3430	.5120	.7290
4	0	.6561	.4096	.2401	.1296	.0625	.0256	.0081	.0016	.0001
	1	.2916	.4096	.4116	.3456	.2500	.1536	.0756	.0256	.0036
	2	.0486	.1536	.2646	.3456	.3750	.3456	.2646	.1536	.0486
	3	.0036	.0256	.0756	.1536	.2500	.3456	.4116	.4096	.2916
	4	.0001	.0016	.0081	.0256	.0625	.1296	.2401	.4096	.6561
5	0	.5905	.3277	.1681	.0778	.0313	.0102	.0024	.0003	.0000
	1	.3280	.4096	.3602	.2592	.1563	.0768	.0283	.0064	.0004
	2	.0729	.2048	.3087	.3456	.3125	.2304	.1323	.0512	.0081
	3	.0081	.0512	.1323	.2304	.3125	.3456	.3087	.2048	.0729
	4	.0004	.0064	.0284	.0768	.1563	.2592	.3602	.4096	.3280
	5	.0000	.0003	.0024	.0102	.0313	.0778	.1681	.3277	.5905
6	0	.5314	.2621	.1176	.0467	.0156	.0041	.0007	.0001	.0000
	1	.3543	.3932	.3025	.1866	.0938	.0369	.0102	.0015	.0001
	2	.0984	.2458	.3241	.3110	.2344	.1382	.0595	.0154	.0012
	3	.0146	.0819	.1852	.2765	.3125	.2765	.1852	.0819	.0146
	4	.0012	.0154	.0595	.1382	.2344	.3110	.3241	.2458	.0984
	5	.0001	.0015	.0102	.0369	.0938	.1866	.3025	.3932	.3543
	6	.0000	.0001	.0007	.0041	.0156	.0467	.1176	.2621	.5314
7	0	.4783	.2097	.0824	.0280	.0078	.0016	.0002	.0000	.0000
	1	.3720	.3670	.2471	.1306	.0547	.0172	.0036	.0004	.0000
	2	.1240	.2753	.3177	.2613	.1641	.0774	.0250	.0043	.0002
	3	.0230	.1147	.2269	.2903	.2734	.1935	.0972	.0287	.0026
	4	.0026	.0287	.0972	.1935	.2734	.2903	.2269	.1147	.0230
	5	.0002	.0043	.0250	.0774	.1641	.2613	.3177	.2753	.1240
	6	.0000	.0004	.0036	.0172	.0547	.1306	.2471	.3670	.3720
	7	.0000	.0000	.0002	.0016	.0078	.0280	.0824	.2097	.4783

TABLE IIIb (continued)

n	S	.10	.20	.30	.40	π .50	.60	.70	.80	.90
8	0	.4305	.1678	.0576	.0168	.0039	.0007	.0001	.0000	.0000
	1	.3826	.3355	.1977	.0896	.0313	.0079	.0012	.0001	.0000
	2	.1488	.2936	.2965	.2090	.1094	.0413	.0100	.0011	.0000
	3	.0331	.1468	.2541	.2787	.2188	.1239	.0467	.0092	.0004
	4	.0046	.0459	.1361	.2322	.2734	.2322	.1361	.0459	.0046
	5	.0004	.0092	.0467	.1239	.2188	.2787	.2541	.1468	.0331
	6	.0000	.0011	.0100	.0413	.1094	.2090	.2965	.2936	.1488
	7	.0000	.0001	.0012	.0079	.0313	.0896	.1977	.3355	.3826
	8	.0000	.0000	.0001	.0007	.0039	.0168	.0576	.1678	.4305
9	0	.3874	.1342	.0404	.0101	.0020	.0003	.0000	.0000	.0000
	1	.3874	.3020	.1556	.0605	.0176	.0035	.0004	.0000	.0000
	2	.1722	.3020	.2668	.1612	.0703	.0212	.0039	.0003	.0000
	3	.0446	.1762	.2668	.2508	.1641	.0743	.0210	.0028	.0001
	4	.0074	.0661	.1715	.2508	.2461	.1672	.0735	.0165	.0008
	5	.0008	.0165	.0735	.1672	.2461	.2508	.1715	.0661	.0074
	6	.0001	.0028	.0210	.0743	.1641	.2508	.2668	.1762	.0446
	7	.0000	.0003	.0039	.0212	.0703	.1612	.2668	.3020	.1722
	8	.0000	.0000	.0004	.0035	.0176	.0605	.1556	.3020	.3874
	9	.0000	.0000	.0000	.0003	.0020	.0101	.0404	.1342	.3874
10	0	.3487	.1074	.0282	.0060	.0010	.0001	.0000	.0000	.0000
	1	.3874	.2684	.1211	.0403	.0098	.0016	.0001	.0000	.0000
	2	.1937	.3020	.2335	.1209	.0439	.0106	.0014	.0001	.0000
	3	.0574	.2013	.2668	.2150	.1172	.0425	.0090	.0008	.0000
	4	.0112	.0881	.2001	.2508	.2051	.1115	.0368	.0055	.0001
	5	.0015	.0264	.1029	.2007	.2461	.2007	.1029	.0264	.0015
	6	.0001	.0055	.0368	.1115	.2051	.2508	.2001	.0881	.0112
	7	.0000	.0008	.0090	.0425	.1172	.2150	.2668	.2013	.0574
	8	.0000	.0001	.0014	.0106	.0439	.1209	.2335	.3020	.1937
	9	.0000	.0000	.0001	.0016	.0098	.0403	.1211	.2684	.3874
	10	.0000	.0000	.0000	.0001	.0010	.0060	.0282	.1074	.3487

TABLE IIIc

Cumulative Binomial Probability in Right-Hand Tail

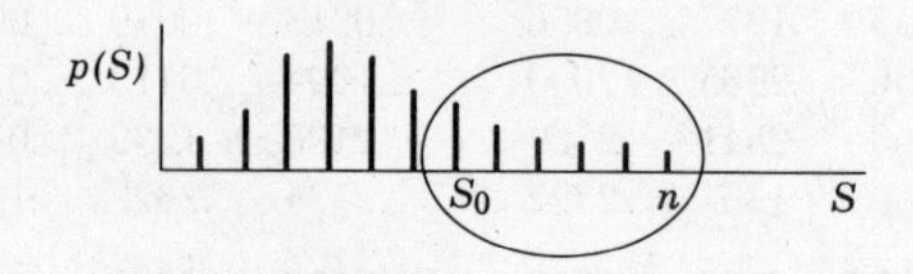

n	S_0	.10	.20	.30	.40	.50	.60	.70	.80	.90
						π				
2	1	.1900	.3600	.5100	.6400	.7500	.8400	.9100	.9600	.9900
	2	.0100	.0400	.0900	.1600	.2500	.3600	.4900	.6400	.8100
3	1	.2710	.4880	.6570	.7840	.8750	.9360	.9730	.9920	.9990
	2	.0280	.1040	.2160	.3520	.5000	.6480	.7840	.8960	.9720
	3	.0010	.0080	.0270	.0640	.1250	.2160	.3430	.5120	.7290
4	1	.3439	.5904	.7599	.8704	.9375	.9744	.9919	.9984	.9999
	2	.0523	.1808	.3483	.5248	.6875	.8208	.9163	.9728	.9963
	3	.0037	.0272	.0837	.1792	.3125	.4752	.6517	.8192	.9477
	4	.0001	.0016	.0081	.0256	.0625	.1296	.2401	.4096	.6561
5	1	.4095	.6723	.8319	.9222	.9688	.9898	.9976	.9997	1.0000
	2	.0815	.2627	.4718	.6630	.8125	.9130	.9692	.9933	.9995
	3	.0086	.0579	.1631	.3174	.5000	.6826	.8369	.9421	.9914
	4	.0005	.0067	.0308	.0870	.1875	.3370	.5282	.7373	.9185
	5	.0000	.0003	.0024	.0102	.0313	.0778	.1681	.3277	.5905
6	1	.4686	.7379	.8824	.9533	.9844	.9959	.9993	.9999	1.0000
	2	.1143	.3446	.5798	.7667	.8906	.9590	.9891	.9984	.9999
	3	.0159	.0989	.2557	.4557	.6562	.8208	.9295	.9830	.9987
	4	.0013	.0170	.0705	.1792	.3438	.5443	.7443	.9011	.9842
	5	.0001	.0016	.0109	.0410	.1094	.2333	.4202	.6554	.8857
	6	.0000	.0001	.0007	.0041	.0156	.0467	.1176	.2621	.5314
7	1	.5217	.7903	.9176	.9720	.9922	.9984	.9998	1.0000	1.0000
	2	.1497	.4233	.6706	.8414	.9375	.9812	.9962	.9996	1.0000
	3	.0257	.1480	.3529	.5801	.7734	.9037	.9712	.9953	.9998
	4	.0027	.0333	.1260	.2898	.5000	.7102	.8740	.9667	.9973
	5	.0002	.0047	.0288	.0963	.2266	.4199	.6471	.8520	.9743
	6	.0000	.0004	.0038	.0188	.0625	.1586	.3294	.5767	.8503
	7	.0000	.0000	.0002	.0016	.0078	.0280	.0824	.2097	.4783

TABLE IIIc (continued)

n	S_0	.10	.20	.30	.40	π .50	.60	.70	.80	.90
8	1	.5695	.8322	.9424	.9832	.9961	.9993	.9999	1.0000	1.0000
	2	.1869	.4967	.7447	.8936	.9648	.9915	.9987	.9999	1.0000
	3	.0381	.2031	.4482	.6846	.8555	.9502	.9887	.9988	1.0000
	4	.0050	.0563	.1941	.4059	.6367	.8263	.9420	.9896	.9996
	5	.0004	.0104	.0580	.1737	.3633	.5941	.8059	.9437	.9950
	6	.0000	.0012	.0113	.0498	.1445	.3154	.5518	.7969	.9619
	7	.0000	.0001	.0013	.0085	.0352	.1064	.2553	.5033	.8131
	8	.0000	.0000	.0001	.0007	.0039	.0168	.0576	.1678	.4305
9	1	.6126	.8658	.9596	.9899	.9980	.9997	1.0000	1.0000	1.0000
	2	.2252	.5638	.8040	.9295	.9805	.9962	.9996	1.0000	1.0000
	3	.0530	.2618	.5372	.7682	.9102	.9750	.9957	.9997	1.0000
	4	.0083	.0856	.2703	.5174	.7461	.9006	.9747	.9969	.9999
	5	.0009	.0196	.0988	.2666	.5000	.7334	.9012	.9804	.9991
	6	.0001	.0031	.0253	.0994	.2539	.4826	.7297	.9144	.9917
	7	.0000	.0003	.0043	.0250	.0898	.2318	.4628	.7382	.9470
	8	.0000	.0000	.0004	.0038	.0195	.0705	.1960	.4362	.7748
	9	.0000	.0000	.0000	.0003	.0020	.0101	.0404	.1342	.3874
10	1	.6513	.8926	.9718	.9940	.9990	.9999	1.0000	1.0000	1.0000
	2	.2639	.6242	.8507	.9536	.9893	.9983	.9999	1.0000	1.0000
	3	.0702	.3222	.6172	.8327	.9453	.9877	.9984	.9999	1.0000
	4	.0128	.1209	.3504	.6177	.8281	.9452	.9894	.9991	1.0000
	5	.0016	.0328	.1503	.3669	.6230	.8338	.9527	.9936	.9999
	6	.0001	.0064	.0473	.1662	.3770	.6331	.8497	.9672	.9984
	7	.0000	.0009	.0106	.0548	.1719	.3823	.6496	.8791	.9872
	8	.0000	.0001	.0016	.0123	.0547	.1673	.3828	.6778	.9298
	9	.0000	.0000	.0001	.0017	.0107	.0464	.1493	.3758	.7361
	10	.0000	.0000	.0000	.0001	.0010	.0060	.0282	.1074	.3487

TABLE IV

Standard Normal, Cumulative Probability in Right-Hand Tail
(For Negative Values of Z, Areas Are Found by Symmetry)

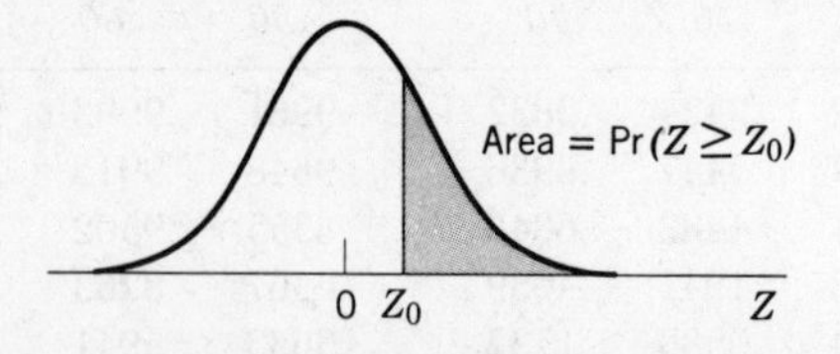

Z_o	Second Decimal Place of Z_o .00	.01	.02	.03	.04	.05	.06	.07	.08	.09
0.0	.5000	.4960	.4920	.4880	.4840	.4801	.4761	.4721	.4681	.4641
0.1	.4602	.4562	.4522	.4483	.4443	.4404	.4364	.4325	.4286	.4247
0.2	.4207	.4168	.4129	.4090	.4052	.4013	.3974	.3936	.3897	.3859
0.3	.3821	.3783	.3745	.3707	.3669	.3632	.3594	.3557	.3520	.3483
0.4	.3446	.3409	.3372	.3336	.3300	.3264	.3228	.3192	.3156	.3121
0.5	.3085	.3050	.3015	.2981	.2946	.2912	.2877	.2843	.2810	.2776
0.6	.2743	.2709	.2676	.2643	.2611	.2578	.2546	.2514	.2483	.2451
0.7	.2420	.2389	.2358	.2327	.2296	.2266	.2236	.2206	.2177	.2148
0.8	.2119	.2090	.2061	.2033	.2005	.1977	.1949	.1922	.1894	.1867
0.9	.1841	.1814	.1788	.1762	.1736	.1711	.1685	.1660	.1635	.1611
1.0	.1587	.1562	.1539	.1515	.1492	.1469	.1446	.1423	.1401	.1379
1.1	.1357	.1335	.1314	.1292	.1271	.1251	.1230	.1210	.1190	.1170
1.2	.1151	.1131	.1112	.1093	.1075	.1056	.1038	.1020	.1003	.0985
1.3	.0968	.0951	.0934	.0918	.0901	.0885	.0869	.0853	.0838	.0823
1.4	.0808	.0793	.0778	.0764	.0749	.0735	.0722	.0708	.0694	.0681
1.5	.0668	.0655	.0643	.0630	.0618	.0606	.0594	.0582	.0571	.0559
1.6	.0548	.0537	.0526	.0516	.0505	.0495	.0485	.0475	.0465	.0455
1.7	.0446	.0436	.0427	.0418	.0409	.0401	.0392	.0384	.0375	.0367
1.8	.0359	.0352	.0344	.0336	.0329	.0322	.0314	.0307	.0301	.0294
1.9	.0287	.0281	.0274	.0268	.0262	.0256	.0250	.0244	.0239	.0233
2.0	.0228	.0222	.0217	.0212	.0207	.0202	.0197	.0192	.0188	.0183
2.1	.0179	.0174	.0170	.0166	.0162	.0158	.0154	.0150	.0146	.0143
2.2	.0139	.0136	.0132	.0129	.0125	.0122	.0119	.0116	.0113	.0110
2.3	.0107	.0104	.0102	.0099	.0096	.0094	.0091	.0089	.0087	.0084
2.4	.0082	.0080	.0078	.0075	.0073	.0071	.0069	.0068	.0066	.0064
2.5	.0062	.0060	.0059	.0057	.0055	.0054	.0052	.0051	.0049	.0048
2.6	.0047	.0045	.0044	.0043	.0041	.0040	.0039	.0038	.0037	.0036
2.7	.0035	.0034	.0033	.0032	.0031	.0030	.0029	.0028	.0027	.0026
2.8	.0026	.0025	.0024	.0023	.0023	.0022	.0021	.0021	.0020	.0019
2.9	.0019	.0018	.0017	.0017	.0016	.0016	.0015	.0015	.0014	.0014
3.0	.00135									
3.5	.000 233									
4.0	.000 031 7									
4.5	.000 003 40									
5.0	.000 000 287									

TABLE V
t Critical Points

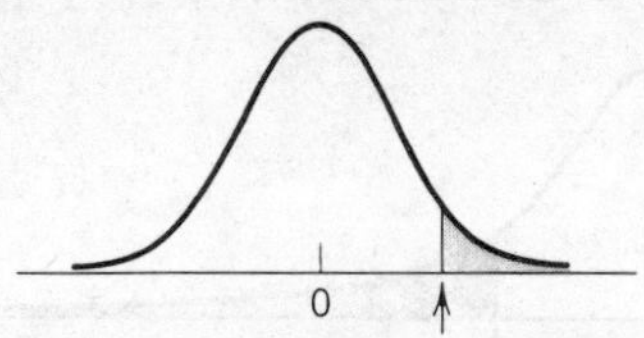

Critical point. For example: $t_{.025}$ leaves .025 probability in the tail.

$d.f.$	$t_{.25}$	$t_{.10}$	$t_{.05}$	$t_{.025}$	$t_{.010}$	$t_{.005}$	$t_{.0025}$	$t_{.0010}$	$t_{.0005}$
1	1.000	3.078	6.314	12.706	31.821	63.637	127.32	318.31	636.62
2	.816	1.886	2.920	4.303	6.965	9.925	14.089	22.326	31.598
3	.765	1.638	2.353	3.182	4.541	5.841	7.453	10.213	12.924
4	.741	1.533	2.132	2.776	3.747	4.604	5.598	7.173	8.610
5	.727	1.476	2.015	2.571	3.365	4.032	4.773	5.893	6.869
6	.718	1.440	1.943	2.447	3.143	3.707	4.317	5.208	5.959
7	.711	1.415	1.895	2.365	2.998	3.499	4.020	4.785	5.408
8	.706	1.397	1.860	2.306	2.896	3.355	3.833	4.501	5.041
9	.703	1.383	1.833	2.262	2.821	3.250	3.690	4.297	4.781
10	.700	1.372	1.812	2.228	2.764	3.169	3.581	4.144	4.537
11	.697	1.363	1.796	2.201	2.718	3.106	3.497	4.025	4.437
12	.695	1.356	1.782	2.179	2.681	3.055	3.428	3.930	4.318
13	.694	1.350	1.771	2.160	2.650	3.012	3.372	3.852	4.221
14	.692	1.345	1.761	2.145	2.624	2.977	3.326	3.787	4.140
15	.691	1.341	1.753	2.131	2.602	2.947	3.286	3.733	4.073
16	.690	1.337	1.746	2.120	2.583	2.921	3.252	3.686	4.015
17	.689	1.333	1.740	2.110	2.567	2.898	3.222	3.646	3.965
18	.688	1.330	1.734	2.101	2.552	2.878	3.197	3.610	3.922
19	.688	1.328	1.729	2.093	2.539	2.861	3.174	3.579	3.883
20	.687	1.325	1.725	2.086	2.528	2.845	3.153	3.552	3.850
21	.686	1.323	1.721	2.080	2.518	2.831	3.135	3.257	3.189
22	.686	1.321	1.717	2.074	2.508	2.819	3.119	3.505	3.792
23	.685	1.319	1.714	2.069	2.500	2.807	3.104	3.485	3.767
24	.685	1.318	1.711	2.064	2.492	2.797	3.091	3.467	3.745
25	.684	1.316	1.708	2.060	2.485	2.787	3.078	3.450	3.725
26	.684	1.315	1.706	2.056	2.479	2.779	3.067	3.435	3.707
27	.684	1.314	1.703	2.052	2.473	2.771	3.057	3.421	3.690
28	.683	1.313	1.701	2.048	2.467	2.763	3.047	3.408	3.674
29	.683	1.311	1.699	2.045	2.462	2.756	3.038	3.396	3.659
30	.683	1.310	1.697	2.042	2.457	2.750	3.030	3.385	3.646
40	.681	1.303	1.684	2.021	2.423	2.704	2.971	3.307	3.551
60	.679	1.296	.1671	2.000	2.390	2.660	2.915	3.232	3.460
120	.677	1.289	1.658	1.980	2.358	2.617	2.860	3.160	3.373
∞	.674	1.282	1.645	1.960	2.326	2.576	2.807	3.090	3.291
	$= z_{.25}$	$= z_{.10}$	$= z_{.05}$	$= z_{.025}$	$= z_{.010}$	$= z_{.005}$	$= z_{.0025}$	$= z_{.0010}$	$= z_{.0005}$

TABLE VI

F Critical Points

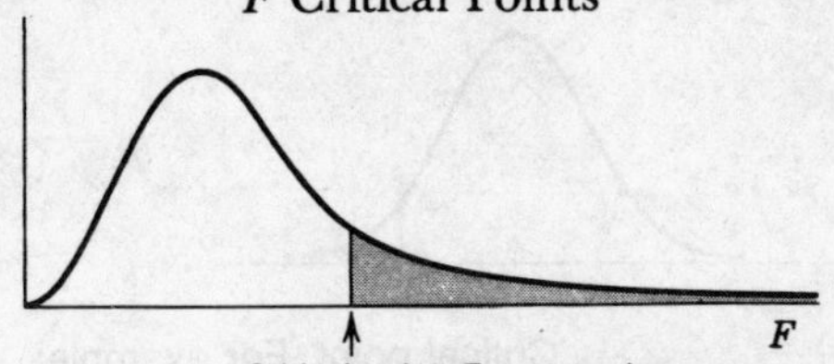

Critical point. For example: $F_{.05}$ leaves 5% probability in the tail.

Degrees of Freedom for Denominator		Degrees of Freedom for Numerator: 1	2	3	4	5	6	8	10	20	40	∞
1	$F_{.25}$	5.83	7.50	8.20	8.58	8.82	8.98	9.19	9.32	9.58	9.71	9.85
	$F_{.10}$	39.9	49.5	53.6	55.8	57.2	58.2	59.4	60.2	61.7	62.5	63.3
	$F_{.05}$	161	200	216	225	230	234	239	242	248	251	254
2	$F_{.25}$	2.57	3.00	3.15	3.23	3.28	3.31	3.35	3.38	3.43	3.45	3.48
	$F_{.10}$	8.53	9.00	9.16	9.24	9.29	9.33	9.37	9.39	9.44	9.47	9.49
	$F_{.05}$	18.5	19.0	19.2	19.2	19.3	19.3	19.4	19.4	19.4	19.5	19.5
	$F_{.01}$	98.5	99.0	99.2	99.2	99.3	99.3	99.4	99.4	99.4	99.5	99.5
	$F_{.001}$	998	999	999	999	999	999	999	999	999	999	999
3	$F_{.25}$	2.02	2.28	2.36	2.39	2.41	2.42	2.44	2.44	2.46	2.47	2.47
	$F_{.10}$	5.54	5.46	5.39	5.34	5.31	5.28	5.25	5.23	5.18	5.16	5.13
	$F_{.05}$	10.1	9.55	9.28	9.12	9.10	8.94	8.85	8.79	8.66	8.59	8.53
	$F_{.01}$	34.1	30.8	29.5	28.7	28.2	27.9	27.5	27.2	26.7	26.4	26.1
	$F_{.001}$	167	149	141	137	135	133	131	129	126	125	124
4	$F_{.25}$	1.81	2.00	2.05	2.06	2.07	2.08	2.08	2.08	2.08	2.08	2.08
	$F_{.10}$	4.54	4.32	4.19	4.11	4.05	4.01	3.95	3.92	3.84	3.80	3.76
	$F_{.05}$	7.71	6.94	6.59	6.39	6.26	6.16	6.04	5.96	5.80	5.72	5.63
	$F_{.01}$	21.2	18.0	16.7	16.0	15.5	15.2	14.8	14.5	14.0	13.7	13.5
	$F_{.001}$	74.1	61.3	56.2	53.4	51.7	50.5	49.0	48.1	46.1	45.1	44.1
5	$F_{.25}$	1.69	1.85	1.88	1.89	1.89	1.89	1.89	1.89	1.88	1.88	1.87
	$F_{.10}$	4.06	3.78	3.62	3.52	3.45	3.40	3.34	3.30	3.21	3.16	3.10
	$F_{.05}$	6.61	5.79	5.41	5.19	5.05	4.95	4.82	4.74	4.56	4.46	4.36
	$F_{.01}$	16.3	13.3	12.1	11.4	11.0	10.7	10.3	10.1	9.55	9.29	9.02
	$F_{.001}$	47.2	37.1	33.2	31.1	29.8	28.8	27.6	26.9	25.4	24.6	23.8
6	$F_{.25}$	1.62	1.76	1.78	1.79	1.79	1.78	1.77	1.77	1.76	1.75	1.74
	$F_{.10}$	3.78	3.46	3.29	3.18	3.11	3.05	2.98	2.94	2.84	2.78	2.72
	$F_{.05}$	5.99	5.14	4.76	4.53	4.39	4.28	4.15	4.06	3.87	3.77	3.67
	$F_{.01}$	13.7	10.9	9.78	9.15	8.75	8.47	8.10	7.87	7.40	7.14	6.88
	$F_{.001}$	35.5	27.0	23.7	21.9	20.8	20.0	19.0	18.4	17.1	16.4	15.8
7	$F_{.25}$	1.57	1.70	1.72	1.72	1.71	1.71	1.70	1.69	1.67	1.66	1.65
	$F_{.10}$	3.59	3.26	3.07	2.96	2.88	2.83	2.75	2.70	2.59	2.54	2.47
	$F_{.05}$	5.59	4.74	4.35	4.12	3.97	3.87	3.73	3.64	3.44	3.34	3.23
	$F_{.01}$	12.2	9.55	8.45	7.85	7.46	7.19	6.84	6.62	6.16	5.91	5.65
	$F_{.001}$	29.3	21.7	18.8	17.2	16.2	15.5	14.6	14.1	12.9	12.3	11.7
8	$F_{.25}$	1.54	1.66	1.67	1.66	1.66	1.65	1.64	1.63	1.61	1.59	1.58
	$F_{.10}$	3.46	3.11	2.92	2.81	2.73	2.67	2.59	2.54	2.42	2.36	2.29
	$F_{.05}$	5.32	4.46	4.07	3.84	3.69	3.58	3.44	3.35	3.15	3.04	2.93
	$F_{.01}$	11.3	8.65	7.59	7.01	6.63	6.37	6.03	5.81	5.36	5.12	4.86
	$F_{.001}$	25.4	18.5	15.8	14.4	13.5	12.9	12.0	11.5	10.5	9.92	9.33
9	$F_{.25}$	1.51	1.62	1.63	1.63	1.62	1.61	1.60	1.59	1.56	1.55	1.53
	$F_{.10}$	3.36	3.01	2.81	2.69	2.61	2.55	2.47	2.42	2.30	2.23	2.16
	$F_{.05}$	5.12	4.26	3.86	3.63	3.48	3.37	3.23	3.14	2.94	2.83	2.71
	$F_{.01}$	10.6	8.02	6.99	6.42	6.06	5.80	5.47	5.26	4.81	4.57	4.31
	$F_{.001}$	22.9	16.4	13.9	12.6	11.7	11.1	10.4	9.89	8.90	8.37	7.81

TABLE VI (continued)

Degrees of Freedom for Denominator		Degrees of Freedom for Numerator 1	2	3	4	5	6	8	10	20	40	∞
10	$F_{.25}$	1.49	1.60	1.60	1.59	1.59	1.58	1.56	1.55	1.52	1.51	1.48
	$F_{.10}$	3.28	2.92	2.73	2.61	2.52	2.46	2.38	2.32	2.20	2.13	2.06
	$F_{.05}$	4.96	4.10	3.71	3.48	3.33	3.22	3.07	2.98	2.77	2.66	2.54
	$F_{.01}$	10.0	7.56	6.55	5.99	5.64	5.39	5.06	4.85	4.41	4.17	3.91
	$F_{.0001}$	21.0	14.9	12.6	11.3	10.5	9.92	9.20	8.75	7.80	7.30	6.76
12	$F_{.25}$	1.56	1.56	1.56	1.55	1.54	1.53	1.51	1.50	1.47	1.45	1.42
	$F_{.10}$	3.18	2.81	2.61	2.48	2.39	2.33	2.24	2.19	2.06	1.99	1.90
	$F_{.05}$	4.75	3.89	3.49	3.26	3.11	3.00	2.85	2.75	2.54	2.43	2.30
	$F_{.01}$	9.33	6.93	5.95	5.41	5.06	4.82	4.50	4.30	3.86	3.62	3.36
	$F_{.001}$	18.6	13.0	10.8	9.63	8.89	8.38	7.71	7.29	6.40	5.93	5.42
14	$F_{.25}$	1.44	1.53	1.53	1.52	1.51	1.50	1.48	1.46	1.43	1.41	1.38
	$F_{.10}$	3.10	2.73	2.52	2.39	2.31	2.24	2.15	2.10	1.96	1.89	1.80
	$F_{.05}$	4.60	3.74	3.34	3.11	2.96	2.85	2.70	2.60	2.39	2.27	2.13
	$F_{.01}$	8.86	5.51	5.56	5.04	4.69	4.46	4.14	3.94	3.51	3.27	3.00
	$F_{.001}$	17.1	11.8	9.73	8.62	7.92	7.43	6.80	6.40	5.56	5.10	4.60
16	$F_{.25}$	1.42	1.51	1.51	1.50	1.48	1.48	1.46	1.45	1.40	1.37	1.34
	$F_{.10}$	3.05	2.67	2.46	2.33	2.24	2.18	2.09	2.03	1.89	1.81	1.72
	$F_{.05}$	4.49	3.63	3.24	3.01	2.85	2.74	2.59	2.49	2.28	2.15	2.01
	$F_{.01}$	8.53	6.23	5.29	4.77	4.44	4.20	3.89	3.69	3.26	3.02	2.75
	$F_{.001}$	16.1	11.0	9.00	7.94	7.27	6.81	6.19	5.81	4.99	4.54	4.06
18	$F_{.25}$	1.41	1.50	1.49	1.48	1.46	1.45	1.43	1.42	1.38	1.35	1.32
	$F_{.10}$	3.01	2.62	2.42	2.29	2.20	2.13	2.04	1.98	1.84	1.75	1.66
	$F_{.05}$	4.41	3.55	3.16	2.93	2.77	2.66	2.51	2.41	2.19	2.06	1.92
	$F_{.01}$	8.29	6.01	5.09	4.58	4.25	4.01	3.71	3.51	3.08	2.84	2.57
	$F_{.001}$	15.4	10.4	8.49	7.46	6.81	6.35	5.76	5.39	4.59	4.15	3.67
20	$F_{.25}$	1.40	1.49	1.48	1.46	1.45	1.44	1.42	1.40	1.36	1.33	1.29
	$F_{.10}$	2.97	2.59	2.38	2.25	2.16	2.09	2.00	1.94	1.79	1.71	1.61
	$F_{.05}$	4.35	3.49	3.10	2.87	2.71	2.60	2.45	2.35	2.12	1.99	1.84
	$F_{.01}$	8.10	5.85	4.94	4.43	4.10	3.87	3.56	3.37	2.94	2.69	2.42
	$F_{.001}$	14.8	9.95	8.10	7.10	6.46	6.02	5.44	5.08	4.29	3.86	3.38
30	$F_{.25}$	1.38	1.45	1.44	1.42	1.41	1.39	1.37	1.35	1.30	1.27	1.23
	$F_{.10}$	2.88	2.49	2.28	2.14	2.05	1.98	1.88	1.82	1.67	1.57	1.46
	$F_{.05}$	4.17	3.32	2.92	2.69	2.53	2.42	2.27	2.16	1.93	1.79	1.62
	$F_{.01}$	7.56	5.39	4.51	4.02	3.70	3.47	3.17	2.98	2.55	2.30	2.01
	$F_{.001}$	13.3	8.77	7.05	6.12	5.53	5.12	4.58	4.24	3.49	3.07	2.59
40	$F_{.25}$	1.36	1.44	1.42	1.40	1.39	1.37	1.35	1.33	1.28	1.24	1.19
	$F_{.10}$	2.84	2.44	2.23	2.09	2.00	1.93	1.83	1.76	1.61	1.51	1.38
	$F_{.05}$	4.08	3.23	2.84	2.61	2.45	2.34	2.18	2.08	1.84	1.69	1.51
	$F_{.01}$	7.31	5.18	4.31	3.83	3.51	3.29	2.99	2.80	2.37	2.11	1.80
	$F_{.001}$	12.6	8.25	6.60	5.70	5.13	4.73	4.21	3.87	3.15	2.73	2.23
60	$F_{.25}$	1.35	1.42	1.41	1.38	1.37	1.35	1.32	1.30	1.25	1.21	1.15
	$F_{.10}$	2.79	2.39	2.18	2.04	1.95	1.87	1.77	1.71	1.54	1.44	1.29
	$F_{.05}$	4.00	3.15	2.76	2.53	2.37	2.25	2.10	1.99	1.75	1.59	1.39
	$F_{.01}$	7.08	4.98	4.13	3.65	3.34	3.12	2.82	2.63	2.20	1.94	1.60
	$F_{.001}$	12.0	7.76	6.17	5.31	4.76	4.37	3.87	3.54	2.83	2.41	1.89
120	$F_{.25}$	1.34	1.40	1.39	1.37	1.35	1.33	1.30	1.28	1.22	1.18	1.10
	$F_{.10}$	2.75	2.35	2.13	1.99	1.90	1.82	1.72	1.65	1.48	1.37	1.19
	$F_{.05}$	3.92	3.07	2.68	2.45	2.29	2.17	2.02	1.91	1.66	1.50	1.25
	$F_{.01}$	6.85	4.79	3.95	3.48	3.17	2.96	2.66	2.47	2.03	1.76	1.38
	$F_{.001}$	11.4	7.32	5.79	4.95	4.42	4.04	3.55	3.24	2.53	2.11	1.54
∞	$F_{.25}$	1.32	1.39	1.37	1.35	1.33	1.31	1.28	1.25	1.19	1.14	1.00
	$F_{.10}$	2.71	2.30	2.08	1.94	1.85	1.77	1.67	1.60	1.42	1.30	1.00
	$F_{.05}$	3.84	3.00	2.60	2.37	2.21	2.10	1.94	1.83	1.57	1.39	1.00
	$F_{.01}$	6.63	4.61	3.78	3.32	3.02	2.80	2.51	2.32	1.88	1.59	1.00
	$F_{.001}$	10.8	6.91	5.42	4.62	4.10	3.74	3.27	2.96	2.27	1.84	1.00

TABLE VII

χ^2 Critical Points

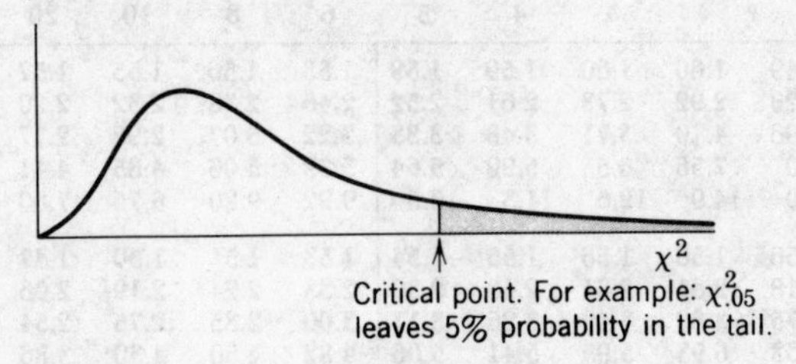
χ^2
Critical point. For example: $\chi^2_{.05}$
leaves 5% probability in the tail.

d.f.	$\chi^2_{.25}$	$\chi^2_{.10}$	$\chi^2_{.05}$	$\chi^2_{.025}$	$\chi^2_{.010}$	$\chi^2_{.005}$	$\chi^2_{.001}$
1	1.32	2.71	3.84	5.02	6.63	7.88	10.8
2	2.77	4.61	5.99	7.38	9.21	10.6	13.8
3	4.11	6.25	7.81	9.35	11.3	12.8	16.3
4	5.39	7.78	9.49	11.1	13.3	14.9	18.5
5	6.63	9.24	11.1	12.8	15.1	16.7	20.5
6	7.84	10.6	12.6	14.4	16.8	18.5	22.5
7	9.04	12.0	14.1	16.0	18.5	20.3	24.3
8	10.2	13.4	15.5	17.5	20.1	22.0	26.1
9	11.4	14.7	16.9	19.0	21.7	23.6	27.9
10	12.5	16.0	18.3	20.5	23.2	25.2	29.6
11	13.7	17.3	19.7	21.9	24.7	26.8	31.3
12	14.8	18.5	21.0	23.3	26.2	28.3	32.9
13	16.0	19.8	22.4	24.7	27.7	29.8	34.5
14	17.1	21.1	23.7	26.1	29.1	31.3	36.1
15	18.2	22.3	25.0	27.5	30.6	32.8	37.7
16	19.4	23.5	26.3	28.8	32.0	34.3	39.3
17	20.5	24.8	27.6	30.2	33.4	35.7	40.8
18	21.6	26.0	28.9	31.5	34.8	37.2	42.3
19	22.7	27.2	30.1	32.9	36.2	38.6	32.8
20	23.8	28.4	31.4	34.2	37.6	40.0	45.3
21	24.9	29.6	32.7	35.5	38.9	41.4	46.8
22	26.0	30.8	33.9	36.8	40.3	42.8	48.3
23	27.1	32.0	35.2	38.1	41.6	44.2	49.7
24	28.2	33.2	36.4	39.4	32.0	45.6	51.2
25	29.3	34.4	37.7	40.6	44.3	46.9	52.6
26	30.4	35.6	38.9	41.9	45.6	48.3	54.1
27	31.5	36.7	40.1	43.2	47.0	49.6	55.5
28	32.6	37.9	41.3	44.5	48.3	51.0	56.9
29	33.7	39.1	42.6	45.7	49.6	52.3	58.3
30	34.8	40.3	43.8	47.0	50.9	53.7	59.7
40	45.6	51.8	55.8	59.3	63.7	66.8	73.4
50	56.3	63.2	67.5	71.4	76.2	79.5	86.7
60	67.0	74.4	79.1	83.3	88.4	92.0	99.6
70	77.6	85.5	90.5	95.0	100	104	112
80	88.1	96.6	102	107	112	116	125
90	98.6	108	113	118	124	128	137
100	109	118	124	130	136	140	149

TABLE VIII

Wilcoxon–Mann–Whitney Two-Sample Test (W Test)

The one-sided prob-value (Pr) corresponding to the rank sum W of the smaller sample, ranking from the end where this smaller sample is concentrated. For $n > 6$ or prob-value $> .25$, see equation (13-16).

n = 2			
m = *1*		*2*	
W	*Pr*	*W*	*Pr*
1	.333	3	.167
		4	.333

Larger Sample Size, n = 3					
Smaller Sample Size, m = *1*		*2*		*3*	
W	*Pr*	*W*	*Pr*	*W*	*Pr*
1	.250	3	.100	6	.050
2	.500	4	.200	7	.100
		5	.400	8	.200
				9	.350
				10	.500

Larger Sample Size, n = 4							
Smaller Sample Size, m = *1*		*2*		*3*		*4*	
W	*Pr*	*W*	*Pr*	*W*	*Pr*	*W*	*Pr*
1	.200	3	.067	6	.029	10	.014
2	.400	4	.133	7	.057	11	.029
		5	.267	8	.114	12	.057
		6	.400	9	.200	13	.100
				10	.314	14	.171
				11	.429	15	.243
						16	.343

Larger Sample Size, n = 5									
Smaller Sample Size, m = *1*		*2*		*3*		*4*		*5*	
W	*Pr*	*W*	*Pr*	*W*	*Pr*	*W*	*Pr*	*W*	*Pr*
1	.167	3	.048	6	.018	10	.008	15	.004
2	.333	4	.095	7	.036	11	.016	16	.008
3	.500	5	.190	8	.071	12	.032	17	.016
		6	.286	9	.125	13	.056	18	.028
		7	.429	10	.196	14	.095	19	.048
				11	.286	15	.143	20	.075
				12	.393	16	.206	21	.111
				13	.500	17	.278	22	.155
						18	.365	23	.210
						19	.452	24	.274
								25	.345
								26	.421
								27	.500

Larger Sample Size, n = 6											
Smaller Sample Size, m = *1*		*2*		*3*		*4*		*5*		*6*	
W	*Pr*	*W*	*Pr*	*W*	*Pr*	*W*	*Pr*	*W*	*Pr*	*W*	*Pr*
1	.143	3	.036	6	.012	10	.005	15	.002	21	.001
2	.286	4	.071	7	.024	11	.010	16	.004	22	.002
3	.429	5	.143	8	.048	12	.019	17	.009	23	.004
		6	.214	9	.083	13	.033	18	.015	24	.008
		7	.321	10	.131	14	.057	19	.026	25	.013
		8	.429	11	.190	15	.086	20	.041	26	.021
				12	.274	16	.129	21	.063	27	.032
				13	.357	17	.176	22	.089	28	.047
				14	.452	18	.238	23	.123	29	.066
						19	.305	24	.165	30	.090
						20	.381	25	.214	31	.120
						21	.457	26	.268	32	.155
								27	.331	33	.197
								28	.396	34	.242
								29	.465	35	.294

Bibliography

Anderson, T. W. (1958), *An Introduction to Multivariate Analysis.* New York: Wiley.

Anderson, T. W. and S. L. Sclove (1978), *Statistical Analysis of Data.* Boston: Houghton-Mifflin.

Berelson, B., and R. Freedman (1964), "A Study in Fertility Control," *Scientific American* (May 1964): 29–37.

Bostwick, Burdette E. (1977), *Finding the Job You've Always Wanted.* New York: Wiley.

Brownlee, K. A. (1965), *Statistical Theory and Methodology in Science and Engineering.* New York: Wiley.

Clark, K. B., and M. P. Clark (1958), "Racial Identification and Preference in Negro Children," in *Readings in Social Psychology,* edited by E. E. Maccoby et al, New York: Holt, Rinehart & Winston.

Clopper, C. J., and E. S. Pearson (1934), "The Use of Confidence or Fiducial Limits Illustrated in the Case of the Binomial," *Biometrika* 26: 404.

Conner, J. R., and K. C. Gibbs, and J. E. Reynolds (1973), "The Effects of Water Frontage on Recreational Property Values," *Journal of Leisure Research* 5: 26–38.

Coopersmith, S. (1967), *The Antecedents of Self Esteem.* San Francisco: Freeman.

Fadeley, R. C. (1965), "Oregon Malignancy Pattern Physiographically Related to Hanford, Washington, Radioistopic Storage" *Journal of Environmental Health* 27: 883–897.

Fraumini, J. F., Jr. (1968), "Cigarette Smoking and Cancers of the Urinary Tract: Geographic Variation in the United States," *Journal of the National Cancer Institute:* 1205–1211.

Freeman, D., R. Purvis, and R. Pisani (1978), *Statistics.* New York: Norton.

Gilbert, J. P., B. McPeek, and F. Mosteller (1977), "Statistics and Ethics in Surgery and Anesthesia," *Science* (18 November): 684–689.

Haupt, A., and T. T. Kane (1978), *Population Handbook.* Washington, D.C.: Population Reference Bureau.

Hooker, R. H. (1907), "The Correlation of the Weather and Crops," *Journal of the Royal Statistical Society* 70: 1–42.

Joiner, B. L. (1975), "Living Histograms," *International Statistical Review* 3: 339–340.

Jones, K. L., D. W. Smith, A. P. Streissgrath, and N. C. Myrianthopoulos (1974), "Outcome in Offspring of Chronic Alcoholic Women," *Lancet* (1 June): 1076–1078.

Katz, D. A. (1973), "Faculty Salaries, Promotions, and Productivity at a Large University," *American Economic Review* 63: 469–477.

Knodel, John (1977), "Breast-feeding and Population Growth," *Science* (16 December): 1111–1115.

Meier, Paul (1977), "The Biggest Public Health Experiment Ever: The 1954 Field Trial of the Salk Poliomyelitis Vaccine," pages 88–100 in *Statistics: A Guide to the Study of the Biological and Health Sciences.* San Francisco: Holden Day.

Miksch, W. F. (1950), "The Average Statistician," *Colliers* (17 June).

Moore, David S. (1978), *Statistics: Concepts and Controversies.* San Francisco: Freeman.

Nemenyi, P., S. K. Dixon, N. B. White Jr., and M. L. Hedstrom (1977), *Statistics from Scratch (Pilot edition).* San Francisco: Holden Day.

Peacock, E. E. (1972), quoted in *Medical World News,* September 1, 1972: p. 45.

Pearson, E. S., and F. N. David (1938), *Tables of the Ordinates and Probability Integral of the Distribution of the Correlation Coefficient in Small Samples.* Cambridge: Cambridge Univ. Press.

Raiffa, H. (1968), *Decision Analysis.* Reading, Mass.: Addison-Wesley.

Rosenweig, M., E. L. Bennett, and M. C. Diamond (1964), "Brain Changes in Response to Experience," *Scientific American* (February):22–29.

Ruffin, J. M., et al (1969), "A Cooperative Double-blind Evaluation of Gastric 'Freezing' in the Treatment of Duodenal Ulcer," *New England Journal of Medicine* 281: 16–19.

Ryan, T. A., B. L. Joiner, and B. F. Ryan (1976), *Minitab Student Handbook.* North Scituate, Mass.: Duxbury Press.

Snedecor, G. W., and W. G. Cochran (1967), *Statistical Methods, 6th ed.* Ames, Iowa: Iowa State University Press.

Salk, Lee (1973), "The Role of the Heartbeat in the Relation Between Mother and Infant," *Scientific American* (May): 24–29.

Tufte, E. R. (1974), *Data Analysis for Politics and Policy.* Englewood Cliffs, N.J.: Prentice Hall.

U. S. Surgeon-General (1979), *Smoking and Health, a Report of the Surgeon-General.* U. S. Dept. of Health, Education, and Welfare, p. 2–12.

Wallis, W. A., and H. V. Roberts (1956), *Statistics: A New Approach.* New York: Free Press.

Wallis, W. A., and H. V. Roberts (1956), *Statistics: A New Approach.* New York: Free Press.

Wangensteen, O. H., et al (1962), "Achieving 'Physiological Gastrectomy' by Gastric Freezing," *Journal of American Medical Association* 180: 439–444.

Wonnacott, T. H., and R. J. Wonnacott (1977), *Introductory Statistics, 3rd ed.* or *Introductory Statistics for Business and Economics, 2nd ed.* New York: Wiley.

Wonnacott, T. H., and R. J. Wonnacott (1980), *Regression: A Second Course in Statistics.* New York: Wiley.

Zeisel, H., and H. Kalven (1972), "Parking Tickets and Missing Women," in *Statistics, A Guide to the Unknown,* edited by J. Tanur, F. Mosteller, et al. San Francisco: Holden Day.

Answers to Odd Numbered Problems

1-1 men, π = .49 ± .0400
women, π = .58 ± .0395* (majority)
under 30, π = .48 ± .0692
over 30, π = .55 ± .0308* (majority)

1-3 (a) Absent housewives, men, and people not in corner homes are all excluded.
(b) The alumni who are most distant, most busy, and most reluctant to face their classmates tend to be excluded.

In each case, avoid bias with a *random* sample of the whole target population.

1-7 Observational studies never prove effects. The given statement is just wild conjecture.

2-1

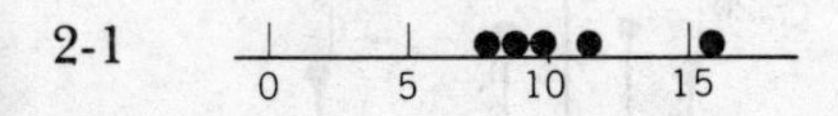

2-3 (a) 60th, 96th
(b) 15 thousand

2-5 (a) mean = \$70.75
mode = \$75

(b)

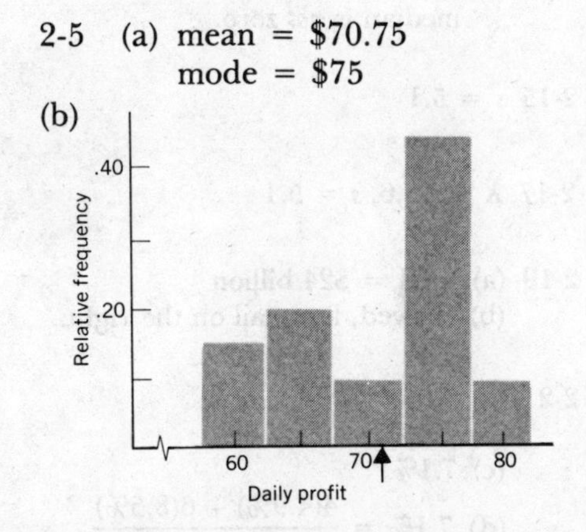

2-7

Mean	Mode
\$70.46	—
70.75	75
71.50	80

(a) The mode depends drastically on the arbitrary degree of grouping.
(b) The fine grouping gives a better approximation.

2-9 (a)

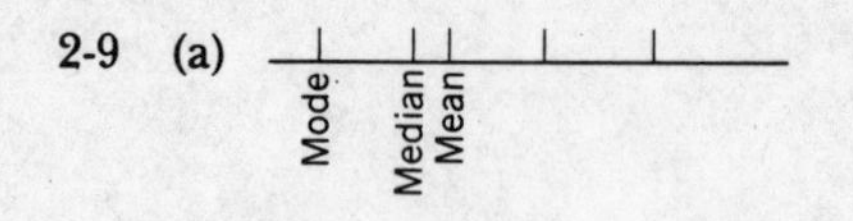

(b)

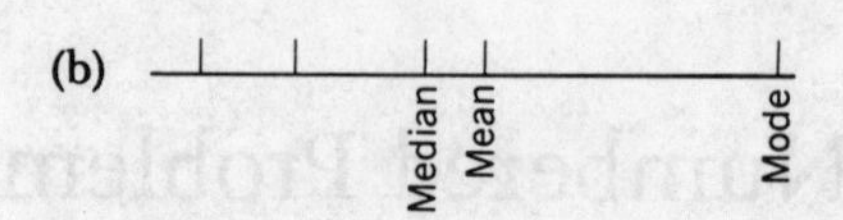

2-11 The mode would switch drastically, from $X = 95\%$ to $X = 5\%$.

2-13 (a) $\overline{X} = 9$
(b) average deviation = 0
(c) average deviation = 0
(d) The average deviation from the median is *not* zero.

2-15 $s = 5.1$

2-17 $\overline{X} = 15.6$, $s = 5.1$

2-19 (a) total = \$24 billion
(b) skewed, long tail on the right.

2-21 (a) 4.9%
(b) 8.5%
(c) 7.1%
(d) $7.1\% = \dfrac{4(4.9\%) + 6(8.5\%)}{10}$

In general,

$$\overline{X} = \frac{n_1\overline{X}_1 + n_2\overline{X}_2}{n_1 + n_2}$$

3-1 (a) The relative frequency distribution will resemble the probability distribution if the class is large (n is large).
(b) .69, .31

3-3 (b) .04, .96

3-5 (a)

X	$p(X)$
0	.48
1	.40
2	.12
	1.00✓

(b)

X	$p(X)$
0	.384
1	.404
2	.176
3	.036
	1.000✓

3-7 (a)

$$\left(\frac{8}{12}\right)\left(\frac{7}{11}\right)\left(\frac{6}{10}\right) = 25\%$$

(b) 9%

3-9 (a)

X	$p(X)$
0	.152
1	.174
2	.248
3	.170
4	.103
6	.153

(b) $\mu = 2.51$
(c) $\sigma = 1.88$
(d)

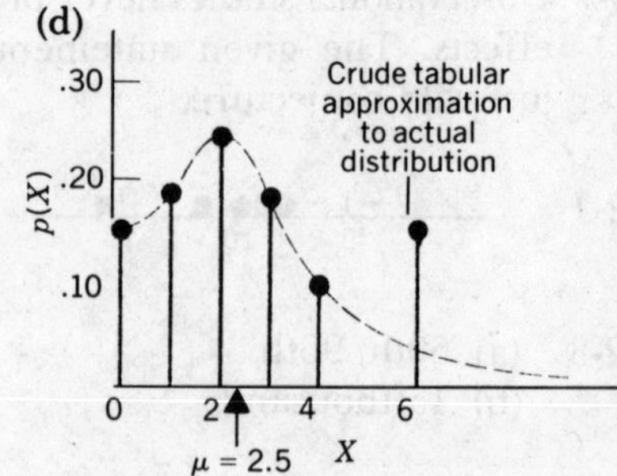

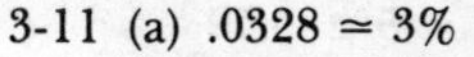

3-11 (a) $.0328 \simeq 3\%$
(b) $.6230 \simeq 62\%$

3-13 Assuming independent missions, $\Pr(\text{surviving}) = .98^{50} \simeq 36\%$.

3-15 (a) $\mu = n\pi$
(b) $\sigma^2 = 25, 4.17, 12.5$

3-17 (a) $.0548 \simeq 5\%$
(b) $.0441 \simeq 4\%$
(c) $.9495 \simeq 95\%$
(d) $.1034 \simeq 10\%$
(e) $.4750 \simeq 47\%$
(f) $.9500 \simeq 95\%$
(g) $.6818 \simeq 68\%$
(h) $.0062 \simeq 1\%$

3-19 (a) $.2743 \simeq 27\%$
(b) $.9332 \simeq 93\%$
(c) $.3830 \simeq 38\%$
(d) $.8977 \simeq 90\%$

3-21 (a) $.2612 \simeq 26\%$
(b) 550
(c) 390
(d) 160

3-23 (a) $.7102 \simeq 71\%$
(b) $.99^7 = .9321 \simeq 93\%$
(c) $\mu = 4.20, \sigma = 1.30$

3-25 (a) 234, 274, 961 millions
(b) 166, 138, 31 millions
(c) Such projections provide interesting possibilities but, even in the short run, need to be carefully qualified.

3-27 (a) $.9772 \simeq 98\%$
(b) 109 minutes

3-29 (a) $.0352 \simeq 4\%$
(b) Yes, we would indeed question the hypothesis of "no effect." For if it were true, a very unlikely outcome actually occurred.

3-31 (a) 2.40
(b) .98
(c)

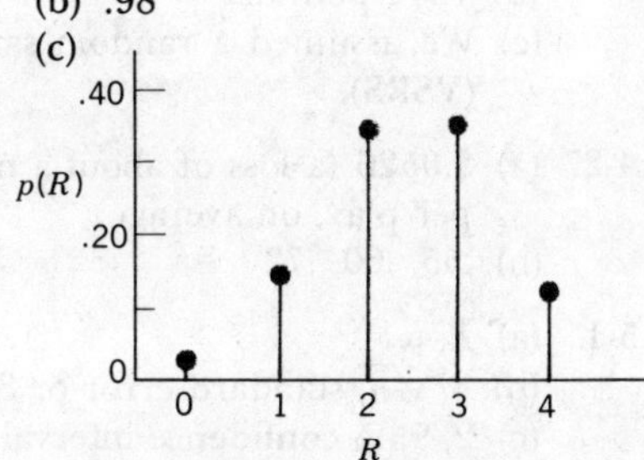

4-5 (a) $\mu = \$20{,}000$, $\sigma/\sqrt{n} = \$1200$, approximately normal
(b) $.0475 \simeq 5\%$

4-7 $.8996 \simeq 90\%$

4-9 (a) 108
(b) $\Pr > .999994$, that is, all but 6 in a million ($Z = \pm 4.8$)

4-11 We assumed very simple random sampling (VSRS), which is a safe assumption only in Problem 4-7. In all the other problems, observations might be so dependent as to make the probability calculations worthless.

4-13 $.0183 \simeq 2\%$ (wcc, .0359)

4-15 $.9207 \simeq 92\%$ (wcc, .9406)

4-17 $.9926 \simeq 99\%$

4-19 μ = population mean
$\bar{X}$ = sample mean
σ^2 = population variance
s^2 = sample variance
π = population proportion
P = sample proportion

4-21 (a) $.2514 \simeq 25\%$
(b) $p < .001$ ($Z = 3.33$)

4-23 $.0823 \simeq 8\%$

4-25 (a) $.0446 \simeq 4\%$
(b) 7912 pounds
(c) We assumed a random sample (VSRS).

4-27 (a) \$.0526 (a loss of about a nickel per play, on average)
(b) .55, .60, .72

5-1 (a) $\overline{X}$, μ
(b) $\sigma/\sqrt{n}$, standard error or SE
(c) 2, 95% confidence interval
(d) wider

5-3 (a) $\mu = 3.30 \pm .392$
$\simeq 3.30 \pm .39$
NOTE: In all confidence intervals, we should round the confidence allowance (.39) to the same number of decimal places as the estimate (3.30). We give the unrounded value (.392) merely as a check on your methodology.
(b) $\mu = 3.30 \pm .514$

5-5 True

5-7 (a) $.83 \pm .0747$
(b) Not at all surprising:
$X - \overline{X} = .15$, which is comparable to $s = .20$.
[The confidence interval in (a) is irrelevant, because it is for the population mean, not for a single observation.]

5-9 (a) $(\mu_1 - \mu_2) = 2.3 \pm 2.49$
That is, the first population is 2.3 inches higher than the second, on average ($\pm$ 2.49 inches, with 95% confidence).
(b) $(\mu_1 - \mu_2) = 2.3 \pm 2.05$

5-11 $\Delta = 40 \pm 57.6$

5-13 (a) $\mu_M - \mu_W = 5 \pm 5.79$
That is, the men at the university earn 5($\pm$ 5.79) thousand dollars more than the women, on average. Or, equivalently,
$\mu_W - \mu_M = -5.00 \pm 5.79$
That is, women earn 5($\pm$ 5.79) *less* than men.
(b) It fails to show discrimination on two counts: (1) This is an observational study, and whatever differences exist may be due to extraneous factors such as men having better qualifications, more experience, etc. (2) We're not even sure a difference exists in the population; the confidence interval includes $\mu_1 - \mu_2 = 0$ (i.e., no difference).

5-15 Seed *B* is better by 6 bushels/acre, (plus or minus 4.50, with 95% confidence).

5-17 (a) $.01 < \pi < .57$
(b) $.06 < \pi < .41$
(c) $\pi = .20 \pm .0157$

5-19 (a) $\pi_1 - \pi_2 = -.19 \pm .077$
That is, the "under 18" population is less likely to believe it than the "over 24,"—less likely by 19($\pm$ 8) percentage points.
(b) Because of the imbalance of 500-100, we would guess the interval is more vague (the interval would clearly be more vague in a 999-1 imbalance, where 1 observation would be terribly unreliable.) Calculation bears this out:
$\pi_1 - \pi_2 = -.19 \pm .101$

5-21 (a) $\pi_M - \pi_W = -.09 \pm .0562$
That is, men are 9($\pm$ 6) percentage points less in favor of Carter.
(b) If men and women are equally numerous in the population, $\pi = .535 \pm .0282$

5-23 $\mu > 65 - 11.06 \simeq 54$
That is, the population mean is estimated to be 65; with 95% confidence, it is at least 54.

5-25 (i) $\mu_B - \mu_A > 6 - 3.57 \simeq 2.4$
That is, we estimate seed *B* to be better than *A* by 6 bushels/acre; with 95% confidence, it is at least 2.4 bushels/acre better.
(ii) mean drop $> 21 - 11.9 \simeq 9$
That is, the drop in mean IQ due to alcoholism in the mothers is estimated to be 21 points; with 95% confidence, it is at least 9 points.

5-27 $\pi_2 - \pi_1 = -.45 \pm .086$
That is, a drop of 45 $\pm$ 8.6 percentage points.

5-29 (b) $\mu = 118.0 \pm 3.16$
(c) $\pi = .13 \pm .066$
Or, from Figure 5-5, $.07 < \pi < .22$

5-31 (a) $\pi = .523 \pm .148$
(b) We agree.

5-33 (a) 28.5, 71.0, 46.2 (per 100,000)
(b) We estimate the vaccine reduces the polio rate from 71 to 29 cases per 100,000, a reduction of 42 cases per 100,000 (with 95% confidence, a reduction of 42 $\pm$ 14).
(c) Such an observational study would have confounded the effect of volunteering with the vaccine. And this is a substantial effect: volunteers have nearly twice the polio rate of non-volunteers (71 vs. 46, per 100,000).

6-1 In each case, the difference is statistically discernible, and H_0 is rejected, at the 5% level.

6-3 (a) H_0: Judge is fair, i.e., $\pi = .29$
(b) $Z = -8.16$, $p \simeq 0$

6-5 (a) mean increase > 9.89
(b) $t = 5.27$, $p < .0025$

6-7 (a) In the population as a whole, men outsmoke women by at least 5 percentage points (4.86, exactly).
(b) $Z = 3.57$, $p < .000233$
(c) Yes, is discernible.

6-9 (a) H_0: the proportion of defective gloves is the old value of 10%, that is, $\pi = .10$.
H_1: the proportion of defective gloves is worse than 10%, that is, $\pi > .10$.
(b) Reject H_0 if $P > 14\%$ (.140, exactly).
(c) Reject all but the first and third shipments.

6-11 I, α. II, β. α, β.

6-13 (a) No, do not reject H_0.
(b) We would follow common sense, because the classical test is narrow-minded, and arbitrarily sets $\alpha = .05$.
(c) Since $\overline{X} = 1245$ exceeds the new critical value of 1235, we reject H_0. So the problem in (a) was indeed inadequate sample size.
(d) Since $\overline{X} = 1201$ exceeds the new and very stringent critical value of 1200.5, we reject H_0; therefore, we find the increase of 1 unit "statistically significant." Conclusion is true—and shows another weakness of classical tests.

6-15 α would decrease, and consequently β would increase.

6-17 (a) H_1: $\pi = .30$
(b) $.1493 \simeq 15\%$

6-19 (a) True
(b) False: For this one-sided alternative, we should use a *one*-sided test, rejecting H_0 when P turns out to be *small.*

6-21 (a) $\mu_{GP} - \mu_P = 2000 \pm 3724$
(b) .2938
(c) No, not discernible.

6-23 The whole 10:30 class really would give a higher rating than the 3:30 class. If the classes were equal, there is less than a $2\frac{1}{2}\%$ chance that sampling fluctuation would produce as large a sample difference as the .8 units actually observed.

We can claim with 95% confidence that a complete poll would show the 10:30 class is better than the 3:30 class, by at least .2 scale points.

Technically speaking, the prob-value for H_0 is less than .025, and the 95% confidence interval for the difference in the population means is

$\mu_1 - \mu_2 = .80 \pm .68$

or for a one-sided claim,

$\mu_1 - \mu_2 > .23$

6-25 (a) $\pi = 82.6\% \pm 4.4\%$, discernible
(b) $\pi_R - \pi_L = 5.0\% \pm 15.1\%$, indiscernible
(c) See Problem 6-26 for an interesting hypothesis.

6-27 If you admit a conceivable population of babies that might have been born in the same place, time, and circumstances, then the confidence interval and prob-value both make sense. Even without such a population, however, the prob-value can be interpreted sensibly because the treatment and control were correctly assigned at random.

7-1 (a) $\hat{Y} = 59 + 7.0X$

(b)

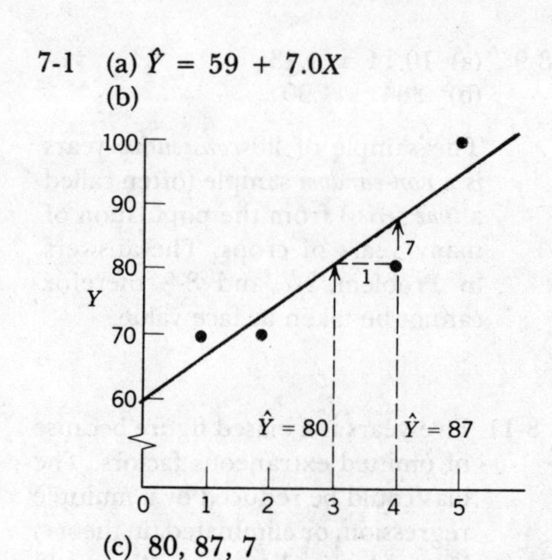

(c) 80, 87, 7

7-3 (a) $\hat{Y} = 119 + 9.0X$

(b) 164, 119

(c)

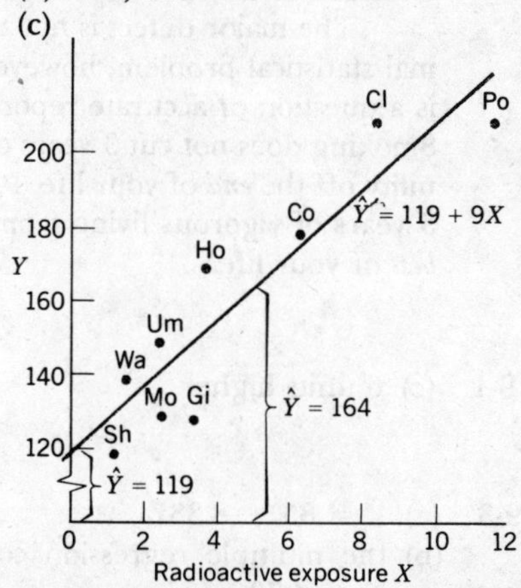

(d) Since this is an uncontrolled observational study, it does not provide proof.

7.7 (a) $\hat{S} = -.78 + .142X$

(b) $\beta = .142 \pm .169$

(c)

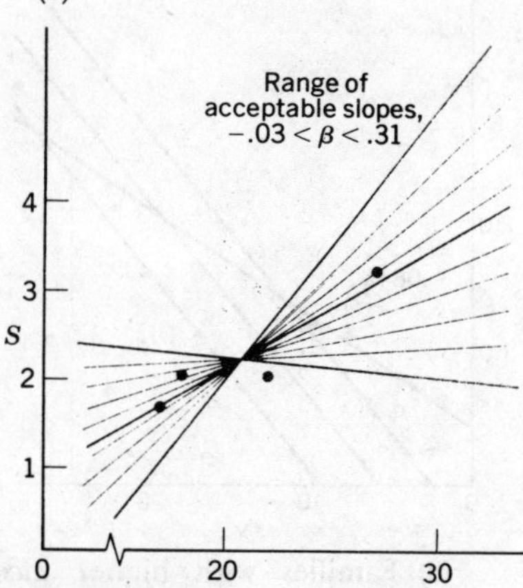

7-9 (a) H_0: $\beta = 0$

H_1: $\beta > 0$

(b) $p < .05$

(c) $\beta > .027$

(d) reject H_0 on both counts

7-11 (a) $9200

(b) $\beta = 800 \pm 492$

(c) Is discernible, since 800 exceeds $\pm$ 492.

(d) No, since this is just an observational study. Perhaps men with higher education also tend to have more ambition, and this may be what produces the higher income.

7-13 $t = 3.48$, $p < .0005$

8-1 (a) $\hat{S} = .413 + .0835X$

(b)

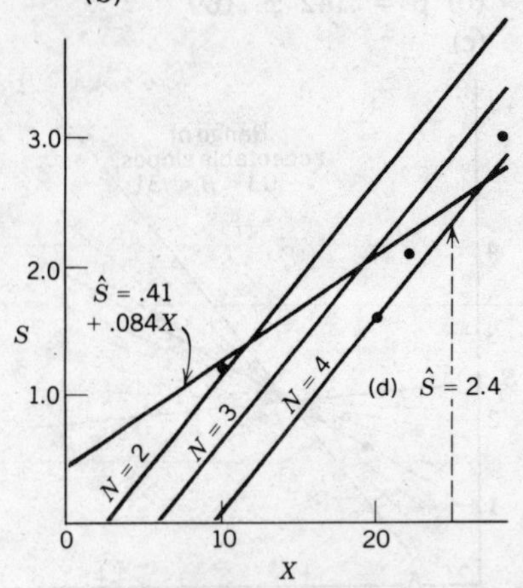

(c) Families with higher incomes tend to have more children, and this depresses their savings. So the simple regression slope is reduced.

(d) 2.39 thousand dollars annually

(e) .148 thousand dollars annually

(f) Solve:

$15.7 = 188b + 23c$

$1.3 = 23b + 4c$

8-3 the simple regression formula,

$$b = \frac{\Sigma xy}{\Sigma x^2}$$

8-5 (a)

±.038	±.068	±.036
1.11	2.21	2.39
<.25	<.025	<.010

(b) We are assuming the 66 nurses form a random sample from a large population. It is for this population that we are calculating confidence intervals and prob-values.

8-7 (a) 10.14

(b) .36

(c) 10.50

(d) 9.66

(e) 9.66

(f) −2.08

8-9 (a) 10.14 ± 4.43

(b) .364 ± 1.90

The sample of 20 *consecutive* years is a *non-random* sample (often called a *time series*) from the population of many years of crops. The answers in Problem 8-7 and 8-8 therefor cannot be taken at face value.

8-11 Five years is a biased figure because of omitted extraneous factors. The bias could be reduced by a multiple regression, or eliminated (in theory) by randomized control. We would *guess* the unbiased figure was about 3 instead of 5.

The major defect is not a formal statistical problem, however. It is a question of accurate reporting. Smoking does not cut 3 years of senility off the *end* of your life. It cuts 3 years of vigorous living from the *best* of your life.

9-1 (c) 6 units higher

9-3 (a) $\hat{Y} = 85 + 8.33Z$

(b) the multiple regression coefficient 4.70

(c) 3.63 too high

9-5 $\mu_T - \mu_C = 8.33 \pm 9.61$

We estimate the effect of the drug is to increase blood pressure by 8.33 ± 9.61. But since this was an observational study without randomization, this estimate may be seriously biased by extraneous variables.

This agrees with the conclusion in Problem 9-4.

9-7 (a)

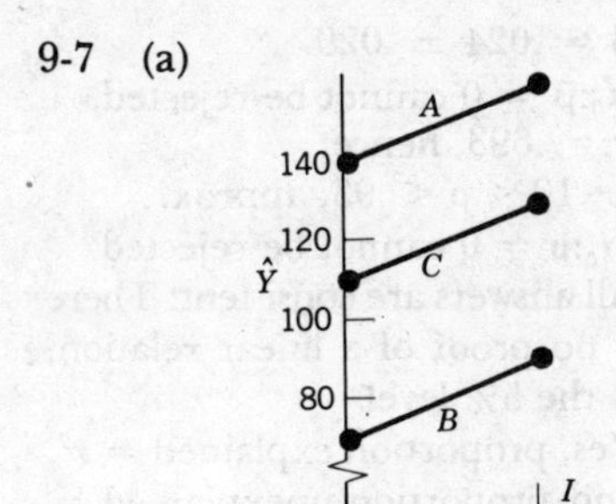

(b)

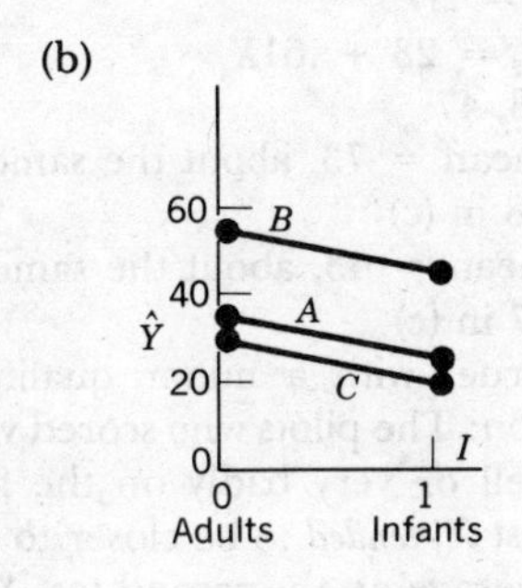

9-9 (a)

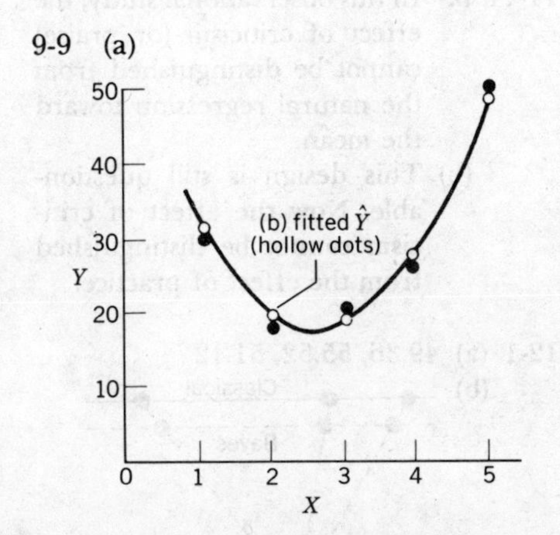

(b) $\hat{Y} = 32.2, 20.2, 19.0, 28.6, 49.0$

(c) $\hat{Y}$ increases by 20.4. Of course, this has nothing to do with the coefficient of X, because this parabola is *nonlinear*.

9-11

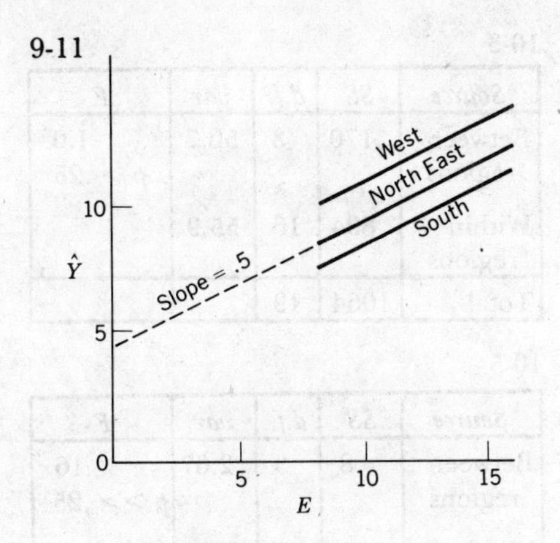

9-13 We disagree. The $2400 figure may be due to discrimination, but it also may be partly due to some more subtle difference between men and women not measured in this study.

9-15

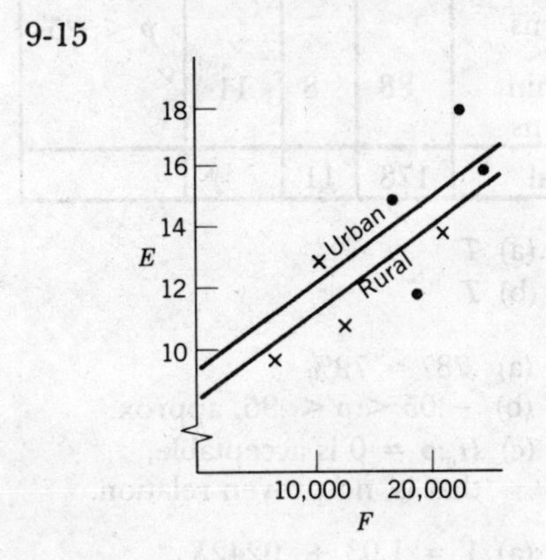

10-1

Source	*SS*	*d.f.*	*var*	*F*
Between fert.	312	2	156	9.0 $p < .01$
Within fert.	156	9	17.3	
Total	468	11		

10-3

Source	*SS*	*d.f.*	*var*	*F*
Between regions	170	3	56.7	1.0 $p > .25$
Within regions	894	16	55.9	
Total	1064	19		

10-5

Source	*SS*	*d.f.*	*var*	*F*
Between regions	8	3	2.67	.16 $p >> .25$
Within regions	114	7	16.3	
Total	122	10		

10-7

Source	*SS*	*d.f.*	*var*	*F*
Between yarns	90	3	30	2.73 $p < .25$
Within yarns	88	8	11	
Total	178	11		

10-9 (a) *T*
(b) *T*

11-1 (a) $.787 \simeq 79\%$
(b) $-.05 < \rho < .96$, approx.
(c) $H_0: \rho = 0$ is acceptable,. that is, no proven relation.

11-3 (a) $\hat{Y} = 1.03 + .0242X$
(b)

Source	*SS*	*d.f.*	*var*	*F*
Expl.	.70	1	.70	4.6
Unexpl.	.76	5	.152	$p < .10$
Total	1.46	6		

Since $p > .05$,
H_0 cannot be rejected.

(c) $\beta = .024 \pm .029$
$H_0: \beta = 0$ cannot be rejected.
(d) $r = .693$, hence:
$-.10 < \rho < .93$, approx.
$H_0: \rho = 0$ cannot be rejected.
(e) All answers are consistent: There is no proof of a linear relation, at the 5% level.
(f) Yes, proportion explained $= r^2$, and proportion unexplained $= 1 - r^2$.

11-7 $r = 1.00$

11-9 (a) $r = .57$
(b) $\hat{Y}_2 = 23 + .61X_1$
(c) 78, 47
(d) mean = 73, about the same as 78 in (c)
(e) mean = 43, about the same as 47 in (c)
(f) True, with a minor qualification: The pilots who scored very well or very badly on the first test X_1 *tended* to be closer to the average on the second test X_2.

11-11 (a) In this observational study, the effect of criticism (or praise) cannot be distinguished from the natural regression toward the mean.
(b) This design is still questionable. Now the effect of criticism cannot be distinguished from the effect of practice.

12-1 (a) 49.36, 55.52, 51.12
(b)

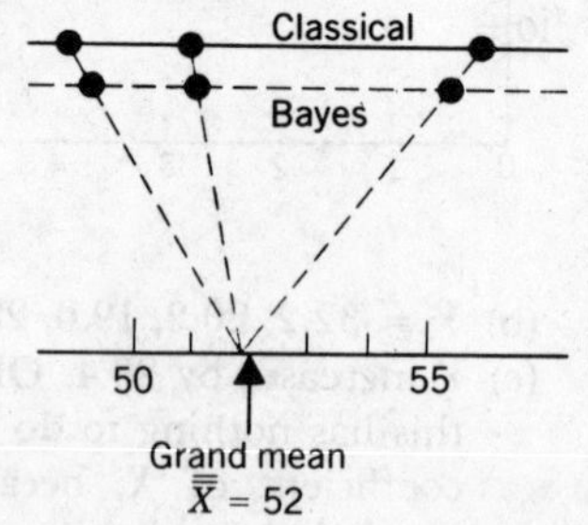

(c) (i) Example 12-1
(ii) the same, Example 12-1

12-3 (a) 37, 37, 37, 37

(b)

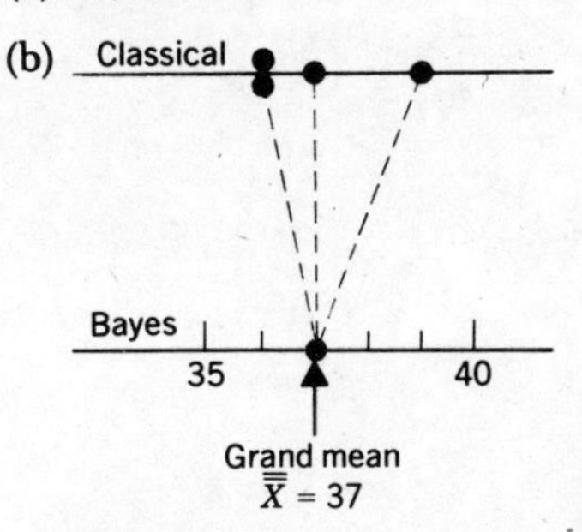

12-5 (a) 13.0
(b) $10.86 \simeq 11$ (girls are estimated to be 11 marks higher than boys, on average).

12-7 (a) 2.8%
(b) 2.48%

12-9 Bayes estimates are 10.66, 6.51, 9.83

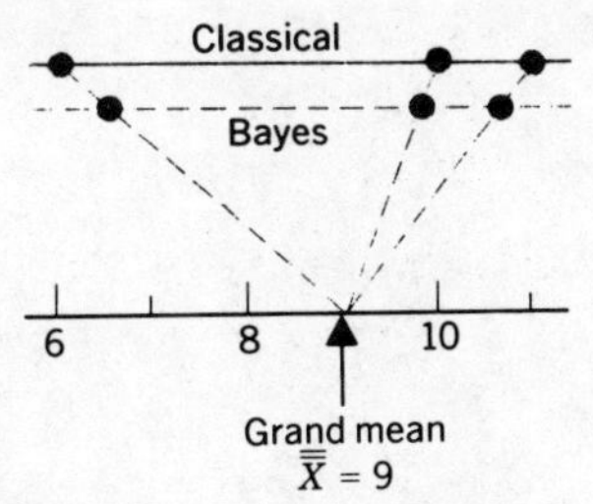

13-1 (a) .0547
(b) .1719
(c) .0547 (for H_0: median difference $= 0$)

13-3 .0047 (wcc, .0082)

13-5 (a) $67 \leq \nu \leq 76$
(b) $63 \leq \nu \leq 69$
(c) $3 \leq \nu \leq 7$

13-7 (a) Yes, approximately normal.
(b) The nonparametric CI is likely less precise.
(c) $67.8 \leq \mu \leq 74.2$, which is indeed more precise (narrower).

13-9 .0392

13-11 for H_0, $p = .0294$

13-13 (a) $< .025$
(b) .035

14-1 (a) $\chi^2 = 4.90$, $p < .25$
(b) No, cannot reject H_0.

14-5 Instead of χ^2, use the proportion of aces P. To calculate the prob-value, convert P to

$$Z = \frac{P - 1/6}{\sqrt{\dfrac{(1/6)(5/6)}{30}}}$$

14-7 (a) Newspaper and social class are independent.
(b) $\chi^2 = 12.4$, $p < .10$, indiscernible.

14-9 (a) Race and education are independent.
(b) $\chi^2 = 8.55$, $p < .025$

14-13 (a) $\chi^2 = 5.76$, $p < .025$
(b) $\pi = .512 \pm .010$
(c) The CI in (b) is better: it says just *how much* π differs from $\frac{1}{2}$.

14-15 (a) $\chi^2 = 6.54$, $p < .025$
(b) $\pi_2 - \pi_1 = .0009 \pm .0007$
(c) The CI in (b) is better: it says just *how much* $\pi_2 - \pi_1$ differs from zero.

Photo Credits

Page 9	Christopher Morrow/Stock, Boston.
Page 28	Cary Sol Wolinsky/Stock, Boston.
Page 64	United Press International.
Page 85	Wide World Photos.
Page 140	Wide World Photos.
Page 158	Frank Siteman/Stock, Boston.
Page 158	Rick Smolan/Stock, Boston.
Page 192	Grant Heilman.
Page 241	Harry Wilks/Stock, Boston.
Page 234	American Cancer Society.

Index

Analysis of covariance, 244
Analysis of variance (ANOVA), 244, 253
  Bayes, 293
  dummy variable regression, 244
  hypothesis test, 258
  regression, table for, 279
  robust, 254
  table, 261
  two factor, 244, 265
  unequal sample sizes, 264
  variation, 253, 261
Answers to odd problems, 357
Average, *see* Mean; Sample mean
Averaging out, 89, 92

Bayes estimates, 291
  ANOVA estimates, 293
  classical methods and, 291, 300
  confidence intervals, 302
  limitations, 301
  mean difference estimates, 298
  posterior estimates, 291
  regression estimates, 295
Bias, 4, 14
  multiple regression reduces, 14, 213, 215
  nonrandom sampling, 4
  observational studies, 14, 159, 213
  simple regression, 213, 215, 228, 282
  two-sample *t*, 243
  unbiased defined, 92
  see also Randomized controlled experiments
Binary variable, *see* Dummy variable
Binomial, 58
  chi-square test, 322, 330
  cumulative, 63
  examples, 59
  independence required, 60
  mean, 65
  normal approximation, 113
  sign test, 305
  variance, 65
  see also Sample proportion
Bivariate distribution, 274

Categorical *vs.* numerical, 327
Center of a distribution, 24
Central limit theorem, *see* Normal approximation theorem
Chi-square (χ^2), 318
  binomial extension, 322
  contingency tables, 323
  goodness of fit, 318
  limitations, 327
Classical hypothesis test, 170, 176. *See also* Hypothesis test
Color convention, 92, 121
Complementary events, 48
Computer output for regression, 217, 223
Confidence interval, 119
  Bayes, 302
  chi-square, alternative to, 326

correlation, 275
difference:
in two means, 133, 136
in two proportions, 146
horseshoe pitching analogy, 124
hypothesis test and, 156
mean, 121, 129
meaning, 5, 123
median, 308
nonparametric, 308
one-sided, 148
paired samples, 136
proportions, 144
random interval, 123
regression coefficients, 206, 224
summary, *inside back cover*
Contingency tables, 323
Continuity correction, 113
Continuous distributions, 21, 66
Control group, 8
Controlled experiments, *see* Randomized controlled experiments; Bias
Correlation, 268
assumptions, 275, 281
confidence interval, 275
determination, coefficient of, 281
hypothesis test, 278, 282
meaning, 270, 281, 282
multiple, 281
population, 273
regression and, 276, 282
spurious, 282
Cross-validation, 301

Deduction and induction, 119
Degrees of freedom, 35, 164
ANOVA, 258, 265
chi-square, 320
contingency table, 325
regression, 206, 207, 224
single sample, 35, 130, 164
two samples, 135
Density function, 66
Dependence, *see* Independence
Descriptive statistics, 19
Determination, coefficient of, 281
Deviations, 32, 33
in regression, 193, 195
Difference:
in means, 133, 136
in proportions, 146
Discernibility, 159
indiscernibility, 159
ordinary significance and, 160
prob-value and, 172
statistical significance and, 160
Discrete distributions, 19, 42
Distribution-free tests, 304. *See also* Sign test; W test
Distribution of regression estimates, 204, 223
of sample mean, 98
see also Probability distributions
Double-blind experiments, 10
Dummy variable, 110
ANOVA, 244
proportions, 110
regression, 236
several categories, 241

Error, rounding, 35
error term in regression, 201, 216
type I *vs.* type II, 172, 176
see also Confidence interval
Estimating (normal) equations, 217
Ethical issues in randomization, 12
Expected value, 54
regression coefficients, 204, 223
sample mean, 99
sample proportion, 110
Experimental design, 8, 204

Extraneous factors, 10, 14
  controlled by multiple regression, 14, 159, 213, 215, 228, 282

F, 258
  ANOVA use, 258, 262
  Bayes use, 293, 302
  distribution, 260
  regression use, 278
  t and, 279
Finite-population sampling, 105
Frequency distributions, 19

Gaussian distribution, *see* Normal distribution
Grouped data, 21

Histogram, 22
Hypothesis test, 156
  ANOVA, use in, 258
  classical testing, 170
  confidence intervals define, 156
  correlation, use in, 278, 282
  critical point and region, 170, 176
  errors of type I and II, 172, 176
  level of test (α), 158, 160, 176
  limitations, 160, 172, 176
  nonparametric, 304
  null hypothesis, 160, 170
  reasons for, 178
  regression, use in, 208, 224
  sign test, 305
  significance, *see* Discernibility
  two-sided, 180
  see also Confidence interval; Prob-value; chi-square; *F*; *t*

Independence (statistical), 60
  binomial assumes, 60
  random sampling assumes, 86
  regression assumes, 201
Induction and deduction, 119

James-Stein estimation, 301

Least squares in regression, 194, 217
Level of Test (α), 158, 160, 176
Lines, elementary geometry, 335
Load limit problems, 104

Mann-Whitney test, 312
Matched samples, *see* Paired samples
Mean, 25
  balancing point, 27
  Bayes estimates, 291
  binomial, 65
  grouped data, 26, 37
  median and, 27
  population, 53, 84
  proportion as, 110
  0–1 population, 110
  see also Expected value; Sample mean; Confidence Interval
Mean squared deviation (MSD), 33
Mean sum of squares (MSS), 263
Median, 25
  confidence interval, 308
  sign test, 305
MINITAB, 217
Mode, 24
Moments, 35
Monte Carlo (Simulation), 87, 89
  computer, 333
  dice, 43
  normal numbers, 332
  regression, 203
  sample mean, 89

Multinomial χ^2 test, 322
Multiple regression, 213
  ANOVA and, 244, 253
  bias reduced by, 14, 213, 215
  computer output, 217, 223
  confidence intervals, 224
  controls extraneous factors, 14,
    159, 213, 215, 228, 282
  determination (r^2), 281
  hypothesis tests, 224
  least squares, 217
  meaning, 216, 228
  model, 216
  normal equations, 217
  prob-value, 224
  ridge regression, 301
  simple regression and, 213, 215,
    228, 282

Nonlinear regression, 246
Nonparametric statistics, 304. *See*
  also Sign test; *W* test
Normal approximation theorem, 101
  for proportions, 111
  for regression, 204, 223
Normal distribution (*Z*), 67
  approximation to binomial, 113
  general, 70
  standard, 67
  t, relation to, 130, 164
Normal equations, 217
Null hypothesis, 160, 161, 170. *See*
  also Prob-value

Observational studies, 11
  biased by extraneous factors, 14,
    159, 213
  multiple regression reduces bias,
    14, 213, 215
One-sided confidence intervals, 148

p value, *see* Prob-value
Paired samples, 136, 306
  advantage, 140
  nonparametric, 306
Parameters of population, 119
Percentiles, 23
Polynomial regression, 246
Pooled variance, 135, 257
Population, 83
  mean and variance, 53
  sample *vs.*, 56, 92, 121
Posterior estimates, 291
Prior information, 179, 291
Probability, 45
  trees, 50
Probability distributions, 42
  binomial, 58
  continuous, 66
  discrete, 42
  mean, 53
  normal, 67
  trees, 50
  variance, 54
Prob-value for H_0, 161
  ANOVA, use in, 258
  chi-square tests, use in, 320
  classical hypothesis test and, 172
  contingency table, use in, 326
  credibility of H_0 measured, 161, 163
  definition, 163
  level of test and, 172
  regression, use in, 208, 224
  t test, use in, 164
  two-sided, 180
Proportions, *see* Sample proportion

Random sampling, 383
  definition, 4, 85
  without replacement, 105
  replacement or not, 86, 105
  simulated, 87

small populations, 105
very simple (VSRS), 86
0–1 populations, 108
Random variable, 44
Randomized controlled experiments, 8
bias removed, 10, 14
double-blind and, 11
ethical issues, 12
multiple regression simulates, 14
social science experiments, 13
Range of sample, 33
Regression, *see* Simple regression; Multiple regression; Regression extensions
Regression extensions, 236
ANOVA, 244, 253
dummy variables, 236
nonlinear polynomial regression, 246
Relative frequency, 20, 37
density, 66
limit and probability, 45
see also Sample proportion
Residual in regression, 193, 206
Ridge regression, 301
Robustness of ANOVA, 254
Rounding error, 35

Sample mean, 25, 98
confidence intervals, 122, 130, 135, 137
distribution normal, 98
expected value, 99
Monte Carlo, 89
standard error, 99
Sample proportion, 108, 144
confidence intervals, 144, 146
continuity connection, 113
distribution normal, 111
expected value, 110
sample mean as, 110, 144
standard error, 111
Sample size required, 129
Sample variance, 34
Sampling, 3, 83. *See also* Random sampling
Scatter diagrams, 194, 273
Sign test for median, 305
confidence interval, 308
paired samples, 306
t test and, 304, 311
Significance, *see* Discernibility
Simple regression, 191
ANOVA, table for, 279
assumptions, 200
Bayes, 295
biased, 213, 215, 228, 282
coefficient distribution, 204
coefficient formulas, 196
confidence intervals, 207
correlation and, 276, 282
error term, 201
hypothesis tests, 208
least squares, 194
meaning, 196, 226, 336
model, 199
multiple regression and, 213, 215, 228, 282
prob-value, 208
regression toward the mean, 286
residuals, 193, 206
see also Correlation
Simulation, 87. *See also* Monte Carlo
Simultaneous linear equations, 217, 338
Skewed distributions, 28, 313
Slope, 335. *See also* Simple regression
Small-population sampling, 105
Social science experiments, 13
Spread (of distribution), 33
Spurious correlation, 282
Standard deviation, 34. *See also* Standard error; Variance

Standard error, 91
  regression coefficients, 204, 206, 223
  sample mean, 99, 105
  sample proportion, 111
  two means, 134
  see also Confidence interval
Statistic, definition, 19, 120
Student's *t, see t*
Sum of squares, *see* Variation
Symmetric distribution, 27

t, 129, 164
  d.f., *see* Degrees of freedom
  distribution, 164
  F and, 279
  mean, use in, 129, 133, 164
  normal, relation to, 130, 164
  regression, use in, 208, 224
  sign test and, 304, 311
  see also Confidence interval
Tables, 339
Test of hypothesis, *see* Hypothesis test
Trees to list outcomes, 50
Type I and Type II errors, 172

Unbiased, *see* Bias

Validity of nonparametric tests, 304
Variance, 34, 54
  ANOVA, 244, 253
  binomial, 65
  grouped data, 34, 37
  pooled, 135, 257
  population, 54, 84
  residual, 206
  sample, 34
  0–1 population, 109
  see also F; Standard error
Variation (explained, unexplained, total), 253
  in ANOVA, 253, 261
  in regression, 277
Very Simple Random Sampling (VSRS), 86. *See also* Random sampling

W test (Wilcoxon-Mann-Whitney), 312

χ^2, *see* chi-square

Z variable, 67, 73; *see also* Normal distribution
Zero-one (0–1) variable, *see* Dummy variable

SUMMARY OF 95% CONFIDENCE INTERVALS

Population Parameter	*95% Confidence Interval*	
MEANS		
One population mean μ	$\bar{X} \pm t_{.025}\text{SE}$	(5-17)
	$= \bar{X} \pm t_{.025}\dfrac{s}{\sqrt{n}}$	(5-13)
Difference in two population means $(\mu_1 - \mu_2)$ (a) independent samples.	$(\bar{X}_1 - \bar{X}_2) \pm t_{.025}\, s_p\sqrt{\dfrac{1}{n_1} + \dfrac{1}{n_2}}$	(5-24)
(b) paired observations	$\bar{D} \pm t_{.025}\dfrac{s_D}{\sqrt{n}}$	(5-28)
PROPORTIONS		
One population proportion π	$P \pm 1.96\sqrt{\dfrac{P(1-P)}{n}}$	(5-31)
Difference in two population proportions $(\pi_1 - \pi_2)$, independent samples	$(P_1 - P_2) \pm 1.96\sqrt{\dfrac{P_1(1-P_1)}{n_1} + \dfrac{P_2(1-P_2)}{n_2}}$	(5-33)
RELATIONS		
Simple regression slope β	$b \pm t_{.025}\dfrac{s}{\sqrt{\Sigma x^2}}$	(7-18)
Multiple regression slope β	$b \pm t_{.025}\,\text{SE}$	(8-15)
ALTERNATIVES		
Bayes shrinkage for any of parameters above, if 0 seemed plausible a priori yet F is substantial	Shrink estimate by $1 - (1/F)$ Shrink allowance by $\sqrt{1 - (1/F)}$	(12-14)
Nonparametric confidence interval for one population median ν	$X_{(q)} < \nu < X_{(r)}$	(13-12)
One-sided confidence interval, for one population mean μ, for example	$\mu > \bar{X} - z_{.05}\dfrac{\sigma}{\sqrt{n}}$	(5-34)

WHERE TO FIND IT

Problem	*Solution Name*	*Solution Formula*	*Worked Example*	*Problem Exercise*
		MEANS		
Estimating one mean	t	(5-13)	Example 5-2	Problem 5-7
Comparing two means (a) independent samples	t	(5-24)	Example 5-4a	Problem 5-13
(b) paired observations	t	(5-28)	Example 5-4b	Problem 5-15
Comparing c means	ANOVA	(10-6) Table 10-4(a)	Example 10-1 Table 10-4(b)	Problem 10-1
		PROPORTIONS		
Estimating one proportion	Normal Z, or Binomial	(5-31) Figure 5-5	Example 1-1 (5-32)	Problem 5-17(c) Problem 5-17(a)
Comparing two proportions	Normal Z	(5-33)		Problem 5-19
Comparing k proportions	Chi-square χ^2	(14-7)		Problem 14-9
		RELATIONS		
Response to one factor (a) numerical factor	Simple regression	(7-5), (7-18) (7-8)	Example 7-2	Problem 7-7 Problem 7-11
(b) categorical factor	ANOVA	(9-13), (10-6) Table 10-4(a)	Example 10-1 Table 10-4(b)	Problem 10-1
Response to k factors (a) all factors numerical	Multiple regression	(8-4) to (8-6) (8-15) (8-36)	Table 8-2 Example 8-3 Example 8-5	Problem 8-1 Problem 8-5 Problem 8-7
(b) some factors categorical	Multiple regression (with dummies where necessary)	(9-8)	Example 9-1	Problem 9-1 Problem 9-7
Nonlinear response	Transformation, then regression	(9-17), (9-18)	Table 9-2	Problem 9-9